Regaining Sanity for the earth

Regaining Sanity for the earth

Why science needs 'best faith' to be responsible
Why faith needs 'best science' to be credible

Klaus Nürnberger

Library of Congress Control Number:		2011903118
ISBN:	Hardcover	978-1-4568-7606-7
	Softcover	978-1-4568-7605-0
	Ebook	978-1-4568-7607-4

Design, typesetting, layout and cover: Klaus Nürnberger

Quotations from the Scriptures are taken from the Revised Standard Version, the New International Version, or they reflect translations of the author.

This book was printed in the United States of America.

To order additional copies of this book, contact:

Xlibris Corporation (United Kingdom)
0-800-644-6988
www.xlibrispublishing.co.uk
Orders@xlibrispublishing.co.uk

Cluster Publications (South Africa)
Tel 033 345 9897
www.clusterpublications.co.za
nelly@clusterpublications.co.za

301678

Contents

Acknowledgements

The manuscript has been peer reviewed and recommended for publication by Prof Vincent Brümmer (Philosophy, University of Utrecht), Prof Ernst Conradie (Theology, University of the Western Cape) and Prof Egmont Rohwer (Chemistry, University of Pretoria). It has been accepted for publication by Cluster Publications (Pietermaritzburg) and published by Xlibris Books (London) for the sake of greater accessibility. I am immensely grateful for all the help I received. I also acknowledge with gratitude the financial support received from the Research Office of the University of KwaZulu-Natal. The views expressed are those of the author and not those of the reviewers, the publishers, or the University of KwaZulu-Natal. All inadequacies of content, language, or format are the sole responsibility of the author.

The gist of the argument

This book is about the relationship between science and faith. But it places this relationship into the context of the perilous direction in which modernity has taken humankind over the last few centuries. Perceptive analysts warn us that current trends lead us blindfolded towards an economic-ecological catastrophe of unprecedented proportions. Collaborative faith-science relationships have become indispensable and urgent.

Regaining sanity

We have reached a threshold in history where the sanity of the earth requires the sanity of human minds. Humankind must develop a new sensitivity for the intricate relationships between the super-competitive and the less competitive; between humanity and nature; between knowledge and commitment; between research and vision; between science and faith.

The word regaining suggests that this sensitivity has been lost. Gratitude has changed into entitlement; needs into wants; contentment into avarice; responsibility into indifference. Humankind is suffering from an obsession that is suicidal for humanity and destructive for all life on earth.

The responsibility of faith and science

Faith and science were co-responsible for the generation of this situation over the last four centuries. They must also be held co-responsible for its resolution. Ever since the Enlightenment faith and science have been drifting apart. In this process, science lost its transcendent foundations; faith lost its credibility. The consequences are catastrophic. Insight needs commitment to be responsible, commitment needs insight to be credible.

Science and faith are foundational and complementary human pursuits. Their rationale is to serve humanity and its natural infrastructure. They must join forces in leading humanity into wholesome directions. The book develops the argument first from the side of science to a point where it can link up with faith. Then it develops the argument from the side of faith to the point where it can link up with science. The aim is the integration of 'best science' with 'best faith' for the sake of humanity and the earth.

Recent books by Klaus Nürnberger

Beyond Marx and Market: Outcomes of a Century of Economic Experimentation (1998)

Prosperity, Poverty and Pollution: Managing the Approaching Crisis (1999)

Theology of the biblical Witness: An Evolutionary Approach (2002)

Zuspruch des Seinsrechts: Verstellt die Lehre die Sache? (On justification by grace accepted in faith) (2003)

Martin Luther's Message for us Today: A Perspective from the South (2004)

Biblical Theology in Outline: The Vitality of the Word of God (2005)

Making Ends Meet: Personal Money Management in a Christian Perspective (second edition) (2007)

The Living Dead and the Living God: Christ and the Ancestors in a Changing Africa (2007)

Die Bybel—verantwoordelik lees is krities lees (2009)

Richard Dawkins' God Delusion: A Repentant Refutation (2010)

Foreword

The current crisis we are facing is multi-stranded, involving warring regions, poverty and hunger, economic disasters and environmental crises, but the common element in all of them is a crisis of values and meaning. This has arisen largely through the processes driven by the rise of science and its effect on belief systems. They led to 'modernity' with its paradoxical devaluation of humanity on the one hand and the dramatically improved ability to deliver material benefits on the other. A reconceptualisation is needed to enable a return to a value-driven system of meaning and behaviour.

In this book, Klaus Nürnberger provides a perceptive integrative view of such a reconceptualisation of the science-faith interface in the context of our current economic-ecological predicament. The book first develops the argument from the side of science to a point where it can link up with faith. Then it develops the argument from the side of faith to the point where it can link up with science. The aim is the integration of best science with best faith for the sake of humanity and the earth.

It can be thought of as having two parts. First, there is a consideration of the relation between the two sources of guidance we have: faith and value systems, to do with meaning; and science, to do with empirical reality. Both need to be integrated to make a viable path into the future. This part of the book can be thought of as a valid and valuable guide in any interfaith dialogue, where "faith" means a meaning position that guides what one chooses to do in life. A value system—a criterion of what should be done—is needed as the basis of any such development. The proposal here is that this criterion is *a dynamically evolving vision of the comprehensive optimal well-being of individuals and communities within the comprehensive optimal well-being of their entire social and natural environments.* This vision translates into concern for any deficiencies in well-being in life.

The second part of the book develops this argument in depth from a Christian, and indeed Lutheran, perspective. This kind of detailed development is needed in order to give real flesh to the proposal, and the author does not disappoint here. In particular, he focuses on the kenotic nature of a deep understanding of Christianity, which is the true deep basis for such development. The task for other faiths—including atheism—is to develop their viewpoint to a similar depth. The author provides a model they could follow.

Just as he emphasises there is no 'pure' disinterested science, there is also no 'pure' spiritual faith. Genuine faith is specific in terms of content, embodied in concrete individuals and embedded in social contexts, and it has concrete outcomes in terms of behaviour. The author asks what the point of the 'science-religion' debate is if the insights gained do not spill over into mainstream scientific and theological work, and if they do not translate into public opinion, political agendas, and personal behaviour.

All this is developed in a careful and well-thought-out way, emphasising that faith and science relate to the same reality we experience daily and in which we are inextricably embedded. But they relate to this reality in two fundamentally different ways. Science explores reality from an immanent perspective, while faith looks at the same reality from a transcendent perspective. Faith is aware of the fact its intuition will remain inadequate, yet it lives within its presence. Science knows that its findings are provisional but tries to extend the boundaries of knowledge. Both views are crucially needed. The author observes reality from both vantage points and sees how they intersect. The outcome is what the author calls 'integrative transcendence'. It is a valuable and worthwhile project.

George Ellis

Professor Emeritus of Applied Mathematics,
University of Cape Town
Cape Town 2010-11-21

Setting the scene

Modern science and Christian faith are formative powers in society rather than private pastimes. They have stood at the cradle of modernity. They are jointly responsible for the direction that modernity will take in the future. Modern civilisation dominates the world of today. It is driven by the quest for mastery, ownership and entitlement. Individual material interests have become paramount at the expense of the comprehensive well-being of the whole of reality. The responsibility engendered by faith has not kept pace with the powers unleashed by science and technology. The present course of modernity may lead humanity towards self-destruction. An appropriate system of meaning, based on the vision of comprehensive optimal well-being, is essential for healthy survival. This demands the knowledge of science and the commitment of faith. The book takes the argument from the side of science to a point where it can link up with faith. Then it takes the argument from the side of faith to the point where it can link up with science. The aim is the integration of best science with best faith for the sake of humanity and the earth.

The overall context

This book deals with the relation between science and faith, or to be more precise, with the relation between the naturalistic assumptions of modern science and the transcendent assumptions of the Christian faith. However, it places this debate into the larger context of the way modern civilisation creates socio-economic imbalances and drives the planet towards potential disaster. Four factors are instrumental in this development:

(a) The human population has grown dramatically over the last few centuries. It is still growing in the most impoverished sections of the global population, while it declines in more developed contexts. Rising discrepancies in life chances and escalating conflict potentials are the result.

(b) Material expectations are soaring. Large 'emerging economies', like China, India, Indonesia and Latin America, are intent on reaching Western standards, while consumption continues to be fired up in richer countries to raise profits and avoid a recession. Now multiply growing numbers with growing needs in a limited world and you are saddled with a time bomb!

(c) Science and technology have unleashed unprecedented power over reality without providing a concomitant sense of responsibility.

(d) Modern humans have shed their accountability to a transcendent, all-embracing authority, appropriating for themselves the role of owners, masters and sole beneficiaries of reality. Overriding obligations and commitments are disappearing in vast sections of the population. The consequences are catastrophic.

Ecologically speaking, the infrastructure of life on earth is in danger of collapsing under the weight of human extraction, production, consumption and waste.[1] Climate change, depletion of scarce resources, over-exploitation of renewable resources, destruction of plant and animal life, pollution of water and air, soil erosion, declining food security in many parts of the world—these are symptoms of a deadly trend that must be reversed if humanity (and much of life in general) is to have a future.

Economically speaking, the current system is leading to widening gaps in productive capacity and life chances between economic centres and peripheries. Growing sections of the world population are marginalised or pushed out of the formal economy altogether. Productive capacity and scarce resources are invested in the luxury consumption of elites rather than the needs of the majority. Rising debt levels haunt individuals and states.

Spiritually speaking, these developments have led to a concentration of human goals on material wants; the artificial creation of a culture of discontent; a narcissistic mindset void of wider horizons and long-term visions; the dismantling of traditional systems of meaning, cultural values and social norms without providing valid alternatives and the concomitant dissolution of extended families and communities.

The emancipatory thrust of modernity

At the root of the looming crisis lie the basic assumptions of modernity. Modernity derives its dynamic from a powerful thrust towards human emancipation and sovereignty. All imposed forms of authority are rejected, including the religious, philosophical, cultural, social, economic and political traditions of earlier times. Modern human beings have 'come of age'. They claim to be the owners, masters and beneficiaries of their life worlds—void of accountability to any authority other than themselves.[2]

Modernity increasingly determines all aspects of life all over the world. While not yet the internalised culture of the majority of the global population, it has become the dominant culture of global elites. Elites exert power and set standards. Through globalising commerce and communication systems modernity radiates its influence into the most remote rural villages and the most intimate personal relationships.

Science, technology, commerce and consumer satisfaction have crystallised out as the major forces in the modern civilisation. Science plays a pivotal role in this regard because it enlightens and empowers the operation of the other three aspects.[3]

In earlier phases of history, faith provided science with basic orientation; science provided faith with knowledge about the world. In rudimentary forms, that has been the case in primary cultures, in antiquity and during the Middle Ages.

Since the Enlightenment, this interaction has practically disappeared. What pre-modern humanity expected from God, modern humanity expects from its own insights and interventions. Divine authority has morphed into human autonomy. Science has displaced religion.

The ambiguity of human intentionality

For secularised people this development is a dream come true. At long last humanity breathes the air of freedom. Why then should it be deemed problematic? Humanity has emerged and evolved as a species with a superior intelligence and a self-seeking intentionality. Like other animals humans are subject to physical causality, genetic determination and environmental conditioning. But human life does not just 'happen'. What makes humans specifically human is their extraordinary scope of intentionality and agency.

Intentionality means that human life is goal oriented. Agency means that such goals can be actively and successfully pursued. Humans are capable of using tools and extracting resources to produce goods useful to themselves. Humans can respond to insights, challenges and opportunities. Humans are also determined by collective attitudes and behaviour, but in ways that are fundamentally different from the herd instinct of antelopes or the organisation of ant heaps. Humans are 'condemned to be free' and freedom implies responsibility.

Human intentionality and agency are highly ambiguous. They can be creative or destructive. They can enhance or obstruct life. They can be self-seeking or community oriented. They can have narcissistic or global horizons. Millennia of human experience have led to overarching systems of meaning that intuit validity, define acceptability and allocate authority. These systems of meaning are fast disintegrating under the impact of modern assumptions.

A system of meaning is an indispensable prerequisite of human life. It forms a "sacred canopy" that circumscribes the identity of individuals, communities, society and nature within the whole of reality. Humans have always distinguished between authentic and inauthentic human existence and social structure in terms of such a system of meaning.

Authentic human existence is defined by acceptable purposes, goals, norms and values. They are acceptable if they are oriented towards a vision of comprehensive optimal well-being—at least within the limited horizons of collective consciousness at any point in time. To be authentically human, humans are expected to act consciously, responsibly, creatively and redemptively. Inauthentic human life, in contrast, is capable of acting mindlessly, selfishly and destructively.

The criterion of an appropriate system of meaning

In this study, I assume that there is an overriding criterion for the validity of a system of meaning, including all forms of Christianity. This criterion is *a dynamically evolving vision of the comprehensive optimal well-being of individuals and communities within the comprehensive optimal well-being of their entire social and natural environments*. This vision translates into concern for any concrete deficiency in well-being in any dimension of life.

I will take this definition for granted in this book. I assume that the criterion of comprehensive optimal well-being is universally applicable. It is also broadly in line with the general thrust of the evolutionary process because the latter displays a powerful thrust towards the realisation of potentials and the fulfilment of life at all levels of emergence. It places all faith communities under the obligation to subject both their own faith assumptions and those of others to critique.

I have argued elsewhere that God's vision of comprehensive optimal well-being is the motivational groundswell of the biblical faith.[4] God is here understood as the transcendent Source and Destiny of reality as such and as a whole. A similar conviction may also inspire a number of religious and ideological alternatives, but this is not necessarily the case. Whether the Christian faith and its alternatives are appropriate reflections of this vision is the question that should underlie Christian theological reflection, philosophy of religion and interfaith dialogue.

Systems of meaning can question, correct, displace and replace each other. They can also complement and enrich each other. This process is essential. The struggle for the truth cannot be suspended. Its goal should be to reach the most appropriate frame of reference available, one that could give the most plausible meaning, the most wholesome directions and the most reassuring authority to the human enterprise as it operates within its total social and natural contexts.

The role of science and faith

Modernity is a product of Western history. Science and faith have both contributed to the genesis of the current impasse and must be deemed responsible for its resolution. At least in the West, the roles of faith and science are inseparably intertwined. In contrast to higher animals, the human species is a creature that cannot do without either knowledge or meaning in its pursuit of a fulfilled, fruitful and responsible life.

Intuiting the transcendent foundations of reality, faith presupposes observation and comprehension of this reality. On its own it does not have the competence to analyse, explain and predict reality. It has to be informed by 'best science'. Conversely, science explores the mechanisms that underlie the processes of reality and does so for a purpose. So it will be guided by some kind of vision and commitment. On its own it does not have the mandate and method to venture into the transcendent foundations of reality. It has to fall back on 'best faith'.

Modernity has unleashed human avarice and extravagance from the religious, cultural and social constraints of earlier times without providing new foundations and criteria of acceptable behaviour. Science and technology have been instrumental in unleashing powers unprecedented in history. Faith did not keep pace with science and technology and failed to instil a sense of responsibility commensurate with these new powers.

As the bearer of life, our planet may be unique in the cosmos. Human behaviour now determines the fate of all other species. Humanity has become capable of destroying, within a few short decades, what nature has built up over the last four billion years. The complementarity between science and faith has always been indispensable. It has now become urgent.

What is 'science'?

Before I move on, I must clarify my concepts. Science, as I use the term, is more than innocent curiosity, a quest for knowledge, a 'value-free' procedure, or the routine experimentation of some scientists or institutions dedicated to a particular research programme. If science were nothing but empirical observation, logical deduction, or mathematical reconstruction, I could not claim that 'science needs faith'. While we all have our respective preconceptions, the shape of bacteria seen under a microscope and the description of the chemical reaction between oxygen and hydrogen do not depend on anybody's convictions.[5]

But there is no 'pure', abstract, disinterested science. Science is geared to specific objectives, embodied in human beings and embedded in society. It operates within particular traditions. It is a motivated human pursuit, a structured establishment, a decisive power in political decision making. It is intertwined with technology. It serves commerce. It co-determines the modern consumer culture. It claims vast amounts of public resources. It profoundly impacts both society and nature. Science is an inextricable part of the culture of modernity that has come to dominate the earth.

When I use the word "science" in this treatise, therefore, I do not refer to any particular method or discipline. I also do not refer to the motivation of individual scientists, or the research programmes of certain institutions. To engage all the sciences represented at modern universities within the constraints of such a treatise is out of the question. It would also explode my sphere of competence.

When I speak of science, I mean the intellectual and institutional movement that is embedded in the structures and processes of contemporary society and that has, as an integral part of modernity, provided this society with the power it currently commands. It is based on a fairly straightforward set of basic assumptions. Underlying these assumptions is the quest of modern humans for freedom, mastery and control.

What is 'faith'?

Much the same can be said of faith. If faith were nothing but the preoccupation of some private individuals and organisations with mystical, sacramental, or spiritual pursuits, I could not argue that 'faith needs science'. I could also leave it to its own devices. But there is no 'pure', abstract, merely spiritual faith. Genuine faith is specific in terms of content, embodied in concrete individuals and embedded in social contexts. It lives because, and as long as, it responds to concrete needs. It is based on tradition and located in community. It has its own history and shape.

Providing a basic frame of reference, it is involved in every aspect of human life. A vibrant faith motivates and guides individual and collective human decision making and action. It can have incredible consequences, whether beneficial or detrimental. It is a multi-faceted social establishment. It makes countless enterprising individuals and NGOs tick. It controls considerable amounts of financial resources. It can become a formidable political power that co-determines the direction of the processes of society and nature into the future.

When speaking of the kind of faith needed by science, I do not refer to a particular conviction, creed, or theology. Faith is trust. To be more precise, it is self-entrustment to what is deemed the foundation, source, driving force, overriding purpose, value, or destiny of reality. Let me call this assumed supra-structure 'the transcendent' because it is not part of immanent reality, but its assumed precondition and valid direction.

There are countless notions of what the transcendent might be. They each form a set of assumptions, a 'sacred canopy', or a system of meaning. They may be more or less appropriate in terms of individual, communal, social and natural well-being. They cannot be left to chance. They can also not be wished away. To ignore them would lead to a truncated view of reality. The struggle for ultimate truth is more fundamental than the scientific endeavour to observe, explain, or predict immanent reality.

When a desperate grandmother cries to God for help because she cannot cope with the crop of children her daughters who died of AIDS have left in her lap, no measure of scientific insight will remove her awareness of ultimate dependence, vulnerability, accountability and mortality. This awareness will normally find expression in a religious tradition in which a concept of the transcendent has crystallised out over the centuries.

A common predicament and a common responsibility

What then is the relation between science and faith thus conceived? Science and faith have acted as catalysts for the emergence and evolution of the modern mindset.[6] They have demystified the world, each in its own way. They have attacked hierarchical social structures. They have propagated human freedom, equal dignity and human rights. However, neither has been able to instil a sense of responsibility commensurate with unrestricted human autonomy combined with vastly increased powers.

Both are integral parts of a civilisation that is running down the planet. They are jointly responsible for finding a way forward. They must complement each other or face the accusation of letting humanity down.[7] Science needs 'best faith' to become oriented and responsible; faith needs 'best science' to become plausible and relevant. They must again find each other if the earth and humanity on this earth are to have a future.

The science-religion debate

If the ongoing science-religion debate has a point at all, *this must be the point.* It is not an academic pastime; it is an urgent task. 'Science-religion' has become a prolific academic discipline in its own right. New books appear by the day. But what is the point if the insights gained do not spill over into mainstream scientific and theological pursuits? What is the point if they do not translate into robust public opinion, political agendas and personal behaviour?

What is the point if students of the natural sciences experience charismatic highs or evangelistic conversions without recognising the significance of science for faith or faith for science? What is the point if popular culture continues on its rootless, rudderless and hedonistic track, oblivious of well-intentioned academic debates? What is the point when scientific information does not challenge popular denial of the dangers humanity faces in the twenty-first century?

The integration of 'best science' with 'best faith' is imperative. Science is about explanation; faith is about meaning. Science is about prediction; faith is about vision. Science is about immanence; faith is about transcendence. Science is about knowledge; faith is about commitment. A fully explained universe can be meaningless. A profoundly meaningful life can be riddled with untenable assumptions.

Faith is not another form of science

The relation between science and faith is saddled with a host of flawed assumptions. We will encounter them on our way through the argument. Here it must suffice to highlight three of the more important:

(a) That faith is all about observation, explanation and prediction, rather than grounding, orientation and vision.
(b) That faith is based on revelation of eternal truth, rather than historically evolving insight.
(c) That there ever was, can be, or will be a perfect version of the reality in which we are embedded. Perfection cannot be a characteristic of a reality in flux.[8]

Many theologians engaged in the science-religion debate tacitly assume that worldview assumptions found in the Bible have a claim to validity similar to those made by modern science. But the pre-scientific worldview of antiquity cannot claim the status of scientific explanations.[9] The world is not flat; the sun does not circle the earth. To investigate the origination and function of reality is the realm of science, not faith. Conversely, the provisional outcomes of scientific explorations cannot claim transcendent validity.[10] That is the realm of faith, not science.

It is not appropriate, therefore, to try and find some 'consonance' between modern science and biblical theology. Doing so one would not compare apples with apples.[11] There is a common denominator, of course, namely pre-scientific human experience of reality as such. There will be, without doubt, some similarities between the observations of the biblical writers and ours today. But that is where the 'consonance' ends. Today we have more powerful tools of observation and explanation.

In the first place, there is a *synchronic* difference. Science is about description, analysis, explanation and prediction of immanent reality; faith is about grounding, meaning, orientation, acceptability, belonging and authority.[12] Both science and faith are valid and indispensable human pursuits. But they cannot be conflated with each other.

This does not make them "non-overlapping *magisteria*" (Steven Gould). The vision of what reality ought to become is based on the experience of what reality ought *not* to have become.[13] One cannot divorce spirituality and ethics from scientific exploration without letting spirituality hover off into the thin air of irrelevance and leaving scientific exploration rootless and rudderless.

In the second place, there is a *diachronic* abyss. The biblical faith emerged and evolved in a phase of history that predates that of modern science by millennia. Modern science began to pick up momentum only half a millennium ago. The biblical authors could not possibly have known what the sciences have unearthed during the last five centuries. They had their own, limited, historically and contextually conditioned views of how reality was put together and how it functioned.

For the same two reasons I do not believe that we should attempt to establish a 'transversal' dialogue between disciplines that supposedly each have their own rationality, validity and dignity in the overall pursuit of knowledge at academic institutions.[14] If we followed this route we would be obliged to grant the claims of Satanism, Astrology or 'Ufology' academic status at such institutions. In terms of an honest interdisciplinary agenda, I believe, this is a dead-end street.

Faith should not claim more than it is entitled to claim. We do not have God in our pockets. We have no better explanations of reality than the scientists. Faith is not empirical certainty but trust. The trust of faith is based on the assurance that God, the ultimate Source and Destiny of reality, is for us and with us and not against us. It is not based on the validity of a pre-scientific worldview.

The only way science and faith can fruitfully relate to each other is to understand the diachronic differences between their sources, the synchronic urgency of their tasks and their respective mandates and functions in relation to the reality that humans experience. That of science is the description, explanation and prediction of immanent processes. That of faith is their grounding, orientation and motivation in terms of transcendent meaning, acceptability and authority.

The approach followed in this book

Faith and science relate to the same reality. This is the reality that we experience and in which we are embedded. But they relate to this reality in two fundamentally different ways.[15] Science explores reality from an immanent perspective, that is, from within. Its mandate and method are limited to entities and processes that, at least in principle, humans can observe, comprehend and manipulate. The ambition of science is to explain the past, master the present and predict the future.

Faith looks at the same reality from a transcendent perspective, that is, from a potential vantage point 'above' reality, as it were, 'with the eyes of God'. A vantage point 'beyond' immanent reality is, by definition, intuitive. But it is not, therefore, arbitrary. It has emerged and evolved over more than a millennium of ancient history in response to changing needs and predicaments. It intuits the ultimate source of reality and the ultimate destiny of reality—the very reality that is experienced in daily life and that is explored by the sciences.

Faith is aware of the fact that such an intuition will remain inadequate as long as we dwell on earth. Yet it lives within its presence. It looks at the past with gratitude and lament. It looks into the future with reassurance, concern and vision. Its life in the present is motivated by a mission. It asks where current developments are taking us, where we are supposed to be heading and how our attitude and behaviour will impact the unique and infinitely precious reality found on this planet.

Assuming that their procedures are appropriate in terms of the reality we experience, the approaches of science and faith are both legitimate and indispensable. But this is not necessarily the case in practice. Science can lose its orientation; faith can get stuck in obsolescence. Science can serve destructive pursuits; faith can hover off into speculation or get trapped by superstition. Science can become self-satisfied in its laboratories and gadgets; faith can withdraw into the cocoon of individual spirituality.

In this book, I try to observe reality from both vantage points and see how they intersect. I look at the same reality, that is, the reality we experience, first from the perspective of science, then from the perspective of faith. Faith can remind science that its method and mandate do not cover the 'sacred canopy' that provides humans with meaning, acceptability and authority. Science can help faith to keep its feet on the ground.

My personal position

Authors are expected these days to position themselves in their spiritual and social contexts. If this section bores you, skip it. Though born and bred in Africa, I am a Westerner. I am a believing Christian and a professional theologian. I am saddened by the fact that increasing numbers of our contemporaries have abandoned any hope of making sense of this faith. What a difference it could make to their personal lives, their communities, society and nature!

I also belong to the academic world. I studied natural sciences before I studied theology. I did most of my research in the social sciences.[16] Throughout my career I have tried to relate the Christian faith to a broad interdisciplinary spectrum. Christians need a conceptualisation of the 'truth' that they can live with without deceiving themselves and others.

My deliberations have taken account of the science-religion debate as far as I am able to overlook and fathom it. But I will not endeavour to duplicate, do justice to, or improve on what my colleagues have done with much greater expertise and dedication. References to the literature are meant to give profile to my own approach in relation to others. Four fundamental theses that will guide my deliberations in this volume:

1.   To rediscover the complementarity of faith and science is not a pious wish or a dispensable luxury but a necessity—not only for individual human existence, but for community, society, humanity and all life on earth.
2.   To be responsible, science does not need just any faith but 'best faith'. Best faith is a commitment to a vision of comprehensive optimal well-being for the whole of reality. This vision must be anchored in ultimate validity, yet responsive to reality as it unfolds. It must be alert to any deficiency in well-being in any dimension of life.
3.   To be relevant, faith does not need just any science but 'best science'. Best science is the most reliable information, explanation and prediction available at any point in time. Scientific insight is in flux. To remain credible, faith must remain critically and constructively abreast with science.
4.   The Christian faith has the *potential* to become 'best faith', but only if it manages to get its act together. It must be motivated by the inherent creative and redemptive thrust of the 'Word of God'. It must be geared to the biblical vision of comprehensive optimal well-being. It must constitute a dynamic response to changing human needs. It must allow itself to be reconceptualised in line with current scientific insight.

Why focus on the Christian faith?

This study is not about interfaith encounters, but about the relation between science as explanation and faith as self-entrustment. Faith emerges, evolves and differentiates in human history. It travels through time and space in the form of traditions. Convictions differ widely in terms of profundity, comprehensiveness, appropriateness and consequence. Convictions defy generalisation.

Because notions of the transcendent are diverse, one cannot deal with faith in the abstract, as if the content of this faith did not matter. For reasons of competence, clarity and brevity, therefore, I will focus on the *Christian* faith on the one hand, and its modern counterpart, *naturalism,* on the other. Both frames of reference are prototypical in terms of the issue to be addressed in this study.

(a) There is an undeniable historical link between the Christian faith on the one hand and the scientific approach that spawned naturalism on the other. Modern science "happened only once", namely in a particular period of the history of Western Europe, from where it spread throughout the world.[17]

(b) Modernity has led to pervasive secularism on the one hand and the typically modern backlash of fundamentalist Christianity on the other. The latter has mustered considerable social, political and financial power, especially in the United States. Therefore it has become the major target of militant atheism among scientists.[18] In secularised Europe, Christianity is largely ignored. In Africa, Asia and Latin America there may be a religious groundswell, but modernity has become the dominant culture among academic and economic elites worldwide.

(c) Obviously the Christian faith is the faith that I know and share. I feel relatively confident to speak on behalf of this faith, both critically and constructively. To conceptualise an appropriate response of a given tradition to constantly changing situations is an existential and intellectual struggle that can be done only from within that particular conviction.

(d) As a Christian theologian, I feel a particular burden to subject the conventional versions of this faith to critique, to restore its plausibility in terms of best science and to liberate its inherent creative and redemptive thrust in the face of the grave dangers we are faced with today. It is the largest and potentially most influential faith community worldwide.

This does not mean that I want to exclude others from our common responsibility. In my career, I have engaged major socio-economic ideologies, ethnocentric nationalism and African religion, which were the most relevant convictions in my social context. I trust that representatives of other faiths can and will speak for themselves. I also trust that interfaith dialogue will increasingly focus on the issues raised by this study.

The failure of the Christian faith

So far my own position. In view of the long-term concerns of humanity, my personal faith is rather irrelevant. But seen in the context of global developments, the eclipse of the Christian faith in modern society is a tragedy. This faith stands, after all, for responsibility, justice, and sacrificial love. It is inherently dynamic, flexible, and responsive. In theory, the Christian faith could have moved with the times. Its biblical original is inherently 'translatable'.

It emerged and evolved over a millennium of ancient Israelite history. It adjusted to ever changing conditions, worldviews, and challenges in biblical times. It crossed the boundary between Palestinian Judaism and the Hellenistic world of the Roman Empire. Its classical doctrines represented contextual reconceptualisations. It followed the broad philosophical movements of the Middle Ages. It could have engaged the various facets of modernity and accompanied science on its adventurous journey. Ever since the Enlightenment, sensitive thinkers and bold activists have tried to do so to the best of their ability.

By and large, however, the original dynamic and flexibility of the Christian message was arrested by institutional entrenchments and doctrinal rigidities. Theology bought into the timelessness of Platonic abstractions. The church adopted Roman law and emulated the feudal system. Christianity proved to be no match for the modern thrust towards emancipation. It remained behind at the air port while the jumbo jet of modernity roared into distant skies. The progress of science, technology, commerce, and consumerism has pulled out the rug underneath its feet.

All this cannot leave a believer unconcerned. Let me put it in crude and cruel terms: If we believers were not engulfed in the enormous Body of Christian traditions that we have inherited from a pre-scientific age, would there be any compelling reason for us to buy into biblical assumptions and doctrinal assertions, situated as we are in the scientific, technological, commercial, and consumerist civilisation of the twenty-first century? Why not simply dump this entire baggage?

Only after we have found a valid answer to this question can we hope to regain our self-confidence and our spiritual authority. This book is an attempt to do precisely that.

Overview

Part I sketches the situation we face in a world dominated by modernity today. Chapter 2 indicates how we got where we are. Chapter 3 indicates the consequences for science and faith. Chapter 4 highlights economics as a particularly revealing case in point.

Then we enter into the sphere of science-faith relationships. Part II looks at reality from a scientific perspective. This includes the Christian faith as an immanent phenomenon. Part IV looks at reality from a faith perspective, and this includes science as a phenomenon. Part II asks, what precisely is the nature of faith in terms of science? Part IV asks, what precisely is the legitimacy of science in terms of faith?

Parts II and IV follow a similar structure. They begin with a reflection on methodology. I differentiate between the approach of *experiential realism* that I deem appropriate for science (Chapter 5) and the approach of *integrative transcendence* that I deem appropriate for faith (Chapter 6).

Then follows an overview of the basic insights of science concerning spiritual phenomena (Chapter 7) and a classical stance of faith over against immanent pursuits such as science (Chapter 11).

Finally, I ask what science can learn from the immanent nature and function of faith (Chapter 7) and what faith can learn from science to update, correct and enrich its worldview (Chapter 12).

Wedged between the perspective from within (Part II) and the perspective from beyond (Part IV), lies a reflection on the feasibility and legitimacy of a perspective from beyond (Part III). Here the affirmation of transcendence by faith (Chapter 8) is pitted against the denial of transcendence by naturalism (Chapter 9).

The conclusion reiterates the basic contention that science and faith are indispensable and complementary pursuits that have to cooperate in trying to change the direction of the modern civilisation towards a creeping economic-ecological catastrophe.

Part I

Modernity —
The overall context of the
problem

Origins and consequences of modernity

The dynamic unleashed by modernity was propelled by the quest for emancipation from authority, including all authoritarian traditions, assumptions and institutions. Modernity aims at unconstrained mastery, ownership and benefit. The conquest of nature, the conquest of foreign lands and the conquest of human beings are all aspects of this drive. Its typical manifestations are science, technology, commerce and consumerism. Modernity has become dominant because of achievements unheard-of in previous stages of human history. Fettered by an obsolete and counterproductive mindset, faith has failed to keep pace with the accelerating modern dynamic. The acquisition of unprecedented powers through science and technology without a concomitant growth of the sense of responsibility has led modernity in exceptionally dangerous directions. The satisfaction of short-lived and trivial human self-interests and desires, regardless of the consequences for nature, society and future generations, has become the dominant motivation. Escalating social discrepancies and ecological devastations are the result. Faith and science have a common responsibility to lead the process in a new direction.

The task of this chapter

This chapter deals with the greater context in which science and faith are embedded. At least in the West, this is the modern civilisation. Modernity has come to dominate the global scene in one form or another, but it emerged and evolved in the West. Today modernity leads humanity in a destructive and suicidal direction.

Christianity and modern science have been powerful spiritual and social factors in the formation of modernity. They are inextricably woven into its fabric. They share in the responsibility to change current economic and ecological trends before it is too late. This is where the gravity and urgency of the relation between science and faith are located.

In what follows, I may seem to indulge in vast, biased and unfair generalisations. People do not like doomsday scenarios. They have a stake in what they have achieved. They resent spoilers. Affluent population groups have no intention of forfeiting their privileges. Marginalised people are yearning for modern living standards. What I have to say in this chapter cannot be popular.

The reader should know that I am an admirer and beneficiary of modernity. Having spent many years of my life in pre-modern contexts, I have been cured of all romanticism in this regard. I have no intention of doing without modern achievements. Nor do I want to denigrate the momentous insights, excellent initiatives and substantive shifts in attitude that characterise modern social reality.

Obviously, positive developments should be supported, rather than discredited. We have no choice but to build the future on these achievements.

But appreciation does not exclude alertness and critique. A diagnosis that highlights what is sound and strong does not lead to a cure. Humanity is moving in dangerous directions. To smother a valid argument under the weight of a thousand qualifications is a luxury we cannot afford. If compared with what professional ecologists depict and demand, my sketch of potential calamities is almost trivial.[19]

In this chapter, I offer a brief overview of the main facets of modernity, its historical evolution and its problematic consequences. My aim is to demonstrate the urgency of a new synergy between best science and best faith.

Section I
Modernity—The Driving Force of Our Times

Current facets of modernity

Modernity, as I use the term, has historical roots that reach back as far as the European Middle Ages, the Renaissance, early capitalism, humanism and the Reformation.[20] It crossed a watershed during the so-called Enlightenment. Its articulation and institutionalisation sucked one dimension of life after the other into its vortex. Today it is accelerating at a breathtaking pace. It has engulfed the global elite and reaches into every rural village. It affects not just humanity, but the natural world.

Modernity has become dominant because it was able to deliver the goods— superior knowledge, technological prowess, enhanced productivity, material need satisfaction, unheard-of wealth, military might and political power. Though still a minority culture in global terms, it renders traditional mindsets of all kinds obsolete, stagnant and irrelevant.[21] The four constitutive manifestations of modernity can be characterised as follows:

The motivation of **science** is to gain insight into how reality operates. Its leading criteria are *evidence, plausibility and predictability*. It looks at unique entities or occurrences in space and time, finds regularities in their nature and operation, determines possible relationships between them, and uses them to make predictions.

The motive of **technology** is to extract utility from reality to serve human purposes. Its leading criteria are *efficiency* and *performance*. It dismantles reality into components, recombines some of them into artefacts that are useful for human beings and discard the rest.

The motives of **commerce** (the liberal economy) are income, prosperity, financial power and status. Its criterion is *profitability*. Its method is to explore consumer preferences, create market demand and organise production and distribution accordingly.

The motive of **consumerism** is to gain as much satisfaction from reality as possible. Its criteria are *utility* and *pleasure*. In the modern economy, consumer demand exceeds the prerequisites for healthy survival by far.[22] Its strategy is to exploit any phenomenon—psychological, physical, communal, societal, biological or material—for status, wealth and sensual satisfaction.

Pursuing reliable insight, **science** eliminates superstition, untested assumptions, ideas without substance and metaphors without demonstrable referents. In its pursuit of performance, **technology** replaces inefficient processes with more productive ones. In its pursuit of profitability, **commerce** outcompetes economically 'irrational' behaviour. In its pursuit of satisfaction, **consumerism** removes obstacles and inhibitions in the way of the fulfilment of needs and cravings.

Science concentrates not on the observing subject, but on the observed object; technology not on the fabricating subject, but on the fabricated object; commerce not on intrinsic value, but on price in the marketplace; consumerism not on obligations and norms, but on personal satisfaction.

Intimate feedback loops operate between the four aspects. Science uses technology to enhance empirical observation. Technology utilises scientific insight to enhance efficiency. Industry uses technology to produce goods in response to market demand. Market demand depends on needs and purchasing power. Purchasing power depends on income, income on productivity, productivity on technology, technology on science. Without consumer demand, commerce would collapse, technology would lose its rationale and science would lose its funding.

Advanced industrial economies depend on the consumer culture for their survival, growth and prosperity. Enterprises have to out-perform their competitors or lose their market share. To increase profits, commerce creates 'needs' and dismantles inhibitions. Marketing has become a highly sophisticated multi-million dollar industry. Every aspect of reality—health, beauty, sex, religion, art, sport, pets, wild animals, forests, beaches, scenery—is being commercialised.

The accumulation of wealth is no longer the central motive. The modern economy is based on 'throughput' of material resources from resource base via extraction, processing, distribution and consumption to waste and the flow of financial resources in the opposite direction. The stronger this flow, the more affluent the society. Productive power is channelled not towards the greatest needs, but towards the greatest purchasing power.

In the end, it is the affluent consumer whose needs and wants keep the economy going. Therefore needs and wants are deliberately created and enhanced among economic elites and the so-called 'middle class'. The 'demonstration effect' of high living standards filters down to the less privileged. To tap this market, easy credit lures consumers into purchases they cannot afford.

That does not mean to say that there are no scientists whose motive is insight per se, no technical experts whose motive is efficient energy utilisation, no commercial ventures that satisfy genuine needs, no consumers that depend on particular products for their healthy survival. All I want to say is that there is a dominant cultural dynamic, motivated by self-interest, that drives the modern economy and that every economically significant pursuit is part of this overall process.

The emergence and evolution of modernity

How did the modern civilisation come about? The quest for mastery was ingrained in the human psyche through evolution in times when defenceless humans were the victims of predators, diseases and natural disasters.[23] "Fill the earth and subdue it!" (Gen. 1:28) was once an injunction on which human survival depended. But the destructive effects of ambition were kept within limits by social conventions and religious constraints.

Modernity is an emancipatory movement. It was sparked off by the dysfunctions of traditional authority. Medieval Europe had a feudal-patriarchal structure underpinned by Christianity. The church was an oppressive feudal institution sustained by a totalitarian ideology. The Christian faith had lost its liberating, transforming and empowering potential. The Enlightenment did not envisage liberation by Christianity, but liberation from Christianity—and from all other oppressive forces.

During the Middle Ages, philosophy had changed from Platonism (Augustine) to Aristotelianism (Thomas of Aquinas) and on to Nominalism (William of Occam). The 'Renaissance' rediscovered the excellence and cheerfulness of ancient Greek and Roman cultures. 'Humanists' among the Reformers, such as Erasmus of Rotterdam and Philip Melanchthon, enthusiastically embraced the treasures of antiquity.

The Reformation undermined Catholic claims to divine truth and institutional legitimacy. Competition between Protestant and Catholic orthodoxies led to bloody religious wars and questioned the validity and credibility of both. The Thirty Years' War (1618-1648) devastated large parts of central Europe and led to large-scale disillusionment with organised and dogmatised religion.

The sun-centred cosmology of Copernicus displaced the earth-centred cosmology of Ptolemaeus. Journeys of discovery and global trade opened up European horizons and led to confrontations with alternative cultures. Encounters with Islam and Eastern religions problematised the Christian claim to divine revelation. Leading thinkers discovered the historical, situational, cultural and epistemological relativity of human insight.

The spiritual revolution

All this caused a spiritual earthquake. Entrenched institutions, accepted doctrines and inherited philosophies began to lose their plausibility and legitimacy. Bloodied by the mutual slaughter of faith communities, encouraged by thinkers to build on reason rather than belief, lured on by the possibilities of scientific research, technological advancement and capitalist production, people became suspicious of inherited claims to validity and authority.[24]

The result was a pervasive loss of emotional stability. Nothing seemed to be certain any more. Life seemed to be floating in a sea of deception and relativity. Because objective certainties disintegrated, one had no choice but to fall back on one's subjective experience. Think of an astronaut detached from the space craft and drifting aimlessly and helplessly into outer space. In such a situation, nothing is left to cling to but the experience of your own self.

Descartes, the father of modern philosophy, is a pivotal example of this mood. He set out to find ultimate truth through systematic doubt. He argued, first, that because he doubted, he had to exist (*cogito ergo sum*). So his personal existence seemed to be certain. This step is a reflection of the modern self-conscious subject that experiences itself in juxtaposition to the outside world.

Second, the outside world had spatial dimensions that could be subjected to geometrical measurements and calculations (*res extensa*). This suggested that the outside world must exist. Descartes was a brilliant mathematician and mathematics began to play an increasingly dominant role in the exploration of physical reality.

Third, his intuition of the existence of a perfect being excluded the possibility that this God could have fooled him into the two previous intuitions. If God existed, they both had to be valid. For Descartes, God's existence provided the bracket between the observing subject and the observed object.

It did not last long before this bridge was deemed superfluous, leaving the experiencing subject and the experienced object juxtaposed to each other. The subject now could be deemed the owner and master of an object that could be explored, dismantled, reconstructed and utilised at will. This was not Descartes intention, who was a devout believer, but he was widely blamed for laying the foundation for the modern subject-object split.

Preoccupation with one's personal survival, identity, self-assertion and well-being is programmed into the brain of the human being, in fact into all living species, by the evolutionary process. But in traditional cultures, it is embedded in, and constrained by, social structures that safeguarded the survival and prosperity of the community. Modernity provided a legitimate reason to break out of this embrace. If dormant human potentials were activated, they would lead to spectacular results, not only for the individual, but for society as a whole. The modern legitimation of the pursuit of individual self-interest was born.[25]

Fettered by irrational and authoritarian assumptions, Christianity gradually lost its plausibility, its authority and its appeal in the wider population. Modernity became the rising star. It was a rebellion against the authority of the institutional church, Christian doctrine, the Bible, the state, inherited social conventions, moral norms, classical philosophy and faith in God—all in the name of human autonomy, scientific evidence, technological prowess, commercial profitability and consumer satisfaction.[26] The respective battle cries of modernity can be summarised as follows:

- Think for yourself (rationalism)
- See for yourself (empiricism)
- Find out what works (pragmatism)
- Pursue your interests (liberal economy)
- Enjoy your life (hedonism)
- Relate to your personal saviour (pietism)
- Assert your personal dignity (human rights)
- Have a say in your government (democracy)
- Claim gender equality (female emancipation)
- Let the youth find its way (anti-authoritarian education)

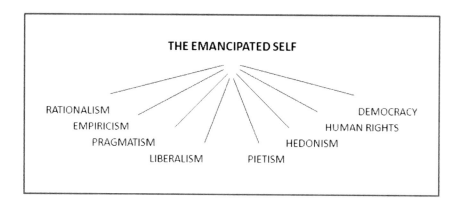

The transition occurred in phases

In modernity, the sovereignty of God made way for the sovereignty of the human being. At the intellectual level, the process began when Nominalism removed God from the Platonic world of ideas and the Aristotelian world of forms that had governed the nitty-gritty of reality, and projected God to the lofty status of an absolute will with absolute power.[27] Now the world became nothing but dumb material. At the end of the process, the human being had emasculated the concept of God and appropriated the claim to absolute power over reality for itself.

The transition from divine omnipotence to human sovereignty was gradual. The thoughts of Descartes, Galileo and Newton were still driven by religious vigour.[28] Kant believed that the concept of God was without epistemological foundation, but indispensable for the ethical dimension of life. However, Kant's ethics was based on reason rather than piety, and soon it was argued that a humanity come of age needed no divine lawgiver, supervisor, or judge.

In response to the emerging dominance of a causally closed worldview, Deism tried to retain the concept of God as prime initiator of the universe, but believed that, once triggered, the world process was left to continue on its own. As Newton had shown, science was quite capable of explaining how it functioned. When the 'God hypothesis' fulfilled no purpose any more, whether in terms of explanation or moral authority, it was first ignored and then consciously abandoned.

The human appropriation of sovereignty over reality and the systematic transformation of reality to achieve human ends replaced God's creative authority and redemptive project as envisaged by the Christian faith.[29] Human self-interest replaced the will of God, however construed, as the "measure of all things".[30] When the excellence of humanity is deemed the culmination of the evolutionary process, it seemed to follow that humanity is entitled to the role of master and owner of reality in all its forms.

Old authorities did not yield without a fight and conflict was the result. When the church continued to exert social influence, oppose scientific insights, legitimate authoritarian rule, condone capitalist exploitation and use spiritual repression to remain in control, indifference changed into hostility across a broad front, not only among philosophers and social activists but also and especially among scientists.[31]

Seen in these wider contexts, atheism was not invented by great thinkers such as Comte, Feuerbach, Marx, Darwin, Nietzsche, Freud, Russel, Heidegger or Dawkins for that matter.[32] These thinkers were merely catalysts that helped to articulate an emerging mood in the modernising population, especially among scientific, technological and commercial elites.[33] Nor is it restricted to the relation between the natural sciences and the humanities at universities. It is a groundswell.

In secular societies, the conflict between Christianity and modernity seems to be over.[34] Religion is no longer attacked by the majority, but simply ignored. The new hankering for spirituality is the psychological dimension of the typically modern quest for gratification. Religion has retreated into the private sphere. With this retreat, it sealed its irrelevance.[35] The aggressive atheism of the last two centuries has made way for a yawn. It is pointless to fight a phantom.

For some people, religion may still provide moral yardsticks, but if morality is based on something non-existent, it does not convince.[36] Seen in this light, the fallout between modern atheists on the one hand and American creationists, Catholics and Islamic fundamentalists on the other may seem antiquated.[37] However, whenever religious intransigence gets in the way of scientific, technological, commercial and hedonistic rationality, it evokes aggressive responses from the latter.

Social, economic and political consequences

Under the impact of exciting new perspectives and initiatives, traditional perceptions and approaches began to crumble across the board. The accumulating achievements of modernity reinforced each other and led to a landslide. Nothing is more successful than success. Once triggered, the process fed on itself. All social processes began to accelerate. Awareness of the potentials of an emancipated humanity engulfed ever more dimensions of life.

In social-structural terms, the demand for equal dignity, freedom and human rights moved from aristocrats to merchants, serfs, slaves, workers, the colonised, people of alternative convictions, the women, the youth, the marginalised and—most recently and yet to become serious—plants, animals and the wider natural environment. Some aspects of life came into fruition earlier than others. Commercial liberalism predated political democracy. Democracy did not immediately lead to female emancipation. The adoption of the liberal-capitalist system was not immediately matched with the democratisation of economic, social and political processes.

In politics, the rising power of the merchant class challenged the feudal system. When the latter disintegrated, princely absolutism tried to hold the fort. But a growing revolutionary fervour dismantled the legitimacy and power of such entrenchments. When they were toppled, dictatorships tended to fill the vacuum. In the end, a democratic social order, based on the rule of law and individual human rights, established itself in virtually all modern societies. It was a major achievement of the emancipatory thrust of modernity.

In economics, the emancipation of individual initiatives from feudal, mercantilist and absolutist constraints did not lead to the kind of equity found in the political realm. Where differential talents and endowments are let loose in a competitive game, a new and dynamic hierarchy of high achievers, average performers and drop-outs has to evolve. Modernising sections of the population quickly outperformed and outmanoeuvred more traditionalist sections. The 'liberal' economy led to vast and growing discrepancies in terms of access to resources, political influence, connections, communications, education, expertise, income, wealth and life chances.

New technologies, such as the steam engine and the 'flying shuttle', led a growing strength of capital as factor of production and the concomitant decline of labour. A growing industrial worker class replaced old professional guilds and peasant communities. A new class structure emerged between the super-rich, the middle class, the workers, the marginalised and the redundant. Incompatible socio-economic ideologies underpinned protracted class struggles. During the 'cold war' these conflicts all but led to a nuclear devastation of the earth as a whole.

Section II
The modern conquest of reality

Control of nature

The new self-image of humanity as owner and master of reality led to new self-confidence and self-assertion. It was directed primarily against nature. Why should the (human) master and owner of nature accept nature in its crude, cruel and undomesticated form? Nature was available to be explored, subdued, exploited, transformed and utilised.

Science provided the insights, technology provided the means and commerce provided the motivation to do precisely that. Science uncovers the secrets of nature; technology transforms nature; commerce turns nature into tradable goods and services, which are consumed and turned into waste. Nature is analysed and taken apart. Useful components are reassembled into marketable artefacts; the rest is discarded.

The destructive impact of human settlement and exploitation on the natural habitat did not begin with modernity or the Industrial Revolution. Ecological deterioration was a constant feature of the human presence since the emergence of sedentary agriculture. Deforestation, degradation of agricultural land and soil erosion followed wherever humans settled in greater numbers. In the long run, entire civilisations were destroyed through the overexploitation of nature.

In pre-modern times, however, the mastery of humans over nature was constrained by a mythology or a system of meaning that insisted on human accountability to a higher authority. According to Genesis 1:28ff, humans were not entitled to mastery, but entrusted with mastery. That changed when modernity jettisoned the assumption of divine authority. Now humans were deemed autonomous masters and owners of reality. Today this assumption is no longer asserted or contested, but taken for granted.

Modernity demolished the traditional awe of nature. Countless creatures and species disappeared under the impact. However, it is the transformation of nature as a whole that matters most. Earthly reality increasingly turns into a world of human constructs—the modern city, the factory, mechanised and computerised agriculture, the synthetic home, electronic means of communication, the financial system, the airways, highways and railway lines traversing a domesticated natural world. "Our children can reproduce dozens of tinny tunes from advertisements, and yet they cannot distinguish between the songs of the meadowlark and the mockingbird."[38]

In fact, these children encounter animals only in the form of animated cartoons and junk food.

In the nineteenth century, the dawning realisation of what humans are doing to nature set off a romantic backlash. The battle cry of the Reformation, 'back to the ', turned into the battle cry, 'back to nature'. The movement articulated a growing awareness that the inexorable pursuit of wealth through industrialisation and commercialisation destroyed both the natural world and the humanity of the human being. It protested against rampant materialism, social inequities and senseless consumerism.[39]

But it could not arrest the momentum. To a substantial extent, the motive underlying romanticism continued to be personal enjoyment—whether in the form of status and power, the beauty of human bodies, ancient ruins, snowy mountains, deep forests, or fascination with the animal world. The adoration of nature led to nature reserves, but they were meant to be utilised by humans for recreation, wild life experiences and—again—profit.[40] Even in this case, nature became a quarry to be mined for the sake of human indulgence.

Towards the end of the nineteenth century, belief in progress towards mastery and utility had become the dominant intellectual and practical motivation in the West. An enormous optimism began to permeate society. The golden age seemed to be dawning. Science would soon unravel the last secrets of reality. Technology would subdue and utilise nature. Modern weaponry provided the means to extend one's power or influence. Economic growth would overcome poverty and want. Socialist revolutions would lead to equality. A sense of opportunity, calling and mission drove countless people—secular and religious—across the seas. At the same time the most advanced economies got ensnared in a deadly competitive conflict that exploded into World War I.

The subjugation of humans

The will to dominate did not stop with nature. Other lands, cultures and ethnic groups befell the same fate. Of course, feudal and imperial tendencies did not begin with modernity, but with the change from nomadic existence to an agriculturally based sedentary lifestyle some six thousand years ago. As in antiquity, modern imperialism was motivated by the quest for social status, economic wealth and political power and legitimated by means of religion and ideology.

Elites have always assumed higher statuses than those of their subordinates. The point is that Christianity was not able to prevent imperialist conquest, economic exploitation and their religious legitimations. While many societies, from the Assyrians to the Romans, to the Muslims, to the Zulus, to the Incas, have embarked on campaigns of conquest and oppression, the European imperial expansion was carried out by the same group of 'Christian' countries, simultaneously with, and aided by, modern science, technology, commerce and lifestyle.[41]

European nations bled each other out in their quest for domination of the rest of the world. Previously autonomous regions were 'opened up' for European emigrants, or turned into resource bases for economic pursuits. Slavery reduced human beings to living machines. Where indigenous populations got into the way, they were subdued, limited to reserves, removed, decimated, or eradicated.

Inside Europe, notably in Britain, the policy of 'enclosures' removed peasants from the land and the Industrial Revolution condemned factory workers to appalling working and living conditions.[42] Humans had become a mere factor of production in competition with machines. In the long run they were unable to compete with the latter.

A century after Columbus, American Indians had been reduced to 10 per cent of their original number. After the war of independence from Britain, their flourishing agricultural economy was deliberately destroyed.[43] By the same time, some 275,000 African slaves had been sent to the "New World" and Europe. During the seventeenth century, that figure rose five-fold to about 1,300,000. It reached 6 million in the eighteenth century.[44]

In South Africa, reserves for black Africans were scheduled to encompass less than 14 per cent of the land and even this quota was never achieved. In the meantime, the imposition of poll tax, the inability of the growing black population to feed itself on deteriorating land and the lure of manufactured goods lured them into the white-controlled economy as cheap labour.[45]

It has become customary to see Christian mission and Western imperialism as two sides of the same coin. There is some considerable truth in these contentions. But it was not only religion that served to legitimate imperialism. One's own culture, administration, political structure, technology, industry and commerce were deemed superior to those of others—and it must be conceded that they were just that at the time! So Europeans deemed themselves not only entitled but called upon to bring the 'light of civilisation' to all of humankind. And this civilisation included Western religion.

It must also be acknowledged that the achievements and powers of modernity enthralled the colonised with their unprecedented possibilities. Dominant elites always become the reference groups for the dominated. Christian schools were instrumental for gaining access to modernity. Soon modernised elites replaced traditional elites in the colonies. In time, they defeated the colonial powers with their own ideological and moral weapons. In both cases, the motives were guided by individual and collective self-interest.

The spike of the twentieth century

These socio-economic developments were not necessarily smooth and linear. The optimism generated by modernity was badly battered by the senselessness, destructiveness and depravity exhibited in World War I. Gloom and apprehension about the direction of Western civilisation spread. But a conservative backlash idealised inherited cultural traits and the excellence of the particular ethnic group or nation.

Immense emotional upheavals led to bitter ideological struggles. Pent-up resentments were swept up by charismatic leaders. National pride, hero worship and pathological power crazes led to the disaster of World War II with its unimaginable loss of life, resources, hardware, infrastructure, buildings, expertise, cultural achievements, moral values and integrity.

The awakening that followed was short-lived. Apart from post-war reconstruction and 'economic miracles', energies were largely channelled into the struggle between competing ideologies and power blocks—capitalist liberalism versus Marxist-Leninist socialism. The so-called 'Third World' became embroiled in the crossfire. Which system would bring greater prosperity? Which would bring greater equity? In fact, however, the dominant motivation among both contenders was to gain political superiority, economic wealth, technological excellence and military power. The competitive game went to the extent of creating weapons of mass destruction that could have turned the earth into a desert a couple of times over.

During the 'cold war' the jury was still out on the merits of the two systems. But the collapse of the Soviet Union put paid to all this. The capitalist option, in Europe somewhat mitigated by various forms of social democracy, had established itself as the only system that could deliver the goods. One could even speak of "the end of history".[46] The fact that Marxism-Leninism was an attempt, however ill-conceived and counter-productive, to overcome the serious failures of capitalism in the first place, was conveniently overlooked.

All along economic growth powered ahead. It experienced a number of bumps, but accelerated nonetheless. In recent times, the profit, utility and pleasure orientation of economic liberalism picked up whatever was deemed exploitable and marketable all over the globe, including all aspects of the human being, and sucked it into its powerful vortex. The small centre-periphery structures of local communities and regional economies are making way for global pyramids whose discrepancies between base and top acquire grotesque proportions.

Religious legitimations

I have mentioned the need for legitimation above. Religion regularly served to legitimate the subjugation of nature by commerce and industry and the subjugation of humans by imperialism. However, it is important to discern the difference between the core convictions of a religion and its use as a legitimating ideology. There are examples for the ideological abuse of religion the world over, including Ancient Near Eastern empires in Egypt, Mesopotamia, Palestine and the Mediterranean. Ancient Israel was no exception.

The kingdom of David and Solomon was a rather modest empire, but it used the Israelite faith to underpin its global aspirations. As Psalm 2, Mica 7: 8-12 and Isaiah 61:5f demonstrate, ancient Israel appropriated classical imperial assumptions found in Egypt and Mesopotamia at the time: the Israelite king was deemed installed by God himself as universal ruler. He was to be surrounded by his own ethnic group as the world elite. He was mandated to subject all nations to his authority and demand tributes from them.

As part of the biblical canon, the model proved to be a convenient legitimation of Christian emperors and popes in medieval Europe. Both claimed to be representatives of Christ, the cosmic king enthroned in heaven. A bitter conflict ensued. While emperors were followed by absolutist rulers who disappeared through the impact of modern assumptions, the ecclesial version never gave up its universal claims and was marginalised as a result.

This is rather ironic, because New Testament texts such as Mark 10: 35-45 and John 13: 1-17 turned the model of Psalm 2 on its head, called it pagan and replaced it with a model of leadership through humble service. It is here were the clash between genuine faith and legitimating ideology presents itself in the starkest terms.

While power, wealth and status always constituted the prime motivation of imperial aspirations, secularisation gradually shifted the legitimating authority from a transcendent God to inner-worldly objectives or 'necessities'. Today imperialism is based less on military and administrative subjugation than on economic clout. It is also largely underpinned by economic ideology, rather than religious commitment.

The Spanish and Portuguese empires still legitimated their conquests with distorted Christian assumptions. The legitimation of the more enlightened British Empire was a mixture of religious, cultural and economic concerns. Marxism and Nazism, though secular, still featured a kind of cosmic, historical, or mythical authority. Hitler legitimated his claim to power with the assumption that 'fate' had ordained him to bring the Germanic master race to world dominance.

The Marxist-Leninist Empire derived its legitimation from the 'historical dialectic' that produced the 'messianic' class of the proletariat. It was to be led by a bourgeois avant-garde that would bring about socio-economic redemption. Marxism did not abandon the claim that humanity was the legitimate master and owner of reality; it just wanted the working class rather than the middle class to be the beneficiary. Once that proved unworkable—precisely because of unredeemed selfishness—most of the 'avant-garde' greedily joined the capitalists.

Many Americans still believe that they are called upon to impose their liberal-democratic values and institutions on the rest of the world. But that is hardly the driving force of American social dynamics. The true motivation of modernity is the pursuit of individual and collective self-interest.

The ideology of liberal capitalism provides the legitimation of an evolving global culture. All transcendent pretensions have disappeared. It is wealth as such, power as such, the 'good life' as such that call the shots. While some humans have become the beneficiaries of science, technology and commerce, other humans and the natural world have become the victims.

Ecological consequences

Underpinned by the first oil crisis in the 1970s, comprehensive ecological studies (such as those of the 'Club of Rome') caused consternation and anxiety. They led to the formation of green parties and ecological movements such as World Watch and Greenpeace. In time, 'green' movements acquired more competence and clout. But such surges of public concern have proved to be rather fickle.

During the Reagan-Thatcher era, public denial, interest-related misinformation and scientific counterclaims have all but swept these early warning under the carpet. The collapse of the communist system and the vitality of capitalist development lulled public opinion to sleep.[47] What remained of economic and ecological consciousness was seriously mellowed down, rendered 'harmless' and amenable to the dominant economic trend.[48]

Influential economists debunked the validity of ecological studies. Instead of refining the methodology and overcoming the imperfections of such early initiatives, as could have been expected from any unbiased science, weaknesses were gleefully pilloried and disturbing scenarios were dismissed as hog wash. It was argued that the earth's resources were unlimited, that the ingenuity of science and technology was infinite and that the market would balance out whatever 'scarcities' may emerge in the future.[49]

Needless to say, these contentions were heavily value-laden, partially true at best and reckless at worst. Whether consciously or not, they legitimated the pursuit of collective self-interests by giant corporations, notably those engaged in the oil and automotive industries, at the expense of less powerful interests—the general public, nature and future generations. Decades have been lost.

In the meantime, the crisis is unfolding. Recent symptoms of climate change and ecological disasters have again agitated the general public, scientists, policy makers and corporations. A consensus begins to emerge among ecologists that humankind is undermining its chances of long-term survival, while diminishing the life chances of masses of other creatures.

But the lure of self-aggrandisement and enjoyment is so deeply entrenched that politicians do not dare to embark on radically new ways of ordering society. Allegations that the danger was exaggerated by incompetent experts and exploited by ruthless money spinners find an eager hearing.

That 'exponential' growth, that is, accelerating growth, poses massive dangers can no longer be disputed.[50] It has developed a global dynamic that cannot easily be stopped. In a finite world, a ceiling is bound to be reached, whether sooner or later, whether gradually or suddenly. Then the process is likely to 'overshoot' and a catastrophe can no longer be averted.[51] Its severity will depend on the magnitude of the overshoot. In view of the massive global structures and processes in place today, the overall effects can be devastating for millions of 'innocent' victims all over the world and for many generations to come.[52]

Humans will not be the only victims. The devastations to the rest of the natural world will be inestimable. According to the worst case scenario, the processes unleashed cannot be reversed, will begin to feed on themselves and make the earth uninhabitable. The crisis is unlikely to unfold in this crass way. It will rather 'creep on us' while we remain largely oblivious of what is happening. I have no doubt that the most privileged will be able to secure a reasonable degree of safety and prosperity for themselves while the least privileged will bear the brunt of the catastrophe. This is already happening on a vast and increasing scale.

After one and a half centuries, ecological movements have not been able to gain a following sufficiently large to make a difference to the general direction of the global economy. Ecological concerns are rare among the super-rich and the impoverished. They tend to arise mostly in middle class population groups that have satisfied their material needs and can afford to be concerned about the quality of life, or whose incomes do not depend on commerce and industry.[53] By implication their attitude would change drastically once real material sacrifices were demanded. Politicians in turn would not take the risk of becoming unpopular.

Section III
The emergence of a self-indulgent civilisation

It should be clear by now that the switch from human accountability to human autonomy had immense spiritual, ethical and social repercussions throughout the world. The quest for knowledge, power, wealth, prestige and pleasure is no longer deemed a manifestation of human selfishness and arrogance to be controlled by divine authority, human ancestry, culture, or an ethical tradition, but a natural phenomenon to be embraced and enhanced.

I do not claim that humans have ever been motivated by less selfish and more circumspect goals. The biblical faith has been exceptionally realistic about the corruptibility of the human being—and that two millennia before the onset of the modern era. What has changed is the assumption of unconstrained human sovereignty over reality, including the self, and the unprecedented power unleashed by science and technology to give clout to this claim.

Without doubt, the reappraisal of human nature was liberating and exhilarating. Humankind seemed to have reached adulthood and the fullness of life. But earlier times had linked the freedom associated with adulthood with responsibility. In modern times, responsibility for the greater whole—including the body, community, society and nature—has made way for individual desire and collective self-interest.

The radicalisation of emancipation

Modernity is based on emancipation, sovereignty, mastery, ownership, utility and pleasure. Recently more radical forms of emancipation began to develop. They can be summarised with the concept of 'popular postmodernity'. Kindly note that I will not deal with the sophisticated school of thought called 'postmodernism' in this book. My aim is to describe the 'average' person who shares the culture of advanced modernity.

Popular 'postmodernity' has abandoned even the criteria and constraints imposed on the modernising population by modernity itself: evidence, rationality, stringency, plausibility, initiative, self-exertion, achievement, precision, reliability, efficiency, productivity, profitability and utility. In industrialised countries, it can afford to dismiss these requirements because it is able to live off the wealth produced by the modern economy. Life has become comfortable enough to relax and enjoy. Social security eliminates risks. Would-be beneficiaries, who do not feel like making a contribution, can easily pose as victims of inequity. They deem themselves entitled to a share in the prosperity generated by the modern economy—as if prosperity rained from heaven and was there for the taking.

This development is particularly detrimental in cases where population groups move from the rigid and debilitating assumptions and institutions of traditionalist cultures straight into the 'postmodern' freedom of glittering cities. They believe that it is possible to bypass the uncomfortable demands of modernity-efficiency, productivity and profitability—and yet enjoy its enticing products. Where a society takes this entitlement for granted, it leads the state into bankruptcy and its citizens into dependency. Autonomy and self-sufficiency fall by the wayside. Most unfortunately this happens on a vast scale in poorer countries.

The point to be made at this juncture is that the postmodern mentality has shed conviction and commitment as such. Nationalist and Marxist kinds of fervour were still driven by a vision—the glory of the nation or the classless society. This vision translated into a mission. Countless people were willing to lose their freedom, their prosperity, even their lives, for what seemed to be a worthy cause. But within a few short decades the grand narratives of nationalism and Marxism had collapsed.

What does this mean? Not only the stories and doctrines of Christianity, but also secular frames of reference, whether liberal or radical, have lost their appeal. Today even the commitments, necessities and restraints of modernity are being jettisoned. Why should social concerns disturb our consciences! Why should Western culture be deemed superior! Why should the 'grand narratives' of the past impress us! Why the preoccupation with truth and validity! Why allow yourself to be bound by a sense of duty! Why strain yourself to attain anything at all apart from enjoyment here and now! I may exaggerate—but I do so for a purpose.

Children become a nuisance. They are placed before tantalising television and computer screens to get them out of the way. A teacher in Germany told me that thirty years ago her task was 20 per cent formation and 80 per cent education; now it is 80 per cent formation and 20 per cent education, because parents have lost control or simply do not bother. She described these children as a bunch of 'ego-monsters'. But they are still lucky enough to know their fathers and mothers and have not been dumped in the lap of their ageing grannies, as millions of children are in South Africa.

The elderly can no longer depend on their offspring. They are expected to fend for themselves, or be taken care of by institutions. Duty, obligation and commitment are no longer part of the vocabulary. Radical freedom and unrestricted enjoyment—that is the agenda of popular 'postmodernity'. Aspirations to attain spiritual excellence through ascetic lifestyles have lost their appeal. Monastic ideals have all but disappeared.

Indulgence is no longer a vice; frugality is no longer a virtue. Puritan self-discipline and modesty have made way for extravagance and the lavish display of status symbols. Platonic idealism has made way for hedonistic pragmatism. The body is no longer considered the seat of temptation, sin and evil, or the temple of the Holy Spirit, but the instrument to maximise pleasure and enjoyment.

Internalised moral inhibitions are considered prudish and hostile to natural impulses. Life must be given the freedom to flourish. Craving for adrenaline rushes, fast cars, dangerous stunts, sexual ecstasy, tempting foods and drinks, intoxication through alcohol and drugs are all taken to be self-evident and legitimate parts of the game.

Popular 'postmodernity' revels in plurality. All systems of meaning are equally valid and, by implication, equally spurious. Everybody is entitled to his/her opinion. The past is gone forever. The future cannot be predicted. The next generation must look after itself. Life is located in the present. Relativity and ambiguity are not deplored or regretted but celebrated. What makes anybody think that you can possess the truth, or that there is such a thing as the truth in the first place! It all happens, after all, in the human brain. Think positively and your life will be positively enjoyable! All points of view and patterns of behaviour are to be appreciated as contributing to the wealth of human spiritual and practical creativity.

One has to concede that systems of meaning have become as diverse as the plurality of plant and animal species. But in earlier times they were still deemed to be valid. Now one finds one's own identity in the encounter with the 'stranger' and the 'other'. While this certainly opens up ingroup-outgroup barriers, enriches all partners and leads to greater tolerance, it can also mean that all convictions and inhibitions are disparaged and abandoned.

Biblical prohibitions saying, "Don't do it!" have made way for the 'postmodern' slogan "Just do it!" Reminders of how problematic some assumptions and patterns of behaviour can become are readily dismissed. Moral appeals are deemed "prescriptive". A critique of untenable positions is deemed "judgmental". Pointing out the devastating consequences of the current sexual chaos is called "prudish". The quest for truth is deemed bigotry and arrogance.

Retrospection, reminiscence, anticipation and foresight begin to disappear. The quest for highs to be enjoyed here and now crowds out the assessment of the past and the options of the future. Spirituality may again be in demand in popular 'postmodernity', but only in the sense that one can freely shop around in the religious marketplace, try this and that kind of spirituality and discard it again without scruples or loss of integrity.

The general mood is powerfully enhanced by technological advancements. All social processes accelerate.[54] Situations change so rapidly that commitment appears to be foolish. One can no longer stick to a profession for life. Whole industries disappear. Careers are suddenly disrupted. Qualifications once acquired become obsolete.

The general atmosphere of flux and relativity permeates all dimensions of life and underscores the general mood of unconstrained freedom. One hops from web site to web site, from job to job, from home to home, from town to town, even from country to country. One enters into multiple or provisional sexual partnerships and kicks up no fuss when they break up. Promiscuity enriches life.

Mass media swamp the human consciousness with processed information and smartly packaged marketing. The electronic media generate shallowness, credulity and distraction. Social media connect people across the continents without forming true relationships, commitments and localised communities with their demands and rewards.

The SMS has reduced communication to the most primitive level. Self-reliant search for the truth has made way for quick fix consultations on the Internet with its insurmountable, unstructured and de-contextualised mixture of trash and truth. While opening up masses of data, the information age inhibits contemplation and critical thought, produces greater confidence in hidden illusions, and renders true knowledge more problematic and fragile.[55]

The artificial creation of dissatisfaction and frustration

Popular 'postmodernity' neatly meshes with the consumer culture. The incessant hunt for emotional highs has led to compulsive patterns of consumption. While affluent societies have reached standards of living that our ancestors of all times could never have dreamt of, the modern human being has become a grumpy, moaning and demanding creature.[56] This mentality is deliberately and systematically enhanced by the marketing, advertising and entertainment industries with the sole object of boosting profits. I cannot put it better than Brian Swimme:

> But at a deeper level what we need to confront is the power of the advertiser to promulgate a . . . mini-cosmology that is based upon dissatisfaction and craving . . .

Advertisers in the corporate world are of course offered lucrative recompense, and, with that financial draw, our corporations attract humans from the highest strata of IQs. And our best artistic talent. And any sports hero or movie star they want to buy. Combining so much brain power and social status with sophisticated electronic graphics and the most penetrating psychological techniques, these teams of highly intelligent adults descend upon all of us, even upon children not yet in school, with the simple desire to create in us a dissatisfaction for our lives and a craving for yet another consumer product. It is hard to imagine any child having the capacities necessary to survive such a lopsided contest, especially when it's carried out ten thousand times a year, with no cultural condom capable of blocking out the consumerism virus.[57]

Rather than leading to fulfilment, the modern civilisation generated ever higher degrees of dissatisfaction with one's material resources and stations in life. Only two or three generations ago it was accepted that life was tough; that you had to work hard to keep your family alive; that duty ranked higher than reward; that enjoyment was linked to festive occasions; that you made do with clothes previously used by older siblings; that you repaired what had broken and mended what was torn; that furniture or cutlery inherited from previous generations were precious. All these certainties have vanished.

The modern civilisation demands constant novelty, constant acquisition, constant trashing. Our grandparents had no electricity, refrigerator, telephone, television set, video-system, washing machine, PlayStation, electronic toy, computer, or cell phone in their houses. They never missed these gadgets because they did not know they would ever exist. Masses of people in poorer countries still have to do without them. Human relationships, community life, creativity and self-investment could provide ample satisfaction. Having grown up in such an atmosphere, I tend to believe that their lives were at least as happy as, and much more balanced than, those of the ever grumbling population of today.

Systemic forces and constraints

This is not just a spiritual problem. The thrust towards economic growth has become a systemic necessity. The Marxist diagnosis could still focus on the capitalist 'accumulation' of 'surplus value' by the bourgeoisie. But mere hoarding of possessions can no longer be considered the prime driver of the capitalist system. The dynamics of modern commerce and industry is generated by the efficiency and the speed of 'throughput' from resource base via extraction, processing, distribution and consumption to waste. It is the strength of the flow within the enterprise that has to outperform the flow in competing enterprises. Wealth has become the speed of a process.

This development is actively and effectively boosted by the advertising and entertainment industries. It is also enhanced by political agendas because the power of the state depends on the power of the economy. The motivation of producers depends on the motivation of consumers because growth in supply depends on growth in demand. The flow of goods and services may not stagnate; otherwise the system grinds to a halt.

Moreover, the artificial generation of needs and wants has to concentrate on people with purchasing power, because the marginalised and impoverished do not contribute to production, have meagre incomes and generate no substantial market demand. Because their potential contributions as consumers or producers are negligible, they have become virtually redundant in terms of the modern economy.

The current value system

The entire package of needs, desires, values and goals has become distorted. The hankering for material possessions and sensual pleasure displaced more profound non-material, social, cultural and spiritual needs and interests. Youth, health, beauty, strength, alacrity and material prosperity have always been deemed desirable. But in earlier times these desires were tempered by a more balanced overall need structure, a sense of undeserved privilege and responsibility for the less fortunate. Today they are glamorised beyond proportion.

Sacrifices for the common good cannot be popular in such a social atmosphere. The 'good life' has become the prime criterion of authentic existence. If one cannot reach it, one indulges in depictions provided by magazines and movies.[58] In uprooted communities, it explodes into violent crime. The inability to compete or perform—whether due to infancy, old age, handicap, disease, illiteracy, bankruptcy, unemployment, or death—has become a nuisance and an embarrassment. A disease or a burial interferes with one's hectic schedule and reminds one of one's vulnerability. Such 'challenges' are psychologically suppressed and tucked away from public notice as far as possible.

One's acceptability and social standing is measured by one's car, home, cell phone, attire, handbag, drink and food. Conspicuous consumption can become more important than the satisfaction of basic needs. A family can go hungry for a television set. Millions have bought into the treacherous 'American dream' that you can easily move from rags to riches. Motivational literature claims that you can always achieve what you desire to achieve. If you fall behind, or drop out of the system, you only have yourself to blame. That this happens with the same kind of probability as winning the great jackpot, is not mentioned by the ideologues, nor considered by their credulous victims.

The media have their own way of making money. On the one hand it is bad taste not to be 'positive'. Radio and television presenters close their shows with the words "have yourself a fantastic day!"[59] 'Problems' have become 'challenges'. Never spoil the party. Never show that your life has become empty and meaningless. On the other hand, the media blow up catastrophes, scandals and sensations, because adrenaline-generating messages sell best. Positive news is in serious short supply.

Magazines portray a fairy tale world of health, affluence and beauty. They do not expose the misery of the underprivileged, nor do they highlight the incredible gifts of ordinary life that should prompt our gratitude and happiness. Even less do they dare to tell uncomfortable truths and call for responsibility. These inconsistencies become perfectly reasonable once you understand that the health and happiness of the consumers are not part of the agenda. Profit is the only consideration.

The time frames of the dominant actors in modern society have become extremely limited. Business leaders and shareholders think no further than the next quarterly results. Investors think no further than the next upturn of the business cycle. Politicians think no further than the next elections. The proverbial man and woman in the street thinks no further than the next salary cheque and what it can buy.

Austerity measures instituted by governments to avert state bankruptcy lead to violent protests. In South Africa, only a tiny fraction of the population bothers to save for unforeseen eventualities or retirement. Most have run up formidable debts that will take years to pay off. Even in the United States, I heard over the grapevine, 72 per cent of the workers over sixty cannot afford to retire.

The younger generation increasingly claims the right to immediate satisfaction of all their needs and desires. Parents have become slaves of their toddlers. Children can no longer be disciplined. They do not learn how to respect and deal with their limits. Teachers are expected to provide 'fun', rather than education and formation. Whatever does not deliver a 'high' of some kind, is not worth pursuing. Peer pressure has always 'forced' youngsters, just as much as adults, to do 'what is done'. But never has 'what is done' assumed the kind of lavishness that is now taken for granted.

Increasingly, young couples are no longer prepared to devote their time and energy to parenting. In culturally uprooted communities, young men leave their pregnant sweethearts for less demanding and more enjoyable relationships. Children are no longer socialised by parents and grannies, or by the encouragement to become inquisitive and creative, but by ceaseless exposures to television programmes, computer games, Twitter and Facebook snippets, or street corner peer groups.

Section VI
Social, economic and ecological consequences

Modernity has produced a 'feudal' hierarchy that dwarfs medieval inequalities. One would have thought that the rationale of the economic enterprise as such is to secure the survival, health and a reasonable level of prosperity of the whole population.[60] This would have to be done in a reasonably balanced way to prevent the unnecessary build-up of conflict potential in the society. If that were the case, one could expect that, sooner or later, a certain threshold of material contentment would be reached beyond which the population would focus on more profound pursuits. Alas, all this is not the case.

Captains of industrial, commercial and financial institutions demand unbelievable remuneration packages. As it happened recently in the American financial sector, they can allocate to themselves huge bonuses even after having run down their enterprises. Trade unions fight for higher incomes as much as shareholders, whether there is a recession or not, whether their demands go at the expense of jobs and work seekers or not; whether kids must do without teachers and patients without nurses or not.

The excessive and wasteful lifestyles of the super-rich defy any sense of proportion. The middle class too has become insatiable. Elites again form the reference groups of underdogs. Excessive consumption, not only by the financial elites, but by the population at large in developed countries, regions and localities is envied and emulated by those who cannot afford it, causing the latter to live beyond their means and end up in debt and misery.

The artificial creation and enhancement of material needs and desires has become an economic necessity. The inherent dynamic of a competitive system does not allow for equilibrium to establish itself. If an enterprise or an entire economy cannot outcompete its rivals in the marketplace, it declines. Expressed in macroeconomic terms, if the capitalist economy does not grow, it contracts. Contraction means loss of profit and employment, thus investment and consumption. The liberal economy thus depends on the growth and diversification of consumption. Contentment cannot be allowed to establish itself because it would undermine the robustness and survival of the entire system as presently organised.

This means that a balanced economy is not on the cards. Because consumption depends on market demand and market demand depends on purchasing power, the creation of needs has to concentrate on the affluent sections of the population. It is they who must be induced to part with their money. So the 'standard of living' is raised disproportionately among the well-to-do. But the demonstration effect of affluent lifestyles filters down to the less privileged, raising their expectations, but not their incomes. When needs rise in relation to incomes, poverty levels are deepened, and the gaps in life chances between rich and poor are widened further.

In sum, modernity espoused personal freedom. There is no doubt that it led to unprecedented wealth, technological superiority and military power. But it also led to the explosive growth of the human population and the equally explosive growth of the expectations of this population; to ever more grotesque accumulations of productive power and financial resources in the hands of the super-productive and the superrich; to growing income discrepancies between upper and lower social strata; to the economic marginalisation and ultimate redundancy of vast sections of the world population; to the prospects of long-term food and water insecurity and to unprecedented ecological destruction.

Change — what change?

The question is whether change will come through planned and controlled human behaviour—in which case humanity will be in charge of its destiny—or through chaotic and catastrophic mechanisms—in which case humanity will be the dumb victim of fate.

The situation is so serious that only a complete turnabout of humanity as a whole can avoid disaster. It is not too difficult to imagine what would have to happen. Four fundamental necessities stand out:

(a) A reduction of average human procreation to a maximum of two children per mother. This would gradually decrease the world population by painless attrition.

(b) A reduction of average human consumption to what is essential for healthy survival. There may be a slight chance that this level could be sustained by the regenerative capacity of planet earth.

(c) A change from capital intensive to labour intensive production. Technology would have to be employed only to facilitate and supplement (rather than replace) the full utilisation of human productive power. This would lead to the incorporation of all able bodied humans in the productive process and do away with both extreme luxury and extreme poverty.

(d) An allocation of resources that reflects the different needs of the productive process and enables all economic agents to make their respective contributions. This would lead to a differentiated but ethically defensible distribution of the common product.

Such a scenario is utopian. The probability that it will happen is virtually nil. But nobody should argue that it is impossible. Human ingenuity, will power and solidarity could bring it about. It would presuppose a drastic reconstruction of the collective mindset of humanity. It is the motivational bedrock that has to change if humanity is to have a future.

As Marxist experiments across the globe have demonstrated, the imposition of an alternative social structure without concomitant changes in the perception and conviction of critical masses in the population will not do the trick.[61] The traditionalist community, the feudal system and the totalitarian state have not presented us with better solutions.

In fact, they have one thing in common: they destroy motivation, initiative and responsibility. Romanticism is out of place. There is no workable alternative to freedom in modern times. Make no mistake, therefore: I do not argue for the superiority of a command economy over a free economy in terms of economic efficiency, social justice and morality. That is a non-starter.

Marxism was a radically *modern* movement, geared to emancipation, human autonomy and mastery—this time demanded ostensibly by the underdog worker classes rather than by social, political and economic elites. Grass roots power could only be generated, it was argued, through the collective action of the proletariat aimed at disempowering the bourgeoisie and taking over the power of the state. The movement was to be led by an 'avant-garde', hailing from the bourgeoisie, who felt called to perform the messianic task on behalf of 'the people'.

The workers lent them initial support because they yearned for the privileges of their bourgeois counterparts. But they were deceived by their bourgeois leaders. To enforce changes in outlook and behaviour, the leaders set up authoritarian structures, which they called the 'dictatorship of the proletariat'. They propagated and imposed a totalitarian ideology that crippled the healthy flow of life. The system was unable to curtail abuses of power. It destroyed personal freedom, responsibility and motivation and led to large-scale hypocrisy and corruption.[62]

In actual practice, Marxist-Leninist states also proved to be even less committed to ecological concerns than capitalist states were. Their focus was on political control, subjugation of nature, productive capacity, economic growth and technological fixes rather than ecological balance.[63]

At the other end of the spectrum we find traditionalist cultures with their heavy emphasis on ancestral, patriarchal and hierarchical authority. Individuals have narrowly defined and tightly controlled statuses and roles in the community based on gender and seniority. Any breach of competence is squashed. Any personal initiative, development of potential and sense of responsibility evokes suspicion and hostility.[64] Both these alternatives to the liberal economy have not been able to deliver the goods.

The modern quest for personal freedom is non-negotiable, even from a Christian perspective. It is the questionable package of assumptions and the distorted value system of modernity that is the issue. Its arrogant claims, its selfish aspirations, its narrow horizons, its short-sighted visions and its ruthless exploitation of humanity and nature need to be overcome.

The relentless and integrated persuasive power of best science and best faith may be the only hope we have. It must be specifically targeted, underpinned by action and aim at institutional transformation. Even small initiatives can trigger 'avalanches' and 'butterfly' effects. Every step in the right direction will at least postpone and alleviate the worst case scenario.

The question is whether human freedom can be integrated into a larger motivational framework characterised by a sense of responsibility for the whole. This applies to individuals, communities and larger structures of society such as political parties, sports bodies, employer and employee organisations and cultural institutions. At the highest level, nation states will have to negotiate common policies. The same is true for multinational corporations. Only a process of consultation and cooperation can lead to a workable and peaceful global system.

Some counter-trends

A cynic could argue that we all survive and prosper because the system is in operation and we do not really want it to change.[65] Consumers have been lured into conformity by the comforts and benefits generated by economic growth. The underprivileged yearn for the higher living standards of their more fortunate contemporaries. It is better not to think than to face the necessary sacrifices.

Of course, this is only one side of the story. Modernity and its effects on humanity, society and the natural world have been observed and described by perceptive poets, novelists, philosophers, psychologists and social analysts ever since the middle of the eighteenth century. They have raised the consciousness of concerned citizens across a fairly broad front. Time and again social critique becomes fashionable.

During the last third of the twentieth century, the student revolt, the Marxist challenge and the oil crisis caused some considerable vexation, excitement and activism. The threat of nuclear annihilation, the first fossil fuel crisis and the threat of 'forest death' have made masses of people wary, especially in central Europe. How can one generation have the cheek of saddling thousands of future generations with the effects of toxic nuclear waste! How will this generation ever be able to give account for upsetting weather patterns and sea levels through global warming, causing untold misery for future generations!

An ocean liner like the Titanic cannot easily be thrust into reverse gear. But its situation is hopeless if the pilot does not realise that it is heading for the iceberg. It would be wrong to belittle the arousal of collective awareness and personal conscience. Wherever freedom is achieved, responsibility has a chance to develop. While modern culture is fundamentally based on self-interest, it has also provided the space for an outburst of practical concern for social justice, the less privileged, the natural environment and future generations.

Numerous non-governmental organisations have sprung up in "civil society". Some of the super-rich, like Bill and Belinda Gates and Warren Buffet, have decided to "give back" to society what they have been able to amass. Giant corporations have realised that they can build up the prestige of their brands by sponsoring recreational, social and ecological projects. Governments have become more sensitive to social and environmental concerns. The international community has moved powerfully in the direction of the peaceful resolution of conflicts and ecological cooperation. Movements among scientists who have become concerned about ecological issues are gaining momentum.

Only time will tell whether these rivulets can flow together into a mighty stream or fizzle out, once again, as happened so often in the past. What we need is the development of a critical mass of those entrusted with power and influence—politicians, business leaders, unions, professional organisations, educational institutions, media chiefs, sports bodies, arts and recreation and, indeed, scientists and religious leaders. My contention is that scientists and religious leaders should be in the forefront.

Let us summarise

Modernity is powered by an emancipatory agenda. It was triggered by a rebellion against imposed constraints, authoritarian mindsets and obsolete institutions. Its primary motivation was the quest for human autonomy, mastery and ownership of reality. This quest brought it into confrontation with a Christian faith that had lost its emancipatory potential. Immense achievements led modernity to universal dominance and the marginalisation of the Christian faith. It expressed itself in the subjugation of nature, foreign lands and the human being as such.

Feeding on itself, its incessant momentum produced a universal attitude of entitlement to extravagant satisfaction of needs and desires irrespective of the consequences for others, society and nature. It is this escalating process that now threatens life on earth in general and the future of humankind in particular. Having been instrumental in the genesis of modernity, science and faith now face the common responsibility of redirecting the modern dynamic towards more equitable, sustainable and responsible goals before it is too late to change course. That brings us to the next chapter.

3 The common task of science and faith

Christian faith and empirical science facilitated the emergence of modernity. Having drifted apart, they were unable to constrain its problematic development. The Christian faith got stuck in an obsolete worldview. It did not shed its authoritarian mindset. It concentrated on the spiritual life of the individual. It lost both credibility and relevance. Science is the epitome of the human quest for mastery. It restricted its mandate and method to immanent reality, was roped in by human self-interest and lost its transcendent grounding and orientation. Science and faith are confronted with a common responsibility and must find a coordinated approach to current realities.

The task of this chapter

Let us briefly recap the last chapter. Modernity is characterised by a relentless thrust towards human autonomy, mastery and ownership of reality. Science and technology led to the acquisition of unprecedented powers. At the same time we witness the eclipse of a sense of responsibility for others, society, nature and future generations. This development is leading humanity in exceptionally dangerous directions.

Science and faith are deeply implicated in the genesis of this situation. The Christian faith has lost its authority, credibility and relevance. Stuck in obsolete assumptions and preoccupied with spiritual needs it has become nothing but a private curiosity. Science, in contrast, embodies the modern quest for human mastery. It commands widespread acceptance, admiration and trust. But it has lost its transcendent grounding and orientation.

This chapter is meant to expose this double failure of faith and science. The two pursuits of faith and science are deemed legitimate and complementary. The aim of the chapter is to open the way for self-critique, reflection on the rationale of the human project and the quest for a common approach to the crisis.

How science and faith intersect

Science is about observation, explanation and prediction. Faith is about existential and social grounding, orientation and vision. There are things discussed by theologians that are not directly relevant for science—for instance whether life is meaningful or meaningless, or whether acceptance should be conditional or unconditional. There are also things discussed by scientists that are not directly relevant for faith—for instance whether the universe will continue expanding forever or implode into a big crunch, or whether subatomic physics is based on probability or regularity.

And yet, because both science and faith are situated within a reality that is multi-layered and multidimensional, science will encounter questions of meaning and value somewhere along its line of reasoning, and faith will encounter questions of structure and regularity somewhere along its line of reasoning.

It is this link, this complementarity that I try to explore in this book. While the scientific methodology precludes value judgements, such judgements cannot be dispensed with as such, not even by scientists. If that were the case, why should pharmacists seek to develop vaccines? Why should lawyers design constitutions that entrench human rights? Why have at least some scientists developed a conscience about nuclear warfare?

Scientists and believers live neither in a quantum world, nor in the vast sweeps of cosmic time and space, nor in a metaphysical world. They live in a macroscopic world, a world where time, space and available energy are circumscribed and inescapable realities. They also live in this world as humans who need meaning, acceptability and authority. This practical circumstance makes the interaction between science and faith necessary and urgent, particularly in view of the looming economic-ecological crisis.

Why science and faith parted ways

The pursuit of scientific insight is much older than modernity. Its most rudimentary forms emerged in pre-historical times and evolved ever since. Its current shape has roots in Chinese, Egyptian, Sumerian and Greek antiquity, to mention just a few. There are parallels in pre-colonial Latin America and elsewhere. In all these cases, science was part of an integrated worldview.

The same can be said of faith. Earlier pursuits of science and faith complemented each other since times immemorial. Together they helped to trigger the modern drive towards emancipation and autonomy. Both were engulfed by a drive towards emancipation. Both are embedded in modern culture and cannot extricate themselves from its embrace. Nor should they even try. Their functions are critically important for the survival and prosperity of humanity.

Ever since the Enlightenment, faith and science have drifted apart. The consequences have been catastrophic. I have been up front with my diagnosis. I argued that the core problem of modernity is the thrust towards human mastery and ownership without a concomitant growth in responsibility. Science and faith are both implicated in this development. Science is the epitome of the modern quest for mastery, and the Christian faith has failed to liberate, transform and orient modern aspirations towards a vision of comprehensive optimal well-being.[66]

For many centuries, a heavily entrenched and oppressive kind of faith underpinned a feudal and patriarchal mindset in Europe. It stifled intellectual, communal, social, economic and political progress. Modernity began as a rebellion against authoritarian mindsets of all kinds. It has gone from strength to strength on that ticket. Being part of an oppressive system, the impact of faith on society dwindled by the same measure. Stuck in obsolete worldview assumptions and authoritarian mindsets, it was unable to regain its future-oriented dynamic. We are saddled with this heritage.

However, even if it could rid itself of its authoritarianism, the Christian faith could not easily figure as part of a modern approach to reality. It assumes that humans are accountable to a higher authority, an assumption that flies in the face of the modern emancipatory thrust. Geared to the transcendent Source and Destiny of reality as a whole, faith insists that the interests of the whole of God's earth are primary and that the valid interests of individuals and groups must be embedded in this greater context. The narcissistic mood of modernity has no appetite for this message.

The clash between the modern quest for autonomy and the Christian insistence on accountability must, therefore, be recognised as the core of the stand-off between modernity and the Christian faith. But there is another critically important factor, namely the perceived obsolescence and irrationality of traditional faith assumptions. Most unfortunately Christian awareness of accountability manifests itself as an irrational subservience to ancient formulations and institutionalisations of this faith. The emancipatory and future-directed dynamic of the Christian faith has been lost.

Untenable perceptions among believers now provide a welcome justification for the dismissive attitudes of the general public towards the Christian faith. Today it is virtually impossible for secularised population groups to take the Christian faith seriously. A public display of religious commitment leads to embarrassment and thinly veiled contempt. Wear a dog collar and you find yourself placed in a box by your social environment! Faith plays virtually no role in all spheres of life that really matter. This book is an attempt to find a way out of this impasse.

Section I
The pathetic performance of the Christian faith

In the last chapter, I have sketched how modernity has led humanity towards a crisis of unprecedented proportions. How could all this happen without both science and faith going on the barricades? Science has lost its transcendent foundations; faith has lost its credibility. If humanity is to have a future, there simply is no alternative to a conviction that is based on a compelling system of meaning, that defines binding criteria of acceptability and that assigns legitimate authority.

If such a system of meaning is to make a difference, it must be oriented towards the dynamic vision of comprehensive optimal well-being, galvanised by any deficiency in well-being in any dimension of life and committed to a sacrificial way of life. The question is why there is so little evidence of that in the modern world. This is a question that must be posed directly and particularly to Christianity, which has been the dominant faith community in the West for centuries.

By its very nature, the biblical faith is meant to be a faith of social justice and redeeming love. A situation of rampant competition, where it is OK for some to get drowned in luxury, while others do not know how to feed their children, is out of character with the spirit of Jesus of Nazareth as depicted in the New Testament. One cannot help but wonder what has gone wrong. Why is social responsibility more developed in secularised Europe than in the 'religious' United States? How could Christianised Rwanda descend into genocide and 'Christian' South Africa into apartheid?

As mentioned in the last chapter, part of the answer is that religion has always been abused to legitimate personal and collective self-interest.[67] The biblical faith is no exception. It began with ancient Israelite kings and the rule of high priests in post-exilic Jerusalem. It continued with Roman imperial power after Constantine, the claim of the popes to absolute authority, the aspirations of kings and princes during the time of absolutism, the 'German Christians' during Hitler's reign and the apartheid regime in South Africa. At present, the biggest failure of Christianity may be that it bought into the liberal-capitalist creed without hesitations, reservations and qualifications.

But that is only part of the story. The assumption of early believers was that the fate of human beings, their communities, possessions and life worlds depended on the benevolence of God, the personal creator and master of reality as a whole. According to common Ancient Near Eastern convictions, the deity had instituted a cosmic order that included natural law, social law and moral law. Transgressions would throw the entire system into disrepair. At least in this sense, Ancient Near Eastern religions still had comprehensive horizons. The individual was embedded in a greater natural, social and moral order.

This potentially inclusive worldview did not evolve to become ever more comprehensive, as one could have expected, but became stunted by a concentration on the spiritual needs of the individual. Why that? To gain a more comprehensive understanding of this sad fact, one has to go back in history—right to the biblical origins of the Christian faith.

In ancient Israel, God was believed to have entered into covenant with his people, decreed a moral code, and expected it to be observed. The emphasis lay on **righteousness**. As Deuteronomy 28 and 30 formulated it, God's blessing would rest upon those who kept the law; God's curse would shatter those who transgressed it. The national catastrophes of 720 BC, when the Northern Kingdom was destroyed by the Assyrians, and 586 BC, when the Southern Kingdom was destroyed by the Babylonians, drove home the point. The prophets and the deuteronomic school insisted that the calamities were Yahweh's punishment for idolatry and transgression. Israel had broken the covenant and had to face the consequences.

On this basis, a form of retributive justice became Jewish orthodoxy. The Jewish preoccupation with the *torah* (= the law of Moses) co-determined its Christian offshoot. The emphasis of the first Christian community lay on the question whether God was a God of incorruptible justice who would have to reject and condemn the sinners, or a God of **redeeming love** who would accept the unacceptable to transform them from within. Jesus proclaimed and enacted the latter version. Paul was its greatest theorist and defender.

In both cases, however, it was the **individual** that mattered. In Old Testament times, the individual was embedded in the relationship between Yahweh and Israel as the 'people of God'. A lack of responsibility would harm the nation and its progeny. In exilic times, it was increasingly felt that to blame children for the transgressions of their fathers was unfair (Jer. 31: 29 ff; Ezek. 18). Collective liability was replaced with individual liability. By the time of the New Testament the focus had shifted overwhelmingly to the healing of individual relationships—between God and humans and between humans and other humans.

At the same time the final reckoning shifted from a future 'Day of the Lord' within ongoing history to a **last judgement** beyond death. This shift was due to the painful experience that people who had done their best to keep the law, suffered, while their oppressors flourished. The Jewish faith clung to the conviction that Yahweh was a God of justice. God's justice could not be frustrated by death. Every single person would rise and answer for what he/she had done. An individualised and spiritualised form of faith was the result. Yet the apocalyptic frame of reference kept universal vistas open. The expectation of "a new heaven and earth" still formed the context within which individual relationships were located.

Subsequent centuries saw further shifts in the direction of **individual spirituality**. Cosmic eschatology made way for the concern of what would happen to the sinner after death. Ecclesial officialdom dispensed means to counter (and abuse) the fear of purgatory and hell. In its response, the Reformation focused on God's redeeming grace rather than good works and ecclesial indulgences. Righteousness was no longer a demand to be met by the human being, but a gift of God, allocated by the proclamation of the gospel in the power of the Spirit.

So far so good. But both parties focused on the individual and the spiritual. The gift was expected to issue in a new life in fellowship with God as well as responsible leadership and citizenship. But salvation was restricted to the **believing community**, while the 'world' needed a harsh regime to prevent it from descending into chaos. Theoretically this worldly task could have included a vision of comprehensive optimal well-being. The main Reformers, Luther and Calvin, still had fairly inclusive horizons. But the Enlightenment on the one hand and Pietism (its religious counterpart) on the other shifted the focus of attention to the inner core of the individual human being and his/her relationship with God.

The synergy between the Christian faith and the **Platonic dualism** between body and spirit enhanced the spiritualising trend. To express the outcome in a simplistic formula, what mattered now was the unmediated and undisturbed relationship between a bodiless soul and a worldless God. Apparently, the body, society, the international scene and nature could be "left to the devil". This is how slavery, colonialism, increasingly bloody wars and the sweat shops of the Industrial Revolution could be tolerated by the Christian conscience.

Of course, there were outstanding exceptions, but they did not change the general picture. One only needs to go through the hymnals of major denominations that originated during the last few centuries to notice the overwhelming emphasis on individualised and spiritualised piety. Social structures and processes no longer seemed to impact the relationship with God. You could be a miserable slave, beggar, or outcast and still enjoy perfect peace with God through Jesus Christ, your personal Saviour.

The need of princely rulers for legitimacy during the era of **absolutism** led to a wrong interpretation of the distinction between the 'two kingdoms of God', one in the heart of the individual and one in the political realm. It suggested that the believer had no mandate to be concerned about social, economic and political structures.[68] Citizens were the personal possession of the ruler. They were supposed to be obedient subjects to the authorities God had placed over them. The latter were free to abuse and exploit their subjects in the interest of power politics, self-aggrandisement and luxury.

Spiritual leaders were state officials and duty-bound to support the rulers. They were trained to cater for the spiritual needs of the individual. Spiritual needs were assumed to encompass nothing but an abstract sin that disturbed the relation between one's soul and God. Pastors had no mandate to deal with 'worldly concerns'. Needless to say they also had no competence to critique untenable social conditions, call for social transformation, or suggest policy directions. Private spirituality and public responsibility had broken apart.

As modernity unfolded, an emasculated faith was relegated to the status of a private pastime. It lost all relevance for the spheres of life that really mattered. Concentrating on the spiritual life of the individual, the 'sacred canopy' of Christianity was not designed to cover the aspects of reality where the greatest imbalances began to develop. In the public eye, the emancipated self had become entitled to pursue its interest. It could ignore overarching concerns and ethical considerations without public censure. The claim of the emancipated human being to unrestricted entitlements was the dominant factor leading to a loss of responsibility.

The impact of faith was crippled, at the same time, by fundamentalist approaches to the Christian tradition. When freedom and democracy finally dawned, Christianity remained tied to a historically obsolete set of assumptions, values and norms. The inherited worldview continued to be equated with the revelation of an eternal truth and inscrutable will of God. Authority—whether biblical, doctrinal or institutional—remained heavily entrenched in the minds of the faithful.

While a new secular discourse developed, theology became an arcane discipline that operated within its own conceptual edifice. With notable exceptions, it failed to respond creatively and redemptively to the insights and challenges presented by modernity and postmodernity. Instead of integrating valid insights of the sciences in its view of reality, it simply ignored them. More fundamentalist circles slated them as ungodly, false and mischievous. When these circles gathered political clout to impose their views, they aroused the anger of the science community, further deepening the gulf between science and faith.

When the self-interested assumptions of the liberal economy had firmly entrenched themselves in the public mind, the basic conservatism of the faith community inadvertently and subconsciously merged with it. Now the individualistic avarice of capitalism was taken to be identical with the Christian way of life. But that was hardly more than window dressing. In fact, faith had long become redundant as a legitimating authority for individual and collective self-interest.

In sum, the Christian faith became implicated in the direction modernity took, and that largely by default. One has to concede that during much of modern history, private and collective interests prevented people from moving beyond their biased perceptions, whether they were believers or not. Ecological concerns were still largely beyond their horizons. Again, there were notable exceptions. But the agenda that faith had adopted was not designed to question the vastly inflated importance of the interests of individuals or collectives of such individuals in relation to society, other living creatures and the inorganic substructures of reality.

Section II
The role of science in gaining human autonomy

Science was much more directly implicated in the development of modernity than faith. The emancipatory thrust of modernity was powered by scientific research and argument. For a time, scientists could claim that science was an objective, non-partisan, factual and disinterested pursuit of insight for the sake of nothing but knowledge. It has often been contrasted with more selfish pursuits like mass marketing or organized sports.[69]

The theory of science has shown that this claim to objectivity cannot be upheld. Although committed much more than other human activities to evidence, careful description, mathematical stringency and utmost precision, it still rests, like all other human pursuits, on assumptions. It produces theories that are model-dependent and that must be substantiated before they can claim to be facts. It is provisional, situational, perspectival and guided by interests.

Seen in the context of modernity, science has always been part of the quest for freedom, power and control.[70] As modernity advanced, the mercenary motivations of scientific and technological endeavours were increasingly exposed by critical social analysts, notably those hailing from the Neo-Marxian tradition. According to these authors, science and technology are not only formidable instruments of domination, but they also provide the legitimation for such domination. The dominated classes in mature capitalist societies accepted this domination because of the material benefits of the system.[71] They were not even aware of their enslavement. Whether these analyses were overdrawn or not, it is hard to refute their validity in principle.

Of course, Marxism itself was a typical outgrowth of modernity. It shared the general quest for emancipation and domination. Indeed it wanted to outperform its capitalist rival in this respect. For its part, neoclassical economics took it for granted that the ruthless pursuit of self-interest by 'economic man' was normal, natural, or 'rational' behaviour. Socialism was dismissed as a devious way of undermining the hard-won freedom of the individual.

Both schools of thought took the emancipatory thrust of modernity for granted. Freedom translated into domination over nature and other humans and their maximal exploitation for private or collective gain. More responsible scientists may want to disown these practices and disentangle the scientific genius from the ravages of modernity. Most scientists may want to be well-intentioned and humble researchers. But the motivations of the science community cannot help but reflect the general mood of the society in which it lives and works.

The dominating and acquisitive spirit unwittingly permeates all aspects of the modern society. Some scientists do tamper with empirical data and manipulate research findings to boost the bottom line of powerful corporations in exchange for kickbacks or research contracts. In fact, these corporations have themselves become dominant conductors of scientific research and technological development.[72] Public research institutes often depend financially on private sector contracts and outsourcing.

Without doubt important and helpful work is being done there. But who pays the piper calls the tune! The point is that research is not just being conducted for the sake of 'pure science', nor to fulfil the needs of society and humanity as a whole. Whether inadvertently or not, science and technology have legitimated and facilitated the thrust of modernity.[73]

The inherent selfishness of modernity has engulfed even traditionally service oriented pursuits, such as the medical profession. Drug companies, hospitals and specialised medical practices have become lucrative enterprises. Many of them have priced themselves out of the market of ordinary citizens. A mechanistic and anthropologically uninformed medical science lowered the death rate in poor societies without attempts to lower the birth rate as well, causing run-away population growth.

Nuclear power generation was adopted on a large scale before the safety of nuclear waste disposal was ensured. Scientists enhanced the rate of extraction and utilisation of fossil fuels. When some scientists became aware of the dangers of global warming, many other scientists laughed them off. Computer science and communications technology vastly increased the competitiveness of powerful actors in the global economy at the expense of weaker actors, leading to large-scale economic marginalisation. It is still not clear what kinds of consequences genetic manipulation and nanotechnology may entail. The list is endless.[74]

Science demands massive amounts of public spending. Are profound cost-benefit analyses made in all these cases, and if so, on the basis of which sets of priorities? Can the priorities of the sciences really face up to a vision of comprehensive optimal well-being? Are opportunity costs taken into consideration?[75] What if the astronomical resources invested in landing a person on the moon, sending a space craft to Mars, or the construction and maintenance of the CERN particle accelerator were spent on education, training, health care, small enterprise development and poverty reduction in the poorest regions of the world?[76]

Who benefits from the marginal increments in knowledge gained from these experiments, apart from a few cosmologists and physicists? Assuming that there are spin-offs for industry and commerce, who will benefit from these spin-offs, given a hopelessly skewed competitive game? Is it responsible to raise expectations that cannot be met, or that will benefit only tiny elites, such as space tourism, the deceptive prospects of universal affluence, or the prolongation of life beyond biologically programmed limits?[77]

Social cost-benefit analyses would have revealed, for instance, the unbelievable squandering of scarce resources in the weapons industry. The havoc that landmines and submachine guns have created in poor countries the world over is hair-raising. Assuming that nuclear weapons were necessary to keep the 'balance of terror' intact between the superpowers during the cold war, how can one justify the overkill in the number and firing power of these weapons during those years?

Why have they not been phased out after the end of the cold war? Why do Russian and Chinese defence budgets again begin to rise? Why has the "safety of America" become such a seductive legitimation of military ambitions, interventions and capital investments? Scientists and technicians develop such devices and leave them in the hands of unscrupulous politicians and military establishments without exercising their clout to ensure minimal and responsible use.

These are random examples that could be multiplied indefinitely. The point is that modern science is not, and never has been, 'nothing but' a disinterested pursuit of knowledge. Science is designed to enhance human mastery and control. The powers intended and unleashed by science and technology are being deployed by fallible and gullible human beings who pursue their individual and collective self-interest.

Human autonomy is the core of the problem

The deliberate utilisation or manipulation of research to serve collective interests at the expense of other such interest, nature and future generations, is a symptom of, rather than the core of the problem. Much more fundamental is the intrinsic thrust of modern science in all its forms towards human autonomy and mastery. By exploring reality, science is designed to get on top of reality, including human reality. Science again serves technology; technology serves commerce;

commerce serves, generates and exploits the consumer culture. The results are highly ambiguous.

On the one hand, nobody will want to do without the incredible achievements of science and technology. The present author is no exception. Science has benefited not only humanity but, at least to a certain extent, also the rest of nature. It is actively engaged in conservation. It discovered the destruction of the ozone layer, the dangers of asbestos, DDT and heavy metals, as well as global warming. It developed early warning systems concerning earthquakes and tsunamis. It hemmed in malaria and eradicated small pox. It boosted the production of food crops to feed a growing population. It works feverishly to find cures for AIDS and cancer. The list is endless.

On the other hand science unleashed powers unprecedented in human history without being able to generate concomitant levels of responsibility. It has often become victim of its own rootlessness. German natural, social and human scientists legitimated and supported Nazi ideology and its pseudo-redemptive agenda. Auschwitz was run along scientific lines. I have mentioned the problematic role played by science in Marxist-Leninist countries. Scientists developed nuclear, biological and chemical weapons without always considering the incredible devastations these weapons can wreak on humanity, nature and the earth. Science and technology have produced means of mass communication that serve dominant media.

It does not have to be that way

Like all other human pursuits, scientific research has a purpose, and its findings will be used accordingly. The question is what this purpose might be.[78] It does not *have* to be self-centred, skewed and truncated; it could also develop more inclusive horizons and concerns if its system of meaning were different. The primary agenda of science could be the attempt to provide tools for getting closer to the vision of comprehensive well-being of all humans within the comprehensive well-being of their total social and natural environments.[79]

Here and there this has indeed been the motivation of responsible scientists. But this is not what normally happens, especially in cases linked to the pursuit of power, wealth and pleasure. Within the general atmosphere of modernity, the purpose of science is clear—it must serve human domination, avarice and self-aggrandisement. Uninhibited self-seeking became not only acceptable, not just fashionable, but determinative for the entire socio-economic system empowered by the sciences.

Science and technology can become tools of human participation in God's creative and redemptive project. They do not have to adopt the theological language of a bygone age to do that. They can even help theology in reconceptualising its message in line with modern insight. They can enhance human awareness of derivation, dependence, ambiguity, guilt, responsibility, vulnerability and mortality. Their knowledge can lead to awe and humility. They can become instruments to move humanity towards God's vision of comprehensive optimal well-being.

Science and technology as such do not contradict true faith. It is the use to which they are put that matters. They can serve true faith and do so with integrity and conviction. The quest for meaning, acceptability and fulfilment through scientific discovery alone, through technological mastery alone, through profitability alone, through material enjoyment alone is a wild goose chase. But science, technology, commerce and consumption can become profoundly meaningful if they are embedded in the greater contexts of divine creativity, benevolence and vision. Science needs faith to be responsible. Faith needs science to be plausible.

Conclusion

The relation between faith and science is not a pastime for intellectuals who enjoy academic entertainment and debate. It is about the hazardous direction in which modernity is taking us. Growing socio-economic imbalances and increasing ecological destruction sound the alarm bells. The future of life on earth in general and humankind in particular are under threat. Rightly understood, both faith and science are critical and emancipatory pursuits. Critique presupposes freedom. Freedom implies responsibility.

Christianity was not able to shed its authoritarian mindset and institutional arrangements. When science forged ahead, faith got stuck in obsolete worldviews and lost its credibility. While social, economic and political thunderclouds gathered, it concentrated on an individualised spirituality and lost its relevance.

Science focused on human mastery. It did not contemplate what humans would do with the new powers it unleashed. It got entangled in the modern thrust towards collective interests and personal desires. In many cases, it became implicated in the modern stampede towards profit and pleasure.

So much for the greater context in which faith and science are embedded. For the sake of the future of humankind and all life on earth, the relation between science and faith must regain an appropriate and compelling expression. They must find a coordinated approach to reality so that they can again communicate with each other, enter into a complementary relationship and tackle the problems humanity is facing in a concerted action.

Economics as a case in point

Neo-liberal economics is the most influential example of a science that has become subservient to narrow commercial interests and unrestrained consumerism. Its models of economic reality, its view of human nature and its perception of the goals of the human enterprise are anything but rational. It legitimates the growth of socio-economic imbalances and ecological destruction that threaten the infrastructure of life on earth in general and the future of humankind in particular.

The task of this chapter

The theory of science has long understood that all science is theory-laden. Theory is value-laden; value depends on conviction; conviction implies a system of meaning. The higher the sciences move in the hierarchy of emergences the more visible the motivational directions of scientific pursuits become and the more devastating the consequences if these directions are self-seeking and destructive.

Economics is a pivotal example. This chapter is meant to demonstrate the subservience of a critically important science to the modern thrust towards collective interests and personal avarice. The point is *not* that economics as an academic discipline has caused these highly questionable developments, but that it has legitimated, facilitated and empowered them, rather than acting as a constant corrective against wrong assumptions and a constant directive towards a more acceptable future.

Is economics a science? Most definitely! It is based, after all, on empirical research and produces elaborate theories. It engages in statistical and mathematical analyses. Though hardly very successful, it tries to predict developments into the future. It is arguably the most respected and influential science in the public arena today. Is it therefore also rational?

Any unbiased observer can see that both the liberal economic system and its academic legitimation are irrational and counterproductive. They are not designed to what should be the prime objectives of a modern economy. The preservation of the natural resource base and the needs of future generations are not part of their traditional agenda. They are strongly biased in favour of the interests of economic and social elites. They lead to the marginalisation of masses of people and, due to increasing globalisation, to the destruction of previously healthy local industries. They do not even serve the genuine interests of affluent consumers, because they do not cater for a balanced satisfaction of human needs.

Before I point out a few of these irrationalities, let me emphasise again that the alternative to liberal capitalism is not a dictatorial, inefficient, corrupt and arrogant state-run economy. As the Soviet and Maoist experiments have shown, such a system is intrinsically counterproductive. There is a difference between a lean, protective, balancing, empowering and efficiently administered set of regulations and a clumsy, corrupt and debilitating imposition of state control on the economy.

The alternative to neo-liberalism is a responsible society, a society that is aware of its embeddedness in the greater network of natural relationships; a society that uses its freedom to construct an equitable and sustainable system; a society in which all citizens have a chance to contribute and benefit equitably; a society where all participants are willing to pull their weight in the productive and administrative processes and to sacrifice for the common good.

Section I
The ideological legitimation of the pursuit of self-interest

Why is there so little evidence of that kind of spirit in formerly 'Christian' countries? One has to understand that there is a constant tussle between convictions and interests. The result is an irrational kind of self-justification. Critical social sciences call it 'rationalisation' when applied to individuals and 'ideology' when applied to collectives. It combines selected facts with clever arguments to argue the legitimacy of the pursuit of self-interest at the expense of others.

The need for self-justification presupposes commonly held convictions, thus some kind of system of meaning. When these convictions have become so shallow and that they are no longer able to question or control self-seeking motivations, the situation seems to move beyond redemption. In modern and postmodern times, interests have all but overpowered, neutralised, or displaced convictions altogether.

Elsewhere, I have analysed the phenomenon of ideology as collective self-justification in quite some detail and do not have to repeat that here.[80] The point to be made at this juncture is that economics as a science has evidently fallen into the ideological trap. Note that rationalisation is not a sign of rationality. It is highly irrational. But its irrationality has a hidden purpose, namely the justification of the pursuit of self-interests at the expense of others, society and nature. In what follows, I only point out a few of the more obvious irrationalities of neoclassical economics.

The free market

It is irrational to assume that the 'invisible hand' of the market coordinates various self-interested pursuits to the greater advantage of the society. In actual fact, every group of economic agents manifestly tries to manipulate the economy in its own favour and the success of such manipulations depends on the command of economic power. The neo-liberal argument fails to take account of the unequal distribution of purchasing power in the case of demand and productive power in the case of supply.

Where an unequal distribution of power leads to escalating differences in productive capacity, income and life chances, a 'hands-off' policy can only benefit the most powerful actors at the expense of the less powerful. The incessant demand for such a policy is, therefore, a transparent ideological weapon to strengthen the positions of more powerful manipulators of the economy at the expense of weaker participants.

On the supply side, immense resources are ploughed into the development of labour-saving technologies, rather than labour-intensive technologies. Such investments may also be tax deductible. There is a built-in bias in favour of capital at the expense of labour. As a result, labour is ever less capable of competing with capital as a factor of production. With machines taking over the functions of labour, the profits of capital owners are artificially boosted, while the army of the unemployed rises steadily in marginalised sectors of the global economy.

In highly developed economies, they can be absorbed in non-productive sectors, while in depressed economies such people simply get into a poverty trap. A rational system would assume that capital and technological sophistication are meant to help humans to produce, rather than pushing them out of the productive process.

On the demand side, the 'free' market is heavily manipulated by the marketing, advertising and entertainment industries. Instead of aiming at a balanced satisfaction of material and non-material needs and a reasonable standard of living, they deliberately and aggressively undermine contentment and peace of mind by means of highly sophisticated psychological techniques.[81] With that they instigate consumers to overspend on non-essentials and prevent an increasingly gullible population from turning to more profound values, once reasonable material requirements have been met.[82] What can be more irrational than that!

Economic growth

Liberal die-hards claim that the resources of the earth are practically inexhaustible or that human ingenuity will always find alternatives once particular resources run out. They claim that economic growth is the only way of creating jobs, while in fact economic growth based on labour-saving technologies has led to increasing unemployment, or the translocation of work to non-productive sectors. Where labour cannot compete with capital, economic growth is, to a large extent, 'jobless growth'.

It is often argued that the production of luxury cars, for example, creates jobs. But the modern production of cars is largely automated. Moreover, the investments made in the production of these luxuries would have the same economic growth effect if it were allocated to the production of goods needed by the infinitely greater number of the less privileged. But in a free market this would presuppose a shift in the allocation of purchasing power, thus greater participation in the productive process, which is simply not meant to be on the cards.

Neo-liberal economics is not bothered by the fact that the benefits of economic growth largely accrue to a minority, rather than leading to greater justice, harmony and healthy social relationships. Such unequal benefits produce envy, frustration and anger among the less privileged. They also lead to the withdrawal of the more privileged into fortified enclaves. They mislead the less privileged into lifestyles they cannot afford because elites always constitute the reference groups for lower social strata.

It is often argued that economic reality is not a zero-sum game. If we grow the pie, the argument goes, the slices of all stakeholders grow. So ostensibly the issue is not about equality, but about productivity. This is a typical rationalisation at the behest of the privileged. Two consequences of such growth without equitable distribution are overlooked:

(a) The rich get a much greater slice of the growing pie than the poor. Say the economy grows 10 per cent; then a person who earns R100,000 per month will get an additional **R10,000**, while a person earning R10,000 per month will get an additional **R1,000**. In this case, the increment for the former is as much as the entire monthly salary of the latter.

(b) As an economic adage goes, there are no free lunches. If both the rich and the poor benefit from economic growth, it is the resource base, thus nature, that has to bear the cost.

Social equity

The power distribution underlying the market can be changed quite easily without falling into the trap of a command economy, for instance by differential taxation and corrective allocations of public resources to research, education and infrastructure. Central European social democracy has demonstrated the superiority of such an approach over the free market—at least until a spirit of entitlement without performance took hold of the population and rendered the system relatively less competitive.[83] However, the recent sub-prime crisis again proved its superiority.

A liberal economy claims to be democratic because every individual has the same chances to develop initiative and prosper. This is simply not true. Economic power is heavily concentrated at the top of a social hierarchy. The advantage of a democratic state is that, at least in theory, the interests of all citizens, whether rich or poor, educated or illiterate, employed or unemployed, shareholders or workers, carry equal weight. Therefore, a democratic state quite naturally shifts the balance of power in the direction of the interests of the less powerful majority. This can lead to a more equitable economy and the fuller realisation of all resources available to the society. As the recent economic history of Sweden or Germany shows, social democracy does not have to lead to a less productive system. But in a liberal-capitalist system, the state is not supposed to interfere in the economy.

Neo-liberal economics also does not censure the fact that enterprises transfer the social costs of cost-cutting measures, social inequity and ecological destruction—unemployment, poverty, family violence, lack of education and training, preventable disease, delinquency and criminality, alcohol and drug addiction, corruption, inefficient administration, deficient infrastructure, polluted and eroded environments, and so on—onto the shoulders of the public. This again goes at the expense of the less privileged majority, nature and future generations.

The satisfaction of needs

Neo-liberal economics is inconsistent even within its own package of assumptions. The law of declining marginal utility says that small increments in the level of consumption among the poor lead to huge gains in satisfaction, while over-consumption among the wealthy does not lead to concomitant satisfaction. While the acquisition of a scooter would be a gift of heaven for the poorest of the poor, the additional satisfaction provided by a second luxury car is negligible if compared with the cost involved. The argument that satisfaction cannot be measured is a typical rationalisation.

This again reflects back on production. An unreasonable amount of research, development and investment is allocated to the production of luxuries for a minority, while the real and crying needs of the majority do not receive the attention they would be entitled to if the system were more rational and equitable. The reason for this imbalance is, of course, that production focuses on consumption, and consumption presupposes purchasing power, which is concentrated in economic elites.

Another irrationality lies in the fact that production is subject to strict cost-benefit analyses, while consumption is left to wanton behaviour. Large-scale wastage und underutilisation of products and resources are not only tolerated, but deliberately and systematically encouraged by aggressive marketing. The reason is simply that people must be lured into behaving irrationally, otherwise consumption cannot grow beyond reasonable levels. That again is required to achieve increasing profits.

This bias shows where the real motivation of the system is located: apply strict efficiency criteria to acquire as much as you possibly can so as to squander as much as you like. This kind of 'freedom' is typical for modernity! The obsession with individual autonomy and mastery leads to gross irresponsibility. And what about the criteria used to measure economic progress and prosperity? I cannot say it better than Christopher Dickey:

> The Gross Domestic Product measures everything except that which makes life worthwhile . . . What isn't bought or sold—housework or caring for your own children, for example—doesn't figure . . . hurricanes and floods push up GDP because the reconstruction gets factored in as new spending. Countries with more prisons look better than those with fewer . . . And if corporations and the very rich are doing well, that can skew the averages to make it look as if everyone is prospering when, in fact, the majority is not. Depletion of finite resources and environmental depredation don't get counted either . . .[84]

A dehumanised concept of the human being

In the modern liberal economy, the human being becomes not only an exploitable object, but an exploiting subject. Neoclassical economics betrays a dehumanising agenda with its mere choice of terminology. The neo-liberal concept of the 'economic man' *(homo oeconomicus)* defines the human being as a profit and pleasure maximiser. The relentless pursuit of self-interest is deemed 'rational behaviour'. By implication, a behaviour that enhances the interests of others, the community, the society as a whole or the natural world is deemed 'irrational'.

Even more dehumanising is the fact that, in economics, 'labour' does not figure as a group of living people with families to feed, but as a marketable factor of production following the laws of supply and demand.[85] A rational system worth its name would assume that humans are something different and something more important than simply less efficient biological machines. But then, although deemed nothing but a factor of production in competition with capital, the owners of capital are represented on the boards of enterprises, while the owners of labour are not—at least not in radically liberal-capitalist economies.

All this presents itself as sober, objective science and is accepted as such by a brainwashed population! In fact, this kind of science represents precious human brain-power and sophisticated methodology put to highly irrational and problematic uses.

A distorted system of meaning

The ideological nature of neoclassical economics has been exposed during the conflict between socialist and capitalist value systems over more than a century. But this 'consciousness raising' has largely disappeared since the collapse of Marxism-Leninism. The failure of communism seemed to have demonstrated the success, thus the validity of capitalism. It is conveniently forgotten that radical socialism emerged as a response to the failure of capitalism to bring about sufficiency and equity in the first place. After its demise, rampant neo-liberalism was allowed to cause even greater social inequities and ecological destruction.

Even moderate intermediate systems, such as European social democracy, have come under siege because they were deemed clumsy, less productive and less competitive. Unfortunately, their initial success stories have been undermined by the gullibility of their own populations. This again demonstrates the power of self-interest in the modern economy. When unreasonable expectations rise above productive capacity, an economic system cannot be expected to compete with a radically liberal one.

The point to be made at this juncture is that the modern system of meaning has been severely distorted and the consequences are becoming ever more dangerous. Adam Smith, the father of liberal economics, could still presuppose a substantial ethical bedrock inherited from a more judicious past. With the transcendent foundations of Western civilisation gone, we are drifting into reckless selfishness and irresponsibility.

Neo-liberal economics has not only legitimated, but actively promoted the acquisitiveness and irresponsibility of the modern mind. It was successful in doing so because it pandered to the human desire to dump moral inhibitions and enjoy life while it lasts. It fell on fertile soil both among those who defended their privileged positions and among those who aspired to catch up with the latter.

It effectively disempowered religious and cultural constraints. It inflated the importance of material consumption at the expense of the satisfaction of every other human need. With that it thoroughly dehumanised a creature sufficiently equipped to reach a higher level of emergence—the capacity to see the comprehensive package of needs of reality as a whole and to place its own needs into this wider context.

Symptoms are the vast budget deficits in developed countries; the recent sub-prime crisis; the exorbitant bonuses claimed by shameless CEOs after they had run their banks into virtual bankruptcy; the deliberate and sophisticated dismantling of all inhibitions by the advertising and entertainment industries; the explosive indebtedness of the general public prompted by easy credit and aggressive lending drives by financial institutions, and the relentless hunt of the younger generation for the next best 'high'.

People are abandoning their self-determination in favour of senseless status symbols, fads and crazes. The privileged are stomping through nature and society in their blind quest for luxury, totally oblivious of the plight of the deprived and marginalised. They destroy the precious treasures of the earth and put the future of their progeny at risk, if they bother to expose themselves to the chores and discomforts of having a family in the first place.

Instead of developing their creativity with self-made toys, and being socialised into the warmth and riches of a vibrant family life, children spend endless solitary hours before television screens and computer games. Humans are carried along by their own technological creations, losing the dignity and wealth of their life worlds, the awareness of history, the treasures of culture, the efficiency and beauty of the natural world, the magnificence of the universe in terms of time, space and energy.

But they also lose their sense that life as such is an incalculable gift to themselves, that they are embedded in large networks of precious and demanding relationships, that they are supposed to transcend the here and now towards wider contexts in space and time, that they could reach for excellence in terms of perception, contribution, maturity, justice, benevolence, sensitivity, concern, mercy, community, solidarity and, above all, self-determination.

Everything is being commercialised. Everything becomes a commodity to be traded—iron ore, soil, trees, antelopes, music, art, education, sport and sex. The human being and its excellences are no exception. Beautiful female bodies boost the 'sex appeal' of beautifully designed luxury cars. Consumers subject themselves seemingly without resistance to artificially enhanced cravings for constant highs—be it in terms of status, power, possessions, food, or ecstasy. The individual is deemed entitled to instant satisfaction. Where the financial resources cannot be earned, one turns to easy credit, corruption, or crime.

Globalised trade leads to steep growth curves in distant economic centres and the marginalisation of millions of people in far flung peripheries. Living collectively far beyond their means, the most advanced economies—the USA, Japan and the European Union—are in danger of succumbing to unprecedented public debts. Of late, lavish bonuses granted to CEOs are no longer connected to achievements. The same is true for professional football players and lawyers.

Of course, economics is only an outstanding example of a mercenary science. The effects of modernity as a system of meaning permeate all dimensions of life, including religious convictions. But economics has a pivotal role to play. When individualised and overblown material satisfaction becomes the overriding motivation of human behaviour, something indispensable and irreplaceable gets lost—something that makes the human being a human being, rather than an animal or a monster.

It has something to do with an awareness of the prerequisites of comprehensive well-being; with the widest possible horizons in terms of space, time and power relations; with the awareness of human derivation and dependence; with family and community; with meaning, acceptability and authority; with commitment, accountability and sacrifice for the greater whole, with human needs that transcend material wants, with respect for a plurality of life that has grown over four billion years and a culture that has grown over twelve thousand years.

Section II
The looming ecological disaster

Economic imbalances are one problem; ecological destruction is another. The devastating impact of human civilisation on the natural world is nothing new. The roots of human acquisitiveness go back to the first beginnings in the history of humankind. The human species is capable of transcending its immediate needs towards higher ambitions. With the emergence of death rituals, symbolic representations and stone tools, humans left the embrace of the animal world and embarked on their cultural, economic and technological ascendancy. They have aspired towards goals lying far beyond their healthy survival ever since.

Reaching such goals implied the acquisition of additional resources. Such resources could be obtained at the expense of fellow human beings, other creatures, or non-living materials. Human civilisations have always led to differences in power and life chances among humans as well as ecological deterioration wherever they emerged and evolved. Historians have shown that at times "such a one-sided exploitation of nature has led to regional environmental catastrophes that have in turn led to the fall of whole civilisations."[86]

As mentioned above, 'economic growth' means acceleration of 'throughput' from resource base via extraction, production, distribution and consumption to waste.[87] It leads to an ever more rapacious exploitation of nature. While ever faster throughput from resource base to waste may be interpreted as an enhancement of the evolutionary process, in fact it accelerates the entropic process.[88] Current assumptions reduce the gifts of nature to material which can be dismantled and exploited at will, leading directly to reckless behaviour towards the needs of coming generations.

Economic growth has become the prime objective of government policy and private enterprise. It can be argued that higher profits gained by the private sector lead to higher revenue for the state—with which it can redress economic imbalances and ameliorate the ecological harm caused by the private sector. But is that the most rational way of organising an economy?

To call the enhancement of the entropic process 'economic growth' must be one of the most serious misnomers in economic theory. Due to the entropic process any construction feeds on a greater degree of deconstruction elsewhere in the system. In ecological terms, even the "creation of wealth" is a problematic concept. Humans remould, dismantle, reassemble and utilise what nature has provided. They seize parts of accessible earthly reality, transform them into useful or desired artefacts and dump the unwanted residue as waste. Often they leave ecological devastations behind, without ever realising what they have done.[89]

Depending on the severity of the consequences, many civilisations have instituted checks and balances, underpinned with appropriate spiritual and legal sanctions. Yet, they were usually meant to secure resources for the members of particular communities, social elites, or ethnic formations, often at the expense of other such groups, but virtually always at the expense of the natural world.

In time, this led to the current division between economy and nature; economic centres and peripheries; capital owners, labour and the marginalised. The heavy hand of human selfishness has rested upon the earth for a very long time. However, our current situation is different because of the following developments:

(a) The immense growth of the human population and its basic survival needs;[90]
(b) The immense growth of material expectations, fuelled by the demonstration effect of affluent patterns of consumption, aggressive marketing and the disintegration of traditional inhibitions and sanctions;
(c) The resultant immense growth of industrial production, dependent on energy derived from fossil fuels,[91] leading to the dramatic depletion of non-renewable resources, overexploitation of renewable resources and the accumulation of waste;[92]
(d) The immense growth of productivity and income discrepancies between economic centres and peripheries,[93] amplifying discrepancies that were still relatively manageable at local or regional level;[94]
(e) The loss of contentment in the modern civilisation, which has not only destroyed potential fulfilment and happiness but has now also become a threat to the survival of the human species itself.

It is foolhardy to imagine that the earth would be able to carry the additional burden involved in raising the consumption levels of 2.5 billion Chinese and Indians—let alone the growing populations in the rest of Asia, Africa and Latin America—to levels already attained in highly industrialised countries without serious and permanent damage. It is already happening all around us and accelerating.

More frightening is the possibility that certain thresholds are reached beyond which they can no longer be stopped or reversed. Such points may trigger an avalanche effect that restructures the very way our earth is organised. Huge quantities of methane gas (a greenhouse gas said to be twenty-three times as potent as carbon dioxide) can be released by melting ice caps; dying forests can release carbon dioxide rather than absorbing it; ocean currents can alter their flows; high temperatures can lead to a proliferation of micro-organisms, and so one can continue.

Quite obviously all this will affect our progeny much more than our less privileged contemporaries. The need for resources will inevitably grow while the availability of resources will just as inevitably decline. Without doubt the rich will be able to secure resources to meet their survival and prosperity needs, by force if necessary. But scarcities will increasingly price these resources out of the reach of the less privileged.

Most critical are food security and the availability of clean water. A growing population with growing appetites for higher quality food will increase food requirements, while the capacity to produce food will decline. Reasons are, among others, declining energy resources, unbalanced soil compositions due to artificial fertilisers, pollution of water and soil, erosion of fertile land, and the law of declining marginal productivity.

The bottom line is that accelerating growth cannot continue indefinitely on a limited planet. There will come a point where the growing nutritional needs of the growing world population can no longer be met, regardless of how generously future generations may want to share their food. The quest for diminishing resources will almost inevitably lead to higher conflict potentials, greater investment in military hardware and expensive security systems for those whose privileges are threatened. As I have argued more than a decade ago:

> The present generation may very well go into history as the most affluent, and also the most peaceful, that the world has ever seen and will ever see. Science and technology have increased our life chances and our standards of living dramatically and may continue to do so until our "natural capital" is exhausted. Then the great "morning after" may dawn and the entire social pyramid may begin to sink. If that happens, the rich may become much poorer on average and the poor may go below the waterline and drown. Needless to say, an increasingly violent scramble for dwindling resources may also ensue, unless humankind undergoes a steep learning process.[95]

In the same work, I suggested the following order of priorities on the agenda of responsible research in economics:[96]

-    The preservation of the resource base of the planet.
-    A modest but healthy livelihood for all.
-    Equity in the distribution of both efforts and rewards.
-    Concern for the weak and vulnerable.
-    Balanced material and non-material need satisfaction.

Is reorientation possible?

Just pause and think of it—within less than a century we have acquired numbers and lifestyles that endanger the unique phenomenon of life that took four billion years to evolve on this unique and precious little planet in cosmic space! In a free society, there are many checks and balances. However, when the entire society is on a wrong course, these checks and balances will be tuned to the interests of the most influential groups, rather than enforce unpopular constraints.[97]

The entire project of the Western civilisation is in danger of collapsing under its own weight. The artificially reconstructed world in which we are embedded develops its own dynamics and sucks the individual, the community, the society and nature into its vortex. The oil industry, to quote just one prominent example, is like a locomotive of old that relentlessly puffs along set rails and pulls everything else with it—entirely oblivious of long-term sustainability and ecological impact. Seemingly nobody is capable of stopping these juggernauts.

Nobody should argue that economic and ecological crises are unavoidable. They are products of the human spirit.[98] They are caused by systems of meaning with narrow horizons, short-term visions and self-centred views of reality. Such systems of meaning could be replaced if humanity became aware of the dangers, decided to change direction, rediscovered true value, gained true freedom and accepted true responsibility. Arguably, all this should be typical of the most highly developed creature in the universe.

Economics should have been the science most intimately concerned with ways to secure the balanced satisfaction of human needs across the social spectrum for the present and the next few generations. In fact, it did the exact opposite. I have offered detailed analyses of these processes in previous publications and do not need to repeat that here.[99] Suffice it to say that three basic switches in the agenda of economics would fundamentally change the operation of the entire discipline:

1. Realising that we live in a limited world, economics can be expected to explore the means and methods of *securing the minimum material prerequisites for the healthy survival of all humans*, present and future, rather than the privileged advantages and exorbitant luxuries of certain sectors and actors in the economy at the expense of others.
2. Realising that the competitive economy is a power game in which differential natural and social endowments lead to runaway differences in productive power, purchasing power and life chances, economics can be expected to explore approaches and methods to *achieve the balanced integration of all economic stakeholders in the productive process and an equitable distribution of benefits and sacrifices* between them. It is abundantly clear that, left to its own devices, the market cannot do that.
3. Realising that economic growth is tantamount to an acceleration of the entropic process that must be contained rather than enhanced, economics can be expected to explore ways of *reducing to an unavoidable minimum the destruction of non-human life, the exploitation of non-renewable resources, the exploitation of renewable resources beyond their capacity to regenerate themselves and the throughput from the resource base to waste* so as to avoid the overload of natural sinks.

Again, it is not all gloom and doom. Apart from outstanding pioneers, there are now many economists and other social scientists who have begun to change course. Ecological awareness has grown dramatically in the last three decades. It has become a factor in national and international politics.[100] In many developed countries, it has led to substantial legislative processes.[101] In response to a shift in policy and the general mood, major oil and car companies have begun to advertise themselves as pioneers in alternative energy generation and utilisation.

While economic growth remains the unchallenged panacea among commercial captains and policy makers, the atmosphere at the biennial Davos meetings of the World Economic Forum has begun to display some unease with an outdated economic orthodoxy, some awareness of the social explosiveness of economic inequity and the destructiveness of ecological deterioration. While during the heyday of the capitalist-socialist conflict social justice for the current generation was the main issue, intergenerational justice has now become a matter that troubles the conscience of countless people. The relation between the two is heavily debated.[102]

Many ordinary people in affluent societies feel the void left by the tyranny of the quest for material wealth and sensual pleasure. There are ecological mass movements such as Bill McKibben's *350.org*, interreligious movements such as Hans Küng's *Foundation for a Global Ethic*, and initiatives of scientists such as the *International Union of Geodesy and Geophysics*.[103] Even seemingly insignificant initiatives such as the recycling of household rubbish must be supported.

Rivulets add up to powerful streams. Small discoveries, inventions and initiatives have changed the world in the past. Such developments may gather momentum.[104] To become effective, however, awareness and responsibility must acquire a critical mass at all levels of human action and intervention. It is the mindset of billions of people that must be redirected away from a deadly course of history.[105] Only then will policy makers dare to change course. This process must be spearheaded by major thinkers and decision makers. We are light years away from such a situation, and there is no time to be lost. Scientists and religious leaders must use all the leverage they have to forge a new awareness and change existing attitudes.

The role of science

Science has a pivotal role to play in changing the direction of technological development, commercial behaviour, ecological awareness among consumers and economic policy. That there is a need for reorientation is not only true for economics. Social equity, balanced need satisfaction and ecological sustainability are not, as they should be, the most self-evident points on the agenda of most disciplines pursued at institutions of higher learning.

As mentioned above, the natural sciences largely serve the profit motive of commercial enterprises, political interests, military power and consumerism. By and large, the social sciences concentrate on human society without taking its impact on the ecosystem into account. The human sciences are, to a substantial extent, oblivious of the imbalances in the material and social world within which they operate. They simply live off the skewed economy without asking further questions. The same holds true for much of academic theology.

The question is not whether science as such is evil. It is not. The question is also not whether we can do without it. We can't. Without science and technology we would have no chance to solve the problems we are faced with in the twenty-first century at all. The question is, rather, which kind of agenda science is willing to serve. And that depends on the 'symbolic universe' or system of meaning within which it operates.

Is it restricted to short-term, collective self-interest and individual indulgence, or is it motivated by long-term social equity and ecological responsibility? Is it built on human mastery or on human accountability? Is it enslaved by an absolutised natural world, or can it sense the dependence of the latter upon a higher Source and Destiny, however conceived?

It is the transcendent foundations of modernity that have gone missing. The needs and wants of human beings have become absolute and are taking countless other species with them on the road to self-destruction.

Science is in a very special position. Because scientists are keen observers, they are often more sensitive to problematic developments than others. Without their research many ills would not be discovered in time and the means to address them would not become available. Because of the immense respect they enjoy in society, they are also in a pivotal position to mould public consciousness. There is no question that science and technology have a kind of influence that is immensely superior to that of the Christian faith and any other faith for that matter.

Ordinary people have no access to highly specialised and sophisticated workplaces and depend on old-fashioned trust. Science is widely and blindly trusted in secularised societies. Unfortunately, scientists can also be bought.[106] I have dealt with that above. Proverbial examples are scientists who denied or played down the consequences of tobacco smoking, global warming, or unregulated credit provisions to serve the interests of powerful sectors of the economy.[107]

Corrupt politicians can be exposed and deposed. Ruthless dictators may cause their own downfall. The work of irresponsible engineers will show cracks. Corrupt business undertakings will lose customers and go bankrupt. Hypocritical religious leaders become targets of ridicule and contempt. Banks that follow irresponsible credit policies may collapse. Consumers that live beyond their means face the heat when they run out of money. But corrupt scientists throw modern society into bewilderment and despair because society invests so much faith in them.

The common responsibility of science and faith

I have argued that the most basic threat humanity faces is the growth of power without a corresponding growth of vision and responsibility. Scientific-technological rootlessness and religious superstition—this is the double impasse that must be faced honestly and boldly. Sam Harris, a militant atheist, has put the modern impasse in a nutshell: "I am terrified of what seems to me to be a bottle neck that civilisation is passing through. On the one hand we have twenty-first century disruptive technology proliferating, and on the other we have first-century superstition."[108]

Well, exactly! Both Christianity and modernity are in crisis and the two crises are linked. While consumerism is the motivational force, science is the ultimate pillar of modernity. For better or for worse, science and faith are embedded in the global embrace of modernity. They have contributed to the rise and the crisis of modernity, each in its own way. Modernity has emerged and matured in the geographical area once dominated by Christianity. Science and faith are confronted with a common problem. They share the same cultural environment.

So science and faith can justifiably be deemed co-responsible for finding solutions. Neither can afford to avoid the question of truth. Truth has to do with the relation between immanent facts, investigated by science, and the transcendent grounding and orientation of the human project, envisaged by faith. The one cannot do without the other. They have to challenge, correct and empower each other. That is the gist of my argument.

Let us summarise

In this chapter, I pointed to the sad fact that a science can easily become subservient to collective self-interest and personal avarice. The assumptions and goals of neo-liberal economics are irrational and counterproductive in terms of the human project and the preservation of life on the planet. Its contention that a free market leads to economic equilibrium ignores the fact that supply presupposes productive power and demand presupposes purchasing power. Its heavy emphasis on economic growth ignores the fact that unlimited growth is not possible on a limited planet. It ignores the growing inequities caused by uninhibited competition. It ignores the fact that the law of declining marginal utility implies massive imbalances in the satisfaction of actual needs between the rich and the poor. Its concept of 'economic man' as a profit and pleasure maximiser dehumanises the human being and leads to a massively distorted system of meaning. Reorientation is possible when a critical mass begins to realise that humanity is on a suicidal course and begins to change its priorities.

Part II

Science —
The perspective from within

5 The approach of science — experiential realism

To cooperate, science and faith must find a common approach to immanent reality. I suggest that 'experiential realism' is such an approach. It is more inclusive than naïve empiricism, but confined to immanent reality. It deems anything real that can be described, critiqued and manipulated and that has demonstrable consequences in this world. This includes mathematical formulae in subatomic physics, hallucinations in psychology, the power of literature, designs in fine art, sequences of events in history, and faith commitments in religions. Experiential realism does not attribute reality to what is not experienced, such as unsupported contentions, fantasy, mere suspicions and superstitions. It concentrates on the observed object rather than the observing subject. It avoids the reification of religious metaphors and Platonic abstractions. It avoids inferences from untested assertions. It does not agree with the 'postmodern' claim that truth is a presumptuous illusion. It does not include what transcends the reality that is accessible to human observation and comprehension, but it does include *notions* of the transcendent as found in philosophy and religion. Such notions have a shape that can be described, critiqued, corrected, or discarded. They also have formidable consequences in real life. Geared to the flow of time, experiential realism is supportive of narrative expressions of human insight.

The task of this chapter

Science is in the business of observation, explanation and prediction. Faith is in the business of grounding, orientation and vision. The criterion of science is evidence. The criterion of faith is validity. For a fruitful dialogue to materialise, faith must respect the mandate and method of science. Science must acknowledge the quest of faith for transcendent foundations.

My procedure is to determine the status of faith in the context of science (Part II) and the status of science in the context of faith (Part IV). An interlude clarifies the relation between the assertion and denial of transcendence (Part III). My aim is to show how science and faith dovetail with each other. The present chapter is meant to find a methodology that is appropriate for doing so. Dialogue between science and theology, it seems to me, depends on an approach that makes it possible to indicate

(a) in how far science and faith deal with the same reality;
(b) where current scientific insights question traditional theological assumptions; and
(c) where theological assumptions necessarily go beyond the mandate and method of science.[109]

When I speak of faith in this treatise, I am referring to the Christian faith. The Christian faith tries to 'become all things to all people' (1 Cor. 9: 19-23). My attempt to become a scientist to the scientists led me to the approach of 'experiential realism'. This is how I perceive the positive sciences to be operating in practice.[110] Similar approaches are called 'critical realism'[111] and 'model-dependent realism'.[112] I will stick to this approach as consistently as I can and see how far it can take us.

In this chapter, I characterise the approach of experiential realism in contrast with other ways of dealing with reality. I also try to explain why scientists are not easily impressed by these alternatives. Then I will reflect on the mandate and method of the scientific enterprise. Which aspects of experienced reality does it cover, and which lie outside its realm? Finally, I will add a few reflections on the merits of a narrative approach to reality.

Methodological considerations tend to become tedious and controversial. The theory of science, epistemology and hermeneutics are extremely complex and contentious disciplines. Not to burden the reader, I will highlight only the most important facets of the proposed methodology. If my argument becomes too involved to maintain your interest, or too inadequate for your level of sophistication, I suggest you simply skip the chapter.

Section I
The approach of experiential realism

Occam's razor

Experiential realism requires methodological asceticism. There is a principle in science called 'Occam's razor' that I intend to apply in this treatise.[113] It calls for the simplest description or explanation that can do justice to the character of the phenomenon to be described and explained. It presupposes fewer assumptions, displays a clearer logic, is easier to grasp and wastes less time than more complicated approaches.

I know that simplicity should not become an idol, that simple explanations are in danger of becoming simplistic and that simplicity cuts down on the richness of possible aspects of the phenomenon and its environment. I endorse the wisdom ascribed to Einstein that "theories should be as simple as possible, *but no simpler.*" Therefore, I shall try to apply Occam's razor in a circumspect, yet bold way.

My reason for exercising methodological asceticism is that I want to maintain the lowest common denominator available between science and faith, which is *actual human experience.* I also want my work to be read by the educated non-scientist and the educated non-theologian alike. I also assume that modern people have a pragmatic outlook on life. They are not impressed by unexplained metaphors, highflying speculations, unnecessary sophistication and impenetrable obscurantism. For communication to succeed, clarity is of the essence.

Unwarranted sophistication is a particularly annoying aspect of academic production. One cannot help but feel that academics sometimes want to impress their peers rather than deal with the issues that confront humanity. Of course, some objects of study demand conceptual and experimental tools that are not readily accessible to the uninformed and untrained. However, Occam's razor demands that descriptions and explanations be as simple and accessible as the object of study allows. There are excellent examples of accessible treatises in the natural sciences.

For its part, academic theology is plagued with sophisticated obscurantism. Much of it is completely indigestible. Much of it does not seem to benefit the community of believers or the society at large. The reason seems twofold. One, the legitimacy of theology at institutions of higher learning is controversial. It would seem that one has to display high levels of inaccessible learning to demonstrate one's academic superiority. Two, competition for scarce top positions produces the need to impress one's peers.

Theologians should subject themselves to a reality test. If their arguments were to be presented to a group of people in a popularised form, whether in a Sunday service, a shop floor gathering, or a sport stadium, would those who attended leave the proceedings with the exhilarating feeling of new insight, disturbing challenge and comforting reassurance? Or would the message create puzzlement, frustration, or derision? In this case, our dialogue partners are scientists—can they make out what we actually want to convey? If not, why talk to them in the first place?[114]

The comprehensive scope of experiential realism

Science demands evidence, logical coherence, mathematical stringency, or at least, plausible conjecture. Note from the outset, however, that experiential realism is a more comprehensive approach than the quest for empirical evidence. It includes experiences of uncertainty and reassurance, pain and well-being, joy and grief, attraction and revulsion, beauty and ugliness, detachment and commitment, guilt and reconciliation, enslavement and liberation, redundancy and indispensability, freedom and dependence, doubt and conviction, assumptions and perceptions, sense of purpose and sense of meaninglessness.

All these experiences can be described, analysed and critiqued. They have consequences in the real world.[115] So they do form part of experienced reality. We must aspire to understand the network of relationships between all dimensions and aspects of experienced reality in their specific contexts, statuses and roles.[116] That is why modern academia includes the natural sciences (such as cosmology, physics, chemistry and biology), the social sciences (such as sociology, law, economics and political science), the historical sciences (that is, the diachronic dimension of all other fields), the human sciences (such as philosophy, psychology, linguistics, phenomenology of religion) and the arts (such as music, painting, drama, literature, poetry).[117]

The assumptions of experiential realism make different disciplines accountable and acceptable to each other as valid academic pursuits. The university offers a platform for interdisciplinary explorations, but we are far from breaking through the isolation caused by the professional specialisations and linguistic traditions of different disciplines. Theology too lives in a symbolic universe of its own. To become relevant, theology has to ask the question in which way faith relates to the entire set of academic assumptions, concepts, theories and findings, thus to the whole spectrum of academic research.

The limited applicability of experiential realism

The criterion of experiential realism can be defined as *accessibility, in principle, to human observation, comprehension and control*. In practice, this access is staggered. That humans become *aware* of something is an indispensable prerequisite of any kind of knowledge. *Comprehension* is a goal that is reached only to a certain extent. *Control* is curtailed by human limitations to a much greater extent. We may see the sun most clearly, have some limited understanding of how it functions, but have no control or influence over it at all. Influence and control do not go all that far beyond the limits of our planet on the one hand and subatomic particles on the other.

The reality we perceive can include fields, waves, particles, or events. It can consist of things, or relationships, or regularities. It can consist of empirical facts, information, direction-giving assumptions or hallucinations. A hallucination is real in the sense that it is a state of mind that shows up in brain scans and that has consequences in the real world.

Whether our comprehension of these aspects is adequate or not, they are all part of a reality out there that does not depend on the consciousness of the observer. There are also areas of human experience that are so elusive that they are not covered by straightforward observation, comprehension and control. They cannot even draw on imagination. This is particularly true for the field of quantum mechanics.

In such areas, we have to depend on mathematical conjecture on the one hand and intuition on the other. Both of these are legitimate but highly problematic and provisional forays into the unknown. They do not clash with the approach of experiential realism, but try to extend its frontiers. In contrast, experiential realism will try to avoid (1) fantasy or fiction that is given out as objective reality, (2) sceptical epistemology, (3) reified Platonic abstractions, (4) deductions from questionable assumptions, (5) reified myths and metaphors, and (6) 'postmodern' relativism. Let us briefly visit each one of them.

1. Fantasy or fiction

As long as fantasy and fiction are recognised for what they are, experiential realism will have no quarrel with them. Fairy tales, poems, novels and science fiction do not pretend to reflect an objectively existing reality out there. They are not illegitimate as works of art. They may have aesthetic, critical, suggestive and motivational value. They are wonderfully creative. They can highlight paradigmatic experience. They can convey profound wisdom. They can entertain. They can enlighten. They can communicate a message. They can stretch one's imagination and open up unrealised potentials—even if one is a hardnosed scientist.[118]

They are neither valueless nor trivial. But they are deliberate constructs of human consciousness. As such they can be described, critiqued, transformed, or abandoned. But they have no objective existence outside the imagination of their producers, owners and consumers. They may fertilise scientific thought and enrich the world of faith. But they are not helpful when it comes to the conceptual precision needed in the interaction between science and faith.

2. Sceptical epistemology

Epistemology tries to establish how the human subject comes to know the objective reality that might or might not exist out there. The epistemological discussion is very involved and highly controversial.[119] Though important, it is not part of the agenda of this book.[120] I will only offer a brief indication of how it fits into the argument of this book.

As I see it, the normal practitioner of science concentrates on the observed object. It can be analysed, critiqued and manipulated. Epistemology, in contrast, concentrates on the observing subject. It tends to focus on the individual self. Some philosophers have described it as narcissistic.[121] Consistent epistemology tends to become sceptical. Can my subjective experiences be trusted to reflect the objective world out there? Many philosophers have come to the conclusion that they cannot.[122]

Experiential realism, in contrast, takes the outside world for real and gets on with the job.[123] It is marvellously self-confident and amazingly successful in doing so.[124] It assumes that the impressions we gain from the outside world through sense perception and subsequent comprehension reflect a reality whose existence does not depend on our consciousness.

That the observing subject is inextricably involved in the observation of the observed object is self-evident. But the epistemological question as such does not provide a helpful basis for practical life or the practice of a science to overcome this problem. The proverbial man and woman in the street take it for granted that entities and occurrences they observe, comprehend and control actually exist. And so do scientists in the normal conduct of their professions.[125]

A few examples may illustrate this point. When I slip on a banana skin and break my hip, when a hailstorm wipes out my crop, or when somebody thrusts a gun into my face and demands the keys of my car, I stop wondering whether there is a world out there or whether it exists only in my mind. Reality is immediate. It demands my attention. It requires decision and action. The same is true for astronomers who discover a new galaxy, economists who follow movements of crude oil prices, and neuroscientists who discover brain waves caused by emotions.

The relation of epistemology to empirical science can be illustrated with a picture. Massive stars begin to look fuzzy and ultimately disappear out of sight the farther they are removed from the observer. But they do not dissolve into nothingness just because we cease to be aware of them. We only have to use more powerful instruments of observation to become aware of their existence.

Science treats even the observing subject itself as an observed object that can be researched by science, manipulated by technology, exploited by commerce and used by the consumer. So in fact, experiential realism turns the epistemological question on its head.

However, the fact that experiential realism focuses on the observed object rather than the observing subject does not imply that science is epistemologically naïve. It only has a more pragmatic approach to the epistemological problem. Let me briefly spell out this contention.

Serious science knows that human knowledge is *partial* in terms of space and *provisional* in terms of time. It tends to stick to an existing paradigm,[126] but in principle science is open to adjust its theories when more compelling insight emerges.

Science knows that knowledge is *perspectival*. The topography of a mountain range looks different from the East as it looks from the West. One can study its geology, its botany, or its entomology.

Science knows that the *capacity of humans to know is limited*, but constantly endeavours to extend its boundaries.

Science is aware of *distorted impressions* and tries to correct them by means of complementary observations and experiments.

Ideally at least, science can know and should know that reality is a *vast, integrated and evolving network of relationships*.[127] Science tends to take reality apart and study the behaviour of the components rather than the whole, but that is a question of methodology, not of principle.[128]

Science knows that knowledge is *based on assumptions* and designs experiments with varying assumptions to find out which of them yield the best results, or how they can complement each other.

Science knows that our access to objective reality is of necessity *mediated through our observational and interpretative apparatus*.[129] But this does not question the existence of objective reality, nor does it remove the importance of criteria such as evidence, plausibility, or reasonable conjecture.

Science knows that *all knowledge is situated in our minds*. But instead of becoming cynical about this fact, it conducts scans to establish precisely where certain emotive, perceptive and interpretative functions are located in the geography of the human brain.

The point is that all the insights mentioned above have been generated, or could have been generated, *from within the scientific approach to reality itself* and without the assistance of a subject-oriented epistemology. In principle, they are all geared to the criteria of evidence, plausibility and reasonable conjecture.

Facts, theories and hypotheses

Modern sciences have a pragmatic approach to the epistemological problem. Empirical facts are established by observing repeated and seemingly regular phenomena. Facts are taken by science to be true because of a 'preponderance of evidence' and can be abandoned if the evidence turns against them.[130]

On that basis, theories are developed. They are provisional models that are meant to extend human knowledge from the known into the unknown. Then inferences are drawn from such theories to form hypotheses. Where possible, empirical tests are conducted either to substantiate or disprove the hypotheses. If a hypothesis cannot be substantiated, the theory is changed or abandoned.

Taking their point of departure from known stellar constellations, for instance, astronomers could calculate that a hitherto unknown object must exist at a given location in space, train their telescopes on that position and find it. In economics, extrapolations of present trends are used to trace probable trajectories into the future if conditions remain constant. Trigonometric curves can reflect both the direction and the acceleration of processes. Fractual formulae are used by chaos theory to depict non-linear processes.

Where for practical reasons such tests cannot be conducted, as in the case in quantum physics, mathematical stringency has to stand in for empirical evidence. Where Mathematics does not lend itself to the object of study, as in the case of historical reconstructions, the most plausible conjectures must suffice. Where the object of study becomes so complex and diffused that evidence is unattainable, as in the sphere of convictions, we have no choice but to fall back on intuition.

Scientific results are subject to intersubjective verification, that is, a scientist in Japan, working on a similar topic under similar conditions, will normally arrive at similar results as a scientist from Brazil.[131] With some exaggeration one can argue that, in contrast with religion and philosophy, scientists deal with the same issues and speak the same language all over the world.

Because insight travels in traditions, one can argue that science always begins with a theory and then tries to integrate new observations into this theory. But no scientific theory is sucked out of the thumb. Obviously, theory and empirical observation form a system of feedback loops—the one cannot do without the other.

Experiential realism must be distinguished from *a blind kind of positivism* that takes the existing world, the existing worldview, the existing law, the existing political dispensation for granted. But the attempt to gain an objective picture of experienced reality makes critique and transformation possible.

3.   The reification of Platonic abstractions

Platonic idealism is based on the abstraction of the idea (or rationale) underlying a phenomenon from the concrete historical phenomena themselves. It is interested in revealing the hidden essence of something rather than its experienced existence. Its most simple form is a generic concept that summarises common characteristics. There are millions of leaves and millions of flowers of different shapes, colours and sizes, but the concept of a leave can be distinguished from the concept of a flower. Platonic idealism is based, therefore, on the observation that certain phenomena can be grouped together.

Abstractions such as *the* flower, or *the* human being, or *the* galaxy give similar phenomena a common name. In theology, world, law, sin, judgement, grace, reconciliation, salvation and death are abstract concepts. In science, energy, wave, substance, probability, causality, evolution, heredity and entropy are abstract concepts. Such concepts are indispensable. They register the existence of similarities, forms, laws, concepts, or universals.

But Platonism treats these abstractions as if they were real entities out there. It assumes that the concept we use for a particular kind of phenomenon defines the *essence* of a phenomenon in contrast with its *existence* or concrete appearance in the flux of time and the relativity of space. Therefore, it deems concepts more fundamental or more real than the concrete phenomena themselves.

Platonism then seeks to establish the authenticity of such an abstraction. It assumes that the pure idea is compromised by the material in which it manifests itself. Conversely, the pure idea exposes the concrete phenomenon as inauthentic. The idealisation of the abstraction then acts as an ethical benchmark. This has far-reaching implications. According to Platonism, the human spirit (soul, idea, essence) ought to extricate itself from the contaminating body (matter).

Platonism ascribes timeless ontology and universal validity to such idealised abstractions. They are deemed 'universals' that apply always and everywhere.[132] They pre-exist their actualisations in the experienced world. They hover above the emergence, evolution, deterioration and decay of concrete phenomena. And because they reflect essence, thus ultimate truth, inferences deduced from them are also deemed reflections of ultimate truth.

We call the practice of attributing ontological reality to conceptual expressions 'reification' (that is, treating concepts as if they were something real out there). This is where the danger lies. Concepts tend to assume a life of their own. They tend to overrule the irreducible uniqueness of entities and occurrences in our consciousness, rather than establishing the dynamic relationships that exist between them.

Once absolutised, they can easily be imposed on the flow of reality. They can lead to preconceptions, prejudices and ideologies. They can distort our dealings with the flow of reality itself. When reified abstractions are imposed on phenomena, anything unique that does not fit into the mould is either cut to size or excluded. Thus the concept of *the* (essential) woman or *the* (essential) African can be imposed on concrete women or Africans whether they fit into these prejudicial concepts or not. Then you are surprised and angry if the woman or African concerned does not behave as expected.

Certain combinations of abstractions can also push a particular social, economic, or political agenda. Prejudice, rationalisation and ideology are based on the generalisation of selected phenomena that are lifted out of their specific contexts and built into a seemingly plausible, situation-independent and timeless argument. Counter-evidence is simply ignored.

Science is great in insisting on factuality. Yet natural laws and mathematical formulae are Platonic abstractions. Whether you count apples or bricks, two times two is always four. The danger is that formulae are deemed to reflect a more fundamental reality than the concrete objects or processes which manifest such regularities. Then they can lead researchers astray. Empirical evidence must remain the ultimate validation of a scientific theory.

I have no expert knowledge in physics or mathematics, but I am suspicious, for example, of the physicist assumption that time is reversible. If you depict time in a geometrical system by means of a line, you have translated it into a spatial dimension. You can easily move to and fro on a geometrical line and express this by means of '+ **t**' or '- **t**', but you cannot move from future to past in real time. The additional 'dimensions' of reality assumed in string theory are also mathematical constructs. Whether there is anything in reality that corresponds to them is not self-evident.

In theological traditions based on Hellenistic metaphysics, relational concepts such as 'sin' and 'grace' can easily assume ontological statuses as unholy or holy 'substances' that can be transferred like water or salt. Endless confusion is the result.

We cannot do without abstractions. But we must be aware that they are drawn from the flow of reality we actually experience. We must be careful not to confuse their status as linguistic tools with the status of the reality from which they have been abstracted.

4. Logical deduction from axiomatic assumptions

The fourth approach under scrutiny is logical deduction from ostensibly axiomatic assumptions. This is an area where science clashes with much of philosophy. The word 'speculation' has pejorative connotations and should perhaps be avoided. As I use the term, however, 'speculation' is not fantasy gone wild or irresponsible behaviour at the stock exchange. It is the deduction of a system of inferences from assumptions that are taken to be axiomatic. Such inferences are deemed reliable reflections of the truth.

As long as one can be certain that the axioms are valid and the logic is sound, deductions may justifiably be taken to be true. But the validity of axioms and the soundness of the logic should not be taken for granted too readily. Mainline economics has taken for granted that the reserves of fossil fuels are virtually inexhaustible. When a non-linear trajectory is taken to be linear in physics, the conclusions may be misleading. The assumptions that demons cause epilepsy or that rituals can cure AIDS should not be taken as axiomatic, because that would obstruct effective treatment and can have fatal consequences.

The history of philosophy is replete with examples. Aristotelian teleology was an assumption that, according to physicist Arthur March, "arrested scientific progress for two millennia" by focusing on purposes embedded in phenomena rather than causes.[133] Hegelian dialectics are logical constructs imposed on reality. They have probably had their most enlightening and most misleading impact in the social, economic and political ideology of Marxism.

In theology, the assumptions of divine perfection, eternity, universality, omnipotence and omniscience are idealised abstractions from biblical reassurances. These reassurances were originally uttered in the context of actually experienced impasses and predicaments.[134] But when pastoral proclamations of God's power and commitment to people in distress were absolutised as axiomatic characteristics of a metaphysically conceptualised deity, they led classical theology into insoluble problems.

The concept of omnipotence, for instance, can mean at least three things. First, it can refer to the idea of an unconstrained will in charge of unlimited power. In this case, both 'will' and 'power' are idealised abstractions from concrete experiences of human will and human power in daily life.

When applied to the concept of God, omnipotence is a speculative postulate. It warranted neither by scientific observation, nor by the existential experience of faith, nor by the biblical texts. It leads to abstruse, misleading and irrelevant conclusions. Being omnipotent, can God kill himself? Can God go back in time and rectify what has gone wrong in the past? Can God reverse the process of decay in a corpse?

Second, it can refer to the total amount of energy in the universe that operates according to set regularities and constrained potentialities. Expressed in theological terms, all power in the universe in all its manifestations is the power of God, the Source and Destiny of reality.

As a believer, one can envision a higher kind of intelligence that is aware of actual and potential processes in the universe as well as the regularities according to which they function and that is willing to utilise these processes to launch a dynamic thrust into a particular direction. Performing a 'miracle' God does not have to suspend natural law! Though this notion is clearly an anthropomorphic metaphor, there is nothing in science that would contradict such an assumption.

Following the same logic, omniscience would refer to the totality of the actual information flow in the current situation; omnipresence would refer to the fact that one can never and nowhere escape the presence of the Source and Destiny of reality (beautifully expressed in Psalm 139).

Third, divine omnipotence can refer to the pastoral reassurance of believers in distress that human limitations are not God's limitations, that there are always potential options we cannot see or access, that God is for us and with us and not against us, and that we are justified to entrust ourselves to God's benevolence and power, whatever happens.

It does not follow from such assurances that God's creative and redemptive intentionality will bypass or suspend the constraints operative in the universe God has created. Such constraints must be seen as manifesting God's benevolent intentionality because without them reality could not function and God's 'action' could not materialise. Again such reassurances do not contradict science. We shall come back to these considerations in Chapter 12.

I do not want to suggest that it is possible to avoid abstractions, assumptions and deductions when trying to make sense of reality. But these must be made explicit. They must also be clear and plausible to any unbiased observer. They must remain hypothetical and provisional. Even natural scientists make metaphysical assumptions when constructing their theories. But they try to avoid supernatural claims, unreasonable guesses and incoherent logic. Their aim is to establish provisional theories that remain subject to the criterion of empirical evidence.

There are instances where evidence is difficult or impossible to come by. In such cases, science falls back on second best procedures. Subatomic physics uses mathematics to establish the nature of fields. History uses conjecture to establish past developments. Philosophy uses intuition to gain a picture of reality as a whole. In all these cases, the goal is to attain as appropriate a picture of objective reality as possible. In all these cases, the underlying assumptions must be exposed as what they are and scrutinised carefully.[135]

As a discipline dealing with *concepts* of the transcendent, which are part of immanent reality, there is no reason why theology should not base itself on experiential realism. Theology should be just as stringent and yet as flexible as science is, ready to question its own assumptions, incorporate new insights, and overhaul obsolete conceptualisations, even though it is located at a much higher level of emergence and therefore less defined and more open than the natural sciences.

When immanent experiences or expectations are projected into the transcendent realm, however, confusion begins to reign. Ludwig Feuerbach has argued that the Christian concept of God is a projection of unrealistic human wishes (e.g. immortality or omnipotence) and idealised abstractions of human experiences (e.g. power or love) into a non-existent heaven. It is remarkable that, against the background of this critique, modern theology has not become more circumspect in its pronouncements.

An example is the projection of the (entirely valid) concerns for egalitarian communal arrangements and female emancipation into an egalitarian fellowship between Father, Son and Holy Spirit. The egalitarian relationships within the 'divine' realm is then used to legitimate the struggle for social equity.[136] But the biblical tradition and classical Trinitarian doctrines cannot be recruited to back up such speculations. Hierarchy and patriarchy were taken for granted in pre-modern times. They only became problematic under modern socio-economic conditions.

Nor is such a procedure necessary. Their abolition must be based on the dysfunctions they cause in contemporary society. The theological legitimacy of equality and community can be deduced quite naturally from the message of God's suffering, redeeming and transforming acceptance of the unacceptable (or the redeeming love of God), which signifies the core of the Christian faith.

5. Reification of biblical myths and metaphors

More serious for the science-faith interface is the case where assumptions deemed unquestionable are based on reified myths and metaphors. This is the area where science clashes with a biblically based theology most fundamentally. Scientists are confident that their concepts refer to concrete realities out there. But they have huge problems with the metaphors used by faith.[137] Such metaphors do not seem to refer to anything real at all.

In a certain sense, all language is metaphorical. It is a set of symbols that is used to refer to certain aspects of experienced reality. When science speaks of a 'field', or a 'particle', or 'evolution', words taken from the language of daily experience are used to refer to less tangible aspects of reality. Because it is metaphorical, human language is polysemous, which means that a word can have many meanings.

The intended meaning depends on its context within a communicative tradition. Traditions have their own immanent history. They always refer to series of prior human experiences and their interpretations. This is also true for the sciences. When we use a metaphor to express divine self-disclosure, therefore, we must be mindful of the history of the meaning of the metaphor within the tradition in which it functions.

If these metaphors are traced back to the historical and existential motives that led to their formulation, however, one may find that they are rooted in the interpretation of concrete experiences. It makes a difference in actual life, after all, whether God, the transcendent Source and Destiny of reality, is seen as an impersonal mechanism, a loving father, or a merciless judge.

By its very nature faith depends on metaphors much more profoundly than science. The word 'God' denotes the transcendent Source and Destiny of reality as such and as a whole. If we call something transcendent, we want to say that it is not directly accessible to human observation, comprehension and control. Strictly speaking, it is not even accessible to human imagination.

We have no language for the transcendent. To give something ineffable a name, we must use the language we have. But such language is, inescapably, metaphorical, and metaphorical language takes its concepts from immanent experience. This makes them sound as if what they refer to is part of immanent reality, which they are not meant to do.

Why then speak about the transcendent at all? Because humans cannot live without meaning, identity, acceptability, reassurance, orientation and authority. These cannot be derived straight from experienced reality as such. Faith is persuaded that God, the transcendent Source and Destiny of reality, manifests 'himself' in the gift of meaning, acceptability and authority expressed in human forms of communication. The result is a *notion* of the transcendent that has a distinct content.

If the transcendent is perceived to have entered into a personal relationship with humanity, as in the biblical faith, it is the language of human relationships that we have to utilise. There is no other way. All the standard Christian concepts are taken from the sphere of personal relationships: father, son, spirit, creator, redeemer, healer, king, judge, law, grace, anger, wrath, goodness, mercy, life, light, power—you name them.

The same is true for myth. A myth is not a fairy tale. It is a symbolic narrative meant to explicate persistent experiences or motives within a particular cultural context. It is projected to the beginning of time to indicate that it is valid at all times. Myth is a pervasive and profound means of communicating transcendent meaning. A similar narrative is the parable. The parable of the prodigal son, for instance, has never happened. It expresses a truth about divine-human relationships that 'happens' all the time.

Although notions of the transcendent are taken by faith to be manifestations of the transcendent, they are part of immanent reality. But such concepts are believed to mediate between the transcendent proper and the community. They are based on communication. Christian theology explicitly admits this to be the case with its doctrine of incarnation. God communicates God's creative and redemptive intentionality (the *logos*) through human communicative channels. The 'Word of God' is proclaimed by humans on the basis of a millennium of human spiritual discernment.

The danger is that these means of communication are reified. Reification means that a mythological story or a metaphorical concept, for instance 'father' or 'son', is interpreted as a first-order reference to some ontologically existing entity out there similar to human fathers and sons. When metaphors and myths are reified, they miss their intended meaning, not only in terms of their content, but also in terms of their character as human conceptualisations of the self-disclosure of the transcendent.

When such reifications are attributed to divine revelation, they are given an ostensibly unassailable status. Used as axioms from which inferences are drawn, they can lead to highly implausible conclusions. Both axioms and inferences can then pose as revealed truths that have to be believed, whether they make sense or not. This kind of problem frustrates a Christian laity that has learnt to think in terms of empirical and historical evidence. How can Christ have a divine and a human nature at the same time? How can God have one essence, yet subsist in three persons?

For the interface between science and faith, this is anything but trivial. As the obstinacy of creationism and the persistence of apocalyptic assumptions in theology show, such 'truths' can begin to compete with scientific insights. The biblical statement that God created the world, whether with his 'hands' (Gen. 2) or by 'imperial decree' (Gen. 1), is an anthropomorphic metaphor. Taken literally, it clashes with big bang cosmology and the theory of evolution. Taken literally, the 'eschatological Kingdom of God' similarly clashes with the law of entropy. We shall come back to that in the last chapter.

When theologians and believers insist that the Bible must be right and science must be wrong, they regularly make a fool of themselves. A body of inferences drawn from reified metaphors that are taken to be axiomatic or 'revealed' is likely to become contradictory, unintelligible and ultimately meaningless. Metaphors and myths are meant to *point to* the transcendent. If the metaphor or myth is taken literally, the intended meaning gets lost and an idol takes the place of God.

More historically enlightened theologians who operate with such abstractions and metaphors must get rid of their obscurantist language and state the meaning they intend to communicate in absolutely clear terms. If we cannot sort out our conceptual chaos, we will not be able to enter into a fruitful dialogue with the sciences. We will not even be able to make sense of our own assumptions.[138]

The metaphors of father and son as an example

The metaphors of 'father' and 'son' are rooted in **first**-order experiences of the relation between a biological father and his biological son. A **second**-order discourse applies them to the relation between an adopting father and an adopted son. In the Ancient Near East, we find a **third**-order discourse that utilises these concepts to express the divine authorisation of a human king.[139] Here the metaphor expresses initiation, legitimacy, authority, justice, even caring love, but certainly it has nothing to do with biological procreation. A **fourth**-order discourse applies this notion to the relationship between God and Jesus, his messianic representative on earth. A **fifth**-order discourse applies it to those who share in the new life of Christ through the power of the 'Holy Spirit'. 'In Christ' we are all 'sons and daughters of God'.

If the difference between first-order referent and metaphor is not taken into account, one ends up in insoluble logical tangles. A confused believer recently asked me: who was the father of Jesus—the divine Spirit or Joseph, the spouse of Mary? If there was a divine father and a divine son, there must have been a divine mother. Alternatively, the human son of a human mother must have had a human father. The unlikely progeny of a divine father and a human mother would have to be a half-god or a superhuman being—both of which are rejected by classical Christian theology.

Moreover, if Mary was the human mother of a divine Son—one even associated with the work of creation—how can a creature of God be the mother of God, as popular Catholicism assumes? If that were the case, could she not claim divinity or at least priority of dignity and authority over her divine son? Again this eventuality was rejected by classical theology. Such conceptual confusions are rife within the faith community and totally incomprehensible for the scientific community.

Biologists can only shake their heads when serious academics engage in such discussions. Biologically speaking, a human male has XY chromosomes. These cannot be derived from a 'spiritual' entity. In scientific terms, we are here dealing with two different levels of emergence. Theologians too should know better. The metaphor refers to the initiation and empowerment of Jesus and his followers, not to biological origination. 'Birth of the Spirit' does not preclude, but presupposes biological birth. When we trace the metaphor back to its origins, the problem disappears.

The meaning of 'revelation'

But has the truth of God not been revealed? Christians are persuaded that it has. But 'revelation' of divine intentionality must not be confused with *information* about the origin of immanent reality or *explanation* of its functions. One should also not confuse revelation with metaphysical *speculation*. 'Revelation' refers to the self-disclosure of the creative and redemptive intentionality of God, the transcendent Source and Destiny of reality, not to the way the world is put together.

Assuming that there is something like revelation, how does it happen? Revelation can be thought of in two ways, top-down or bottom-up. The top down approach assumes that God has disclosed God's eternal and immutable truth once and for all through supernatural channels. The biblical are identical with the 'Word of God' as they stand. Believers are called upon to hear it, accept it and obey it in their dealings with immanent reality, whatever the circumstances.

The bottom-up approach begins with ordinary human experience. Dealing with immanent reality, humans are confronted with mysteries and calamities they cannot master and cannot avoid. They try to make sense of the situation. Confronted with what ought not to be, they gain the assurance that the foundations of reality are committed to what ought to be. God, the transcendent Source and Destiny of reality, is not hostile but benevolent. They also understand that well-being depends on prudent behaviour.

They exhort and strengthen each other with this dual message. Because it is deemed foundational they consider it to be the 'Word of God'. Gradually their horizons widen and their insight into reality becomes more profound. As the 'Word of God' responds to changing situations of need, it undergoes an evolutionary process that culminates in the Christ-event, yet continues to move towards a more acceptable future.

In the first case, God reveals 'himself' directly and unmistakably from heaven, as it were, without historical and situational mediation. It is not subject to time and space. It is eternally and universally valid. If experienced reality contradicts the 'Word of God', it is just too bad for reality. Reality must have fallen into sin and evil from which it must be redeemed. The Word of God stands; it is the world that totters. In the second case, God is taken to be the transcendent Source and Destiny of the very reality we experience. God reveals God's intentionality precisely through the evolutionary flux of reality.

Looking at the scriptural evidence without doctrinal bias, the first alternative is not tenable. It has Platonic rather than biblical roots. The biblical tradition is inextricably embedded in Ancient Near Eastern history. It explicitly acknowledges that God's revelation responds to changing situations of need. It clearly reveals a progression of insight into God's creative and redemptive intentionality.

Insight into the self-disclosure of God emerged and evolved in a historical process, was mediated by a series of redemptive responses to changing needs within a millennium of biblical history, and handed down in the form of a package of traditions. Theologically speaking, there is nothing otherworldly about this process. God always uses normal immanent processes to disclose God's creative and redemptive intentionality. Anthropomorphic metaphors are no exception.

6. 'Postmodern' relativism

Popular postmodernity and philosophical postmodernism, its more sophisticated counterpart, are extremely diffuse phenomena. I cannot possibly do justice to either within a few paragraphs.[140] For current purposes it suffices to refer to the idea that all convictions, assumptions, procedures, values, norms and goals are equally valid and therefore equally spurious.

Obviously, this is an overstatement, even a caricature of postmodernity. A good friend and ardent postmodernist told me that I am setting up a straw puppet that does not exist, with the sole purpose of shooting it down. But I formulated it in this crass form to make a point that can be overlooked only at our peril. Call it something else if you will.

That 'postmodernity' is such a diffuse development is symptomatic for its rejection of foundational assumptions. Some observers distinguish between three different interpretations of the phenomenon.[141] There are those who would speak of *'late modernity'*, rather than postmodernity, arguing that the emancipatory project of modernity has remained an unfinished task. They challenge us to move forward in the same direction. In this case, the emancipatory thrust of modernity is in full operation. Jürgen Habermas would fit into this category.

Then there are the *'true postmodernists'* who reject all inhibiting structures of meaning and social organisation, including those of modernity. They believe that all of reality is subject to historical flux, situational relativity and choice of perspective and want to enjoy life as it comes in its endless wealth and beauty. They argue that the constraints of modernity are oppressive, debilitating and hostile to life. They attribute the current social, economic and ecological crisis to the unrealistic assumptions and aspirations of modernity. In this case, the emancipatory thrust of modernity is radicalised, rather than channelled into responsible behaviour.[142]

Finally, there are those whose convictions and patterns of thought are based on *pre-modern* assumptions. Having been forced into a defensive and apologetic stance by modernity, many believers and conservatives in general are relieved that postmodernity seems to give them a new kind of legitimacy because it embraces all kinds of worldviews and institutions as manifestations of the riches of human cultural creativity.

The third variety is not truly postmodern. It is pre-modernism taking advantage of the postmodern rejection of the constraints of modernity. It is a 'cheap way out' of the dilemmas posed by modernity and a return to pre-modern commitments. It is deceptive because postmodernity is not pre-modern but ultramodern. 'Postmodernity' is not willing to buy into the absolutised assumptions of religious convictions. It is one thing to say that all faith communities are entitled to their own ways of looking at reality, however unfounded, ill-informed, or misleading. It is quite another thing to take seriously the claim that established faith assumptions about God, the spiritual realm and the earth are eternally and universally valid, as pre-modernists assume.

As far as I can see, practising scientists are at a loss how to deal with the phenomenon of postmodernism.[143] Fantasies, utopias, paradises and heavens have a nasty habit of suggesting that they actually exist or could exist while they don't. Conversely, to deem demonstrable natural and social phenomena 'linguistic constructs' or non-existent figments of our imagination, located only in our brains and nowhere else, is both silly and dangerous. "If the real world does not exist, then the claim of science to be describing the real world is nonsense . . . Nothing, but nothing, drives scientists crazy like these postmodernist claims."[144]

Being experiential realists, scientists are not able to live in the happy-go-lucky skies of relativism where people of all cultural extractions and religious persuasions float aimlessly and effortlessly from cloud to cloud, enjoying the riches of different perspectives, constructs and approaches in an unconstrained and non-hierarchical fellowship. True believers cannot either. Both serious science and serious faith feel the urge to establish the truth as far as human limitations allow them to do so. This is not just a pleasant pastime. What humans deem true has consequences for individuals, communities, society and nature. Truth simply cannot be left to chance or whim. It is way too serious.

Scientists can get very angry when you tell them that science and technology are based on the same kind of unsubstantiated assumptions that religions claim as their foundations. The real world contains gravity that pulls you down, physical walls through which you cannot move, chemical reactions that can flatten a suburb, biological processes that feed on carcasses, genes that make you male or female, mental conditioning that determines behaviour, assumptions that are far-fetched, procedures that do not work, collective interests that are brutally pursued at the expense of others, convictions that lead to bloody conflicts. The idea that all convictions, assumptions, cultures, procedures, values, norms and goals have the same dignity and validity is wishful thinking. It is also misleading in practical terms. Wooden hoes cannot compete with agricultural machinery. Magical spells of diviners will not cure AIDS. Traditional culture cannot be allowed to legitimate female subjugation.

This is not meant to discredit all 'postmodern' thought as such. Apart from radicalising the emancipatory thrust of modernity, postmodernism can also be considered a critical and corrective intellectual movement that has much in common with basic Christian motivations. Here are a few items on the 'postmodern' agenda with which I can identify.

1. Reality is in flux, contextually determined and embedded in a system of energy differentials. The whole of reality, with all its parts, emerges, evolves, deteriorates and decays. 2. The observation and interpretation of complex systems is model dependent. Insight is provisional, partial and perspectival. We must be open for different points of view, new discoveries and changing goals, rather than bound to supposed revelation, inherited traditions, fixed paradigms and preconceived objectives. Insight can never reach closure. 3. True knowledge is gained from participatory and interdisciplinary research and ongoing dialogue. 4. Absolutised discourses, paradigms and frames of reference must be exposed and dismantled as idolatrous, particularly when they serve hidden power interests. 5. All structures and processes must continuously be subjected to scrutiny and transformation. 6. Relations with other persons and one's place in the community co-determine one's identity and dignity. 7. Openness towards the other and the community is the prerequisite of true democracy, while the reckless pursuit of individual and collective self-interest is essentially

undemocratic and inhuman. 8. Hierarchies not based on indispensable functions within the whole and accepted by the lowest strata are not legitimate. And so on.

I have difficulties with the following assumptions that one sometimes encounters in postmodernist discourses. Wear the cap if it fits—if not, ignore it!

1. There is no objective reality. What we know represents nothing but constructs and prejudices located in the human mind. Language refers only to language that refers only to language. There is no objective referent. My response: Since Kant we know that sense impressions of the outside world are processed by our faculties of understanding. This does not imply, however, that there is no such objective reality, or that our understanding cannot reflect such reality, or that it cannot reach closer approximations to objective reality, or that there are no regularities that govern this reality.

2. All assumptions and convictions are equally valid and equally spurious. My response: Assumptions and convictions can be less or more appropriate, first because they can approximate reality to a greater or lesser degree, and second because they have consequences in the real world that can be beneficial or detrimental to the fullness of life.

3. A pre-modern version of 'postmodernism' argues that liberation is an illusion because we cannot escape historicity, contextuality and power relations. Old enslavements inevitably make way for new enslavements. My response: There are qualitative differences between authorities; there are degrees of freedom and enslavement; there is a difference between commitment based on responsibility and involuntary enslavement to authority.

4. The consciousness of individuals cannot create or change mental and social structures. Structural change only happens when cracks and flaws in the old paradigms and systems can no longer be patched up. My response: This is a fatalistic and unwarranted assumption. History is replete with great innovations in science, technology, commerce, social construction, charity, political organisation, worldview and religion. There is a dialectical relationship between personal consciousness and impersonal social structures.

5. All generalisations are metaphysical constructs without substance. My response: We must indeed be alert to flux, variability, contextuality, relativity and the ideological legitimation of collective self-interest. But generalisations are indispensable. There are millions of flowers in nature, but it is possible and necessary to distinguish a flower from a seed. There are billions of men and women in the world, but it is possible and necessary to distinguish a man from a woman. There are many religions and worldviews around, but there is a difference between Hinduism and Islam.

6. Because all human beings (in fact, all phenomena) must be accorded equal dignity, all hierarchies are evil. My response: There is a qualitative difference between a stone and an organism. There are differences in complexity, volatility and transience between emergent levels of reality. Social hierarchies based on superior knowledge, expertise, or function are indispensable. There is a difference between an elite that uses its resources to uplift the less privileged and one that tramples the latter under its feet. Criteria of acceptability based on whether attitudes and actions are life enhancing or life threatening are appropriate.

7. All systems of meaning and 'grand narratives' are oppressive, ideological and deceptive. My response: The deconstruction of structures essential to individual, communal and social life without reconstruction is counterproductive. Oppression must indeed make way for freedom, but freedom without structure, authority and accountability is a dangerous illusion.

We have to distinguish between various kinds of discourse

I do not want to suggest that the six approaches mentioned above are necessarily evil in themselves or that those who engage in them are dumb. Within their own limitations, they can lead to valuable insights. Particular frames of reference follow their own internal logic that may be persuasive and display considerable coherence and beauty. The metaphysical edifices constructed by early theologians up to the Middle Ages are impressive feats of intellectual rigour. But they are totally out of sync with modern scientific thought.

Fiction, poetry, myth and metaphor each have their place in human creativity and communication. But we must know what we are doing when we employ them as linguistic tools. Once we mistake them as references to empirical fact or theory in the scientific sense of the word, we misunderstand their rationale. In the science-faith interface, they regularly produce more confusion than clarity, more impasses than solutions. Science and faith want to enlighten, not to obfuscate. They must use linguistic tools that are able to transmit clear messages.

Modernity is characterised by a pervasively experiential, pragmatic, utilitarian and hedonistic frame of mind. Notwithstanding all epistemological provisos, science is geared to a particular kind of story, the story of what actually happened, happens now and will happen, whether humans are aware of it or not. If faith is to enter into dialogue with science, it has to respect this necessary self-confinement.

As part of my engagement with this down-to-earth orientation, I will be critical of the approaches mentioned. On the other hand, experiential realism is comprehensive and versatile enough to include all kinds of experience. The approach of experiential realism is, therefore, much more inclusive and much more profound than naïve empiricism or positivism. The theory of a hierarchy of emergences (to be discussed in the next chapter) provides a plausible rationale for this inclusiveness and profundity.

Experiential realism also raises the awareness of specific natural scientists to levels above their closely confined and truncated agendas. It explains why the social sciences, the human sciences and the arts must be taken seriously as legitimate pursuits at institutions of higher learning.

History of thought, phenomenology of religion and theology have a place in academia not because they claim to possess arcane knowledge of otherworldly realities, but because they deal with assumptions and perceptions as historically grown mental structures or 'symbolic universes' (Peter Berger). These notions can be described, critiqued, transformed, replaced or abandoned. They have consequences in the real world. It would be arbitrary and irresponsible to exclude this dimension of experienced reality from critical analysis and scrutiny.

Pragmatic hedonism—a friendly reminder

Before we continue, I have to remind the reader of an important distinction. My approach of choice is experiential realism. I try to show how this approach can help faith to clean out its baggage and engage science in a fruitful dialogue. We must realise, however, that experiential realism is an epistemological approach, rather than a conviction based motivation. I will have to say a lot about the motivation of the Christian faith in Part IV. Here it suffices to say that it is very different from the motivation of popular modernity.

The motivation of popular modernity is *pragmatic utilitarianism and hedonism*, that is, the hedonism of the consumer culture, the profit motive of commerce, the efficiency orientation of technology and the quest for mastery in science. In the population at large, it was the popular movement of pragmatism and hedonism that pushed the concept of God into the realm of fantasy, rather than philosophy or science.

Pragmatic and hedonist utilitarianism can appeal to science, technology and commerce when they seem to deliver the goods. But when things go wrong, it may turn to other sources of help and comfort. It is committed neither to religious truth nor to scientific fact. The recent upsurge of an appetite for mystery, spirituality and fundamentalism proves that. Conversely, science, technology and commerce follow a mercenary agenda as far as they serve the utilitarian and hedonistic motivation.

In popular postmodernity, 'belief' no longer means "perceiving the truth of something", not even "taking this as the basis for life",[145] but grasping an uprooted tree that happens to float by in want of something more dependable or satisfying. Both science and faith have to come to terms with this general cultural atmosphere.

People governed by desire can become like beach sand swept in all kinds of directions by advertising, entertainment, drugs, fashions, fads, the ecstasy of sport events, ideologies and populist demagogues. Such people may even turn to religious kinds of spirituality. Modern society is replete with cults and spiritual experiments. They may also turn to ecstatic, pragmatic and hedonistic forms of Christianity.

Some churches provide spiritual 'highs' as part of their package of offerings; some priests promise release from the eternal fires of hell; some ecstatic preachers offer participation in 'heavenly glory'. The 'prosperity gospel' of some communities generates mass appeal by overtly legitimating the pursuit of material self-interest. But all that is very different from what I wish to stand for in this book.

Section II
The mandate and method of science

In this section, I want to delimitate the mandate and method of science in relation to the mandate and method of theology—theology being the analytical, critical and reflective pursuit of faith. My aim is to show that the mandate and method of science are confined to the limits of immanent reality, while faith is located in the intuitive awareness of the transcendent foundations of reality.

Of course, there are also aspects within immanent reality that fall outside its method and mandate. This distinction is important because it precludes an absolutisation of scientific knowledge. Let us consider a few examples of aspects of reality that are covered by science and aspects that are not.

Aspects of reality covered by science

1. **Particularities**. As I want to use the term, a particularity refers to a unique phenomenon, entity, or occurrence.[146] It has some kind of impact on human consciousness. It can be described as such. This description can be intersubjectively verified, refined, or corrected.[147] Examples are a mosquito, a nuclear explosion, a political decision, a line of poetry, a metaphysical assumption.

Particularities are constituted by energy conglomerations that have a particular shape or profile, that undergo a particular process, that have a specific location in space and a limited endurance in time. Every particularity is unique in its origination, its characteristics and its consequences. However, it is embedded in a vast and dynamic network of relationships that includes a virtually infinite number of other particularities.

2. **Similarities or 'concepts'**. Similarities are characteristics that a number of particularities have in common. Similarities are expressed by means of concepts or generalisations. Examples are the concept of a chair, a storm, a mammal, a song, a triangle, an ancestor, a species, or a war.

3. **Regularities**. Regularities are typical spatial structures, temporal sequences and logical progressions in the relationship between particularities. Examples are geometrical relationships in three-dimensional space; the constant movement and irreversible sequence of time; causality, gravity, entropy, fractals and the behaviour of complex systems that can, at least in principle, be expressed in mathematical formulae.

Regularities deemed universal are formulated in laws. Regularities are combined and extended by inferences to form theories. Assumed regularities that need to be substantiated are formulated in hypotheses. Regularities abstracted from their contexts and applied to new contexts to enhance comprehension, generalisation and prediction are called paradigms.

4. **Information**. Encoded information lies at the heart of the entire process of emergence from the quantum level through biological evolution to human communicative systems. Although information only manifests itself as far as it is embodied, it cannot be dismissed as something unreal.[148]

5. **Random behaviour**. Random, chance and contingency indicate phenomena discussed and explored by the sciences, but they are difficult terms to define. Do they refer to events that happen "out of the blue" without causal antecedents whatsoever? Or are they too complex to be expressed by means of mathematical formulae? Or do they defy prediction because of the 'chaotic' behaviour of complex systems? Or do they appear to be random because we are unable to fathom all the causative factors that impact on their origination?

For my purposes, three phenomena are worth noting in this connection. One, whether there is something like pure chance in the universe is a contested issue. Perhaps humans are just not able to register the multitude of hidden causes that underlie apparent chance events. This question is important in the discussion of the theory of "non-interventionist direct divine action", which assumes that God acts specifically in situations that are not determined by causality. As far as I am concerned, this is another case of a "theology of the gaps" mentioned above. I will discuss this in Chapter 8.

Two, there are situations that are so finely balanced that any impulse from anywhere in the cosmos can determine the direction of the ensuing process. Chaos theory refers to such situations as "high sensitivity to initial conditions". Such impulses can set off processes with non-linear trajectories that have exponentially growing effects.[149] In Chapter 6, I utilise this phenomenon to explain "free will" and "switches" that lead processes in particular directions.

Three, wherever the probability of an occurrence can be established, there must be some kind of regularity that governs it, however diffuse and complex, otherwise there would be pure chance rather than probability. In this case, the lack of predictability is due to a degree of complexity or immensity that goes beyond the threshold of human observation and comprehension. This consideration is important for my take on entropy and derived phenomena such as biological death.

A few empirically based concepts[150]

A law of nature is a sequence of cause and effect that has been observed repeatedly and no exception has ever been encountered.[151]

A model is a schematic representation of reality that abstracts only the most important properties of the particular relationships under review.

A *hypothesis* is an unverified scientific conjecture that combines all known factors with intelligent guesses concerning unknown factors and uses the combination as a provisional assumption that awaits empirical substantiation.

A *theory* is a concise and coherent but provisional system of observations, laws and hypotheses that selects aspects deemed significant, that is meant to reach a more comprehensive understanding and that is capable of explaining and predicting (or 'retrodicting') certain phenomena.

The procedure that moves from facts to theories is called *induction;* the procedure that moves from theories to predictions is called *deduction*. The two processes form a dialectic, or a 'feedback loop', from theory to research, from research to theory.[152]

A *paradigm* is a model of reality that is taken as a point of departure by a particular research community and that defines and constrains the parameters for research.

A few non-scientific procedures

Intuition is based not on facts, but on an awareness of greater networks of relationships that are not immediately demonstrable or self-evident. Intuition plays an important role in scientific discoveries, because it explodes the rigidity of current assumptions.

Fantasy, poetry and fiction are based on pure imagination without the need for empirical verification, logical coherence, or plausible conjecture. Science fiction has become popular among people fascinated by science and technology. It may occasionally prompt serious science to venture into uncharted territory.

Speculation is not based on facts, plausible conjectures, provisional theories, or pragmatic paradigms, but on inferences from untested assumptions.

Aspects of experienced reality not covered by the natural sciences

The natural sciences are not designed to cover the entire spectrum of experienced reality. As mentioned above, the object of experiential realism is more comprehensive than the object of empirical science. Some phenomena fall outside the mandate and method of the sciences, but can be deemed part of experienced reality.

1. Subjective experiences and objective reality

To begin with, experiences come into play at a personal level that are not intersubjectively verifiable in terms of science or operationally efficient in terms of technology. Examples are repulsion, attraction, aesthetic appreciation, sensual enjoyment, ambition and personal judgement. Such 'subjective' experiences are called *qualia*.

Emotions are based on the hormonal and neurological infrastructure of the brain. They interact with the higher cortex and even with the immune system.[153] However, the fact that these experiences involve electrical and chemical processes does not explain their experiential character. To subject them to empiricist criteria would miss or distort their nature. Yet they are real. Humans have similar experiences of this kind, not only with regard to the formal capacity of having such experiences, but also in terms of their content and their consequences. Let us look at a few examples.

When I am at the beach, I do not want to analyse the spectrum of sunlight or the composition of seawater, but bask in the heat and battle with the waves. I share this experience with countless others. When I am struck with the beauty of a painting, I do not notice that it is made up of canvass, linseed oil and artificial stains. These attributes of the painting are irrelevant unless they produce special artistic effects. A female model appeals to a great number of men.

When falling in love, scientists may legitimately interpret my preoccupation with a young lady as part of the functions of my hormone system, the hard wiring of my brain or a neurosis. And yet, by doing so, they would miss the existential significance of the phenomenon altogether. The moment I reduce my romantic moments to the operation of my hormones, they cease to be romantic moments.

Ethical, aesthetic, romantic, ecstatic and mystical experiences cannot be dismissed as baseless fantasies; they are real for those who experience them; they are shared by many others; they have definite consequences. There is something in objective reality that triggers such experiences, and there is something in our internal experiential capacity that responds to these triggers.

The fact that such attitudes and experiences are 'subjective' does not make them illusory. An abandoned lover can commit suicide in desperation. The pangs of hunger are 'subjective', but no starving person would deem them unreal, inconsequential, or insignificant. The causes and consequences of the discomfort can be described in physiological terms; the discomfort itself cannot.

Feelings of satisfaction, attraction, pleasure, or beauty represent distinct phenomena that have objective preconditions, causes and consequences in this world, from personal relationships through market behaviour to politics. They determine the impact of events and actions on the quality of life. They can trigger reality-transforming actions. They can mobilise powerful social movements. They underlie the pragmatic and hedonistic drives of the consumer culture, commerce, industry and science. So the presence of fine art, music, or drama at institutions of higher learning is entirely appropriate.

2. Historical flux and the laws of nature

The natural sciences look at particular phenomena, find similarities and establish regularities that can be built into theories with assumed universal validity.[154] They are interested in patterns that can be explained and generalised. Patterns allow predictions or extrapolations of current trends into the future.

According to science, natural laws are recurrent regularities whose validity can be ascertained by repeating particular procedures. A single deviation from the regularity that defines a law of nature would amount to a falsification of the underlying theory. In a certain sense, science is akin to Platonic metaphysics, which abstracts 'essences', ideas, or forms from the flow and plurality of reality.

Fair enough. But humans experience the 'flow' of time. The flow of human life, especially, cannot be described in terms of generalisations.[155] Human beings experience reality as sequences of events that are unique in personal, situational and historical terms. They cannot ever be repeated. They cannot ever be reversed. Scientific results, in contrast, are "universal and hence ahistorical".[156] Theoretically, such a sequence can be run backwards towards its origination.

Above, I have defined the word 'fact' as used in scientific terms. However, we have to distinguish natural laws from facts. Facts are specific outcomes of historical developments. The Latin root of the word (*factum est*) refers to what 'has been done' and cannot be undone. That I was born a male is a fact. That the Americans have dropped a nuclear device on Hiroshima is a fact. A fact is not a law. It cannot be generalised. There is no second instance of such an event. It is unique. This divergence in the approach to phenomena is fundamental.

Say I am watching a soccer match. If I focused on regularities, I would discover that all football fields have the same size, all balls have the same shape and all teams consist of eleven players wearing similar outfits. But these commonalities in themselves are boring. The excitement is provided not by the regularities, but by the unpredictable and unrepeatable. The outcome of 3:1 in a match between two particular teams at a particular time and place is a fact. As such it is unique and significant. It is this uniqueness and significance that prompts the gush of adrenalin.

The flow of life consists of sequences of unique occurrences. They are triggered by historical switches at points of high sensitivity to initial conditions, whether through natural causes or human agency. Each one of these switches changes the direction of the world process. They set off causal networks that continue to impact reality at least up to the end of life on earth. At least potentially, faith is highly conscious of this responsibility.

To ignore or deprecate the study of this basic character of reality as 'unscientific' would be foolhardy. In fact, there is no reason why science should do that in the first place. Platonic abstractions from the flow of time can lead to a seriously distorted or truncated view of reality. The theory of evolution is geared to time, although there are persistent regularities that guide the processes. The same is true of big bang cosmology or the development of foetuses and infants.

3. The dignity of unique phenomena

All real phenomena are unique. All common characteristics are abstractions from reality. We cannot do without them, but we must not lose the irreducible historical and substantive uniqueness of all of reality out of sight. Because any future is open within the parameters set by the preceding past, every given state of cosmic reality is contingent, thus unique. According to economic theory, scarcity determines value. Applying this principle to the realm of human experience, it stands to reason that awareness of the uniqueness of entities and occurrences is of fundamental importance both for humans and for their relation to the rest of reality.

Biological species are valuable beyond their potential utility for humans precisely because they are unique. Genuine ecological concerns go beyond maintaining the available gene pool for future human utilisation. Genuine insight into history goes beyond the lessons we can learn from the past. Historical phenomena have a truth and a significance of their own. Individual human beings are unique in spite of their common gene pools. Their infinite dignity does not only lie in the assumption that they are all "created in the image of God", whatever that may mean, but that every single individual is totally irreplaceable.

4. What ought to be

Science is not competent to engage in value judgements. But they cannot be avoided. They can also not be suspended while a scientific investigation is in process. A non-human scientist could probably say 'Let the oil spill in the Gulf of Mexico continue and see what happens!' or 'Let us manipulate the human gene pool in such a way that babies are born without solid spines and see how they will cope!' Such investigations would be scientifically appropriate, but inhuman and irresponsible.

Naturalists such as Dawkins, Kauffman and Swimme quite obviously appreciate the aesthetic and ethical dimensions of life. They are struck by awe when perceiving the natural world in all its majesty, beauty and horror. They are upset by what they deem inappropriate interpretations of reality. They ascribe the origin of this wonderful world to the equally wonderful but comprehensible process of evolution.

Fair enough, but evolution as such does not tell us why we should appreciate anything at all. Value judgements are not part of the scientific mandate and method. But they do fall into the realm of convictions. Faith can become the custodian of the dignity of the specific phenomenon.

5. Specialisation and the multidisciplinary coverage of reality

It is ever more difficult to gain a unified picture of reality. Very few scientists bother to 'view all fields together' in a holistic way and draw out the consequences. Because faith is aware of the transcendent Source and Destiny of reality as a whole, it has the potential of gaining comprehensive horizons. It is unfortunate that for political, financial, social, psychological and practical reasons attempts to reach a holistic view of reality and an interdisciplinary methodology are scarce and rudimentary. Increasing specialisation has led to what some unkind observers have called "professional idiots". Such specialists know everything that can be known about a tiny aspect of reality, but next to nothing about other aspects, let alone a picture of the vast network of relationships that determine their overall context.

Perceptive and enlightened scientists know that the compartmentalisation of fields of knowledge, as well as their different languages and incompatible methodologies, present major challenges. The interaction of social, psychological and biological factors, for instance, is increasingly being accepted within the scientific community. Socio-psycho-somatic medicine is a case in point. There is no reason in principle why science could not develop a comprehensive approach to reality. However, there are at least four reasons why this does not happen in practice.

First is the complexity of experienced reality. Each level of emergence displays an evolutionary (and entropic) dynamic that causes reality to spread out into immense variability. Practical necessity dictates that every science focuses on a particular dimension of reality. The sciences are isolated from each other and fragmented within their own domains. There is very little interdisciplinary cross-fertilisation, augmentation and correction.

Each one has its own history, tradition and language. Within each science, there are further specialisations dealing with ever more circumscribed areas of research. Each research project deals with a particular abstraction from reality as a whole. Each follows a particular approach. As a result, every science has a truncated view of reality.[157] When sciences never take note of each other's insights, their findings may become misleading and their prescriptions may become dysfunctional.

Second, a reductionist bias has led conventional science in a particular direction. It takes wholes apart into their components. The assumption is that a careful analysis of each of the ingredients can explain the operation of the whole. The most elementary entity or the most fundamental formula can provide the key for a comprehensive understanding of reality. Their dynamic interaction within a greater whole is not necessarily taken into consideration. Concentration of the specific at the expense of the comprehensive yielded spectacular results—and equally spectacular failures. As we shall see in the next chapter, the theory of emergence overcomes this impasse.

Third, we must take note of the typically modern value system. Science is all about mastery. Technology has an instrumentalist attitude towards reality. Commerce has a self-interested valuation of reality. The consumer culture is geared to utility and pleasure. It is the usefulness of the part that matters, not the beauty and dignity of the whole.

Fourth, particular sciences tend to extrapolate the assumptions and rules of their particular fields to aspects of reality beyond their fields. Physicists tend to interpret reality in terms of physical causality, biologists in terms of evolution, economists in terms of market forces, Freudian psychologists in terms of sexual drives, theology in terms of personal intention and purpose.

Liberal economics, for instance, tended to ignore the insights of cultural anthropology, social psychology, political science, social ethics, ecology and other related fields. As a result, it produced massive failures in the fields of economic sufficiency, social equity and ecological sustainability.[158] Similar failures happen routinely in the medical field when the body is seen as a machine rather than an infinitely more complex whole. The interface between biological, nutritional, psychological and sociological dimensions of human life is frequently not taken into consideration.

6. The knowable unknown and the unknowable

Human observation, comprehension and control have limitations. Even our imagination has limitations. The window through which we observe reality is tiny and opaque because the human capacity to observe and comprehend is limited and compromised. The facts and regularities that the positive sciences have unearthed may be spectacular and entirely valid. We can also expect more such facts and regularities to surface and existing findings to be refined and redefined.

But all that taken together will not and cannot possibly tell the whole story. The reason is that we are human and not divine. Even if we were able to set up a streamlined interdisciplinary and multidimensional enterprise that would share a common discourse, that would function optimally in all respects and that took all the dimensions of experienced reality, their dynamic developments and their location within the entire network of relationships into consideration, we would not remove the mystery of reality as such. Faith is exceptionally aware of this mystery.

Humility demands that we acknowledge our limitations. The greatest scientists have been very humble people. The realm of reality that is, in principle, accessible to human observation, comprehension and control is embedded in a much larger whole. This whole certainly includes whatever is accessible to our comprehension and manipulation, but it also goes far beyond what we can ever know or influence.

7. Explanation and meaning

Our deliberations so far were confined to the immanent realm, the realm accessible, in principle, to our observation, comprehension and control. Whatever we know, whatever we could know if we had the tools, and whatever we will never know even under the best of circumstances because we are human and not divine, is still part of the same immanent reality that we experience—the reality that naturalists call nature. Because of its restricted mandate and method science is not able to go beyond immanent reality and answer 'ultimate questions'. These questions are not about observation, explanation and prediction, but about meaning, acceptability and authority.[159]

What is the most basic foundation of reality? What is the identity of the individual, the community, the society and humanity within the whole of reality? Do humans as such have a legitimate place within the reality they experience? Is the human being as a species meant to be an outcast, a foreigner, a custodian, or a dictator? In how far is their subjugation and exploitation of reality legitimate? The answers to all these questions depend on the particular system of meaning within which humans operate.

We shall come to problem of transcendence in Part III. Here it suffices to state that in terms of an immanentist perspective, worldviews, ideologies and religions are formidable spiritual and social powers that lead to massive consequences. A system of meaning derives its power from the fact that it responds to the human needs of stability, reassurance, belonging, identity, acceptability and authority.

A sense of belonging, for instance, is of critical importance for human existence. Foreigners, outcasts and dictators do not belong. They stick out like a sore thumb. Being integrated in the patriarchal hierarchy in African religion, belonging to the 'chosen people of Israel', being part of the 'messianic' proletariat or the revolutionary avant-garde in Marxism, being ready to sacrifice your life for king and country, having earned a Nobel prize in science, keeping up with fashions and fads in the consumer culture, falling in line with the expectations of a peer group, platoon, or gang—these are fundamental and explosive motivations that have massive consequences.

For all their perceived autonomy and sobriety, even scientists cannot escape the power of their need to belong, to be acceptable, to have authority. They cannot operate unless they are recognised as legitimate members of the scientific community. They share the collective ethos of this community. They need to prove their excellence. They need an institutional infrastructure, public funds, recognition of their legitimacy and respect for their status.

More profoundly, however, scientists must be sure in which way they can legitimately belong to the very reality they investigate. Are they an integral part of it, accountable to an authoritative supra-structure, or are they detached and sovereign masters over it? Although very few modern people ever reflect on this question, sovereign mastery over reality cannot simply be taken for granted. In fact, it has proved to be destructively dangerous.

Intriguingly, Stuart Kauffman, a sensitive naturalist, assumes that humans can feel "at home in the universe" once it can be shown that they are probable, expected and thus legitimate products of the evolutionary process.[160] Expressed in anthropomorphic terms, humans are nothing 'foreign' to reality, nothing 'unwanted' by nature, nothing 'illegitimate' in terms of evolutionary regularity, nothing 'imposed' by fate or chance. But does belonging not also imply obligation? Will a person who belongs to a context not treat this context in a more circumspect and responsible way than a person who deems him/herself its sovereign master and owner?

Kauffman's concern that the human being really 'belongs' provides a striking parallel to the Christian assurance that God, the ultimate Source and Destiny of reality, accepts the human being in spite of its depressing finitude and hair-raising ambiguity. The difference is that in Kauffman's scheme of things, humans belong to a self-generated, self-sustaining and self-referential 'nature'. The evolutionary process here functions as the foundation of human existence. There is no transcendent affirmation of meaning, legitimacy and authority.

But why restrict the need of meaning, belonging, identity, acceptability and authority to humanity? One can easily extrapolate these needs and reassurances beyond humanity to other creatures, even to reality as such, of which humanity is just a miniscule part. Awareness of the right of reality to exist is the prerequisite of the awareness that humans have a right to exist. But then other creatures also belong, are acceptable and have a right to exist.

Reality itself 'belongs', is no stranger, has a legitimate place within its wider context, has a right to function as it does, has dignity, value and meaning, has not simply happened, but is expected and treasured by—well, by whom or by what? This is what the awareness of transcendence, as found in faith, is all about. In contrast, the 'nature' of naturalism does not belong to anything and owes nothing to anybody or anything. It simply exists. It is a self-generating, self-sustaining, self-sufficient, self-contained and self-destructive ultimate. There is nothing beyond. I will draw out the consequences of these alternatives in Chapters 8 and 9.

Section III
The role of narrative in science and faith

The postmodern concept of 'narrative' gives us a useful handle to define the relation between faith and science, conviction and knowledge, orientation and explanation. But then we must introduce a few distinctions. As we have seen, science, though entirely incapable of operating outside interpretative frameworks, assumes that there is a reality out there whose flow does not depend on human observation and interpretation. Contrary to common belief, subatomic indeterminacy is no exception. As we have seen, science is undeterred by the fact that total objectivity cannot be reached, but concentrates its energies on gaining as objective a picture of reality as possible.

Conversely, faith, though entirely incapable of escaping the realities of this world, concentrates its energies on valid, meaning-giving interpretations of experienced reality. Such interpretations are not constrained by the demand for empirical evidence. Faith narratives express significance and validity rather than evidence and historicity. Significance depends on the assumption of transcendent validity. Food, for instance, is only significant because the life it sustains is deemed important. Hailing from prehistoric times, this kind of story often takes the form of myth, metaphor, parable, ritual and symbolic representation. Let us characterise the two kinds of narrative.

The scientific narrative

Both science and faith are *embedded* in narrative.[161] The reality that science investigates is a process in time, thus a story. Big bang cosmology is a story. The operation of the 'law of entropy' is a story.[162] Emergence theory is a story. Complexity theory is a story. The evolution of life is a story. Chaos theory, sensitivity to initial conditions, fractal mathematics—all these exciting new insights reflect a reality in flux. Even the accumulation of scientific insight itself is a story. Even more pronounced is the narrative character of decision making processes that make up actual human lives and ongoing human history.[163]

Such narratives are indispensable in any effort to comprehend and interpret experienced reality as explored by the sciences. As such they fall solidly within the ambit of experiential realism. They make us aware of developments in the past that have led to present situations. They define the parameters within which the present opens up a spectrum of future options. They make us sensitive to the kind of consequences different options would entail. They reveal which options have been selected and realised. They show us how processes unfold over time.

Narrative as descriptive and interpretative tool has been neglected in Western thought for a number of reasons. Platonic thought abstracted "essence" (which is timeless) from "existence" (which is in flux), drew inferences from seemingly incontrovertible axioms and developed a syllogistic logic that operates with static concepts. Under its influence theology aimed at eternal, universal and harmonious truth. Science similarly tried to establish laws of nature that can be expressed in mathematical formulae and are universally applicable regardless of circumstances and historical flux.

Abstraction from the flow of the cosmic process has been one of the most powerful tools of Western thought and accounts for much of its success. However, something incredibly important gets lost if one focuses primarily on abstract concepts and formulae rather than on the reality from which they have been abstracted. Reality only exists in the form of emerging, evolving, deteriorating and disappearing 'bodies' and the dynamically unfolding relationships between them. Abstraction 'disembodies' reality.

In terms of science, the merit of 'narrative' is that the parameters of time, space and power relations obtaining in concrete situations are taken seriously. There is no static, timeless, universal truth that is independent of a reality in flux. Once this is recognised, it becomes imperative to take 'narrative' as the **first**-order description of reality. The establishment of regularities that determine or guide these developments, characteristic of the sciences, is a **second**-order pursuit. It takes the first order as its 'material' and is only true as a property of the latter.

The Platonic assumption that the idea, which gives material reality its shape, pre-exists and can subsist without the latter cannot be upheld. No natural law, mathematical formula, or geometrical shape—the triangle, the parallelogram of forces, or the bell curve—exists apart from concrete, situation-bound and changing things and relationships. The question what actually happens is primary to the question of which rules it follows when it happens. For science this should be self-evident.

There are at least two kinds of narrative in science. For both of them historical precision is critically important. First, the fundamental ambition of science is to expose the objective narrative of the emergence and evolution of reality and its possible continuation into the future. It is taken for granted that there is such a process out there that does not depend on human observation, interpretation, conviction, or interest. The epistemological concern that humans can never observe and describe the 'thing in itself' (Kant) in its entirety, profundity and complexity is not a deterrent but a challenge to come to as close an approximation to what happened as possible.

Second, there is the narrative of the emergence and growth of human insight into this objective narrative. The history of science is the history of blunder and discovery. Scientists are not ashamed to own this history, but tenaciously press on towards better approximations of insight to objective reality. Part of this endeavour is the narrative of the ways and byways of how individual or discipline-specific insights contradict or support each other and how they mesh with the collective insight of the science community as a whole. Again this vibrant discussion is part of the process.[164]

The narratives of faith

Modern hermeneutics has come up with the immensely fertile exploration of 'narrative' as the location of truth claims that anchor human existence in transcendent validity. In contrast to the narratives of science, the narratives of faith do not have to be based on history. Legends, myths and parables are narratives that provide meaning, identity, acceptability, direction and authority, whether they have a historical core or not.

They define the location of individuals in communities, communities in societies, societies in humanity, humanity in the rest of reality. They are culture specific. They differ vastly. They can be life enhancing or life obstructing. They can be destructive of individual life, community, society and the natural environment. The validity of their messages as criteria of acceptability must be subject to intense scrutiny.

We can distinguish various kinds of narrative found in the biblical faith. *Myth* anchors experienced reality in primeval origins. The projection of the story to the beginning of times indicates that it is valid for all times. *Legend* may contain some historical memories, but these are recounted solely to interpret experienced reality. *Parable* has no historical referent, yet unlocks existential truths. Historicity is not part of the agenda of either these forms.

The creation story found in Genesis 2 and 3 is an example of mythology. *Adam* is the Hebrew word for human being, *adamah* for the soil. We are all Adam; we are all taken from the soil and return to the soil. That women are subject to men, yet closer to the latter than their flock, that humans fall into temptation and mess up their potential well-being, that they have to make a living and face frustrations in doing so, that they suffer pain when raising progeny—these are common experiences of an ancient peasant life knit together in a significance story.

Personified animals and plants are typical for such stories—the snake, the tree of discernment of good and evil, the tree of life. The divine too is depicted in crudely anthropomorphic terms—God moulding the human being with his hands, planting a garden, calling Adam while enjoying a walk in the evening breeze, making clothes for them, and so forth. The subsequent story of Cain and Abel in Genesis 4 displays legendary rather than mythological characteristics. It reflects the conflict between settled agriculturalists and land-seeking nomads. In this story, the nomads claim priority in the eyes of their God in the face of the hostility of settled communities.

The motive behind the creation narrative found in Genesis 1 is neither mythological nor legendary, but theological. It depicts the onset of a cosmic history that, according to the Priestly Source of the Pentateuch, culminates in the temple cult in Jerusalem. The cosmos has been constructed according to a master plan; humans, the final species to emerge, are created in the image of God, that is, as representatives of God; they are entitled to utilise the rest of creation, but also entrusted with the task of ruling over, and caring for, the latter. Most important, the narrative anchors the sanctity of the Sabbath, one of the most basic identity markers of post-exilic Judaism, in primeval history.

'Prophetic' and 'eschatological' narratives do not project legitimating origins but challenging outcomes. Both these stories provide visions of what ought to be. *Eschatology* envisages the comprehensive well-being of the human being in a reconstructed cosmic context. Its projections to the end of times indicates that it provides a vision for all times. Such a vision generates the urge of a mission. It posits values, norms and goals. It arouses motivation and galvanises action.

The eschatological story of the apocalyptic movement deliberately and pervasively uses mythological, metaphorical and symbolical language taken from earlier traditions. This shows that its intention is to communicate validity and normativity, rather than to predict historical futures. It provides reassurance and opens up horizons of hope in desperate situations.

Biblical *'prophesies'* have a similar agenda, but they are much more down to earth. They announce alternative futures within ongoing history—either blessing or curse. Such futures have rarely ever materialised as originally envisaged, but that was not their rationale either. They warn secure believers that unacceptable attitudes and behaviour have dire consequences and reassure the downtrodden that God is about to open up the future for them again. Both prophesy and eschatology are fundamentally different from oracle, divination, futurological extrapolation of current trends, or scientific predictions.

Faith can also attach transcendent significance to particular *historical sequences* of events. Such events are then deemed prototypical or foundational for the community concerned. They can define individual or collective identity, as in the case of the Israelite patriarchs, or the exodus from Egypt. They can legitimate social, economic and political power structures, as in the case of the kingship in Israel, or the establishment of the post-exilic theocracy in Jerusalem. The crucifixion of Christ is the most foundational significance story of the Christian faith.

While such a story may be rooted in history, its rationale is not historicity but validity. It constitutes a system of meaning, defines criteria of acceptability and grants authority. It defines identities and locates individuals and communities within their wider contexts. It provides the 'anchor' of human existence in cosmic history and the 'compass' for human action in the flux of time and the multiplicity of phenomena. Its importance for human life is similar to the force of gravity without which life and its material infrastructure could never have emerged and evolved.

For the biblical faith, this kind of story is fundamental. It picks up distant memories of historical incidents and developments but only on a very selective basis.[165] These events are given a particular interpretation. The aim is not to reach historical precision but to reinforce existential and communal assurances and challenges.

Therefore, the authors and communicators of this kind of story do not feel bound to historical facts and their previous interpretations. They merely utilise them as tools to convey their message. They can augment, change, or transform inherited narratives in response to changing circumstances. They are not ashamed of the ancient forms of the tradition they inherited but freely adapt them to enhance their communicative power.

The symbols and metaphors used in the significance story can be ancient. Biblical examples are the Creator as potter, emperor, shepherd, or king, the heavens above, a hierarchy of angels, spirit as the breath of life, the Kingdom of God, the age to come, or the New Jerusalem. Negative examples are the snake, Satan, the prince of this world, demons, apocalyptic monsters, or Hades, the sphere of death.

One of the major handicaps of theology is that it feels obliged to maintain the continuity with an ancient and long obsolete imagery. The sciences do not feel bound to respect previous stages of their history. Being geared to mastery, they try to understand objective reality as it unfolds before their eyes. They can simply abandon the flawed perceptions of their predecessors. When it comes to meaning, identity, acceptability and authority, in contrast, it is much more difficult to dump the tradition.

However, this does not mean to say that there are no significance stories in modernity—whether in terms of origins or visions. Examples are the idea of the nation, the state, the classless society, the liberal-democratic order, the American dream "from rags to riches"; the vision of a world without injustice, violence and poverty; an ecologically sustainable economy; or a possible colonisation of outer space.

Historical reconstructions and futurological projections of current trends often express indignation, raise awareness of potential catastrophes and call for commitment to try and avoid the worst. But the criteria of acceptability they apply are not rooted in science as such.

How are the narratives of science and faith related?

The significance story of faith is linked to 'factual' history in various ways. The narrative of intended historicity and the narrative of intended validity have to complement each other. Because there is only one reality, the vision of what this reality *ought* to become must dovetail with the knowledge of what this reality *has* become and *may* become if present trends continue. What ought to be cannot be said to be real in experiential terms. But perceptions, concepts and visions of what ought to be can.

There are aspects of the significance story that fall squarely into the realm of experiential realism. First, there is the 'objective' history of how these narratives have emerged and evolved in response to the needs of concrete historical situations. Second, there is the narrative of the impact of the significance story on the life history of individuals and communities. Finally, there is the larger historical context within which the history of concrete needs and the responses of the validity story have occurred.

All three belong to objective history. Objective history is the sequence of events that actually happened, whether humans are aware of it or not, whatever they believe, and however they interpret their traditions. The fact that, given the sources, it is difficult to reconstruct this history does not mean that it never happened.

There are also substantive reasons why convictions fall into the realm of experiential realism. First, in an emergentist perspective, the scientific narrative covers the whole of experienced reality. Religious convictions and their cultural expressions are inescapably part of this reality. They can be explored, described, critiqued, replaced, or abandoned.

The assumptions and rationales of religious and cultural narratives are investigated by historical-critical research into ancient texts, cultural anthropology and phenomenology of religion. These academic disciplines are 'scientific' in the sense that they describe and explain what concrete people out there assume and accept, while withholding their judgement concerning the validity of these assumptions.

Second, there are inescapable prerequisites of human existence that fall within the sphere of experiential realism. Humans are inextricably embedded in reality. To operate at all, they have to try and understand how it functions. It is the task of science to place 'common knowledge' on more reliable foundations. Humans also have to become aware of the possible consequences of their actions. They cannot avoid the fact that they are accountable for these consequences. They have to establish to which authority they are ultimately accountable.

All this leads to a dynamic system of meaning. A credible system of meaning cannot simply be constructed. It impresses its validity on human consciousness. It travels through time in the form of a tradition. A tradition has a social shape and presence in which the individual participates and to which it contributes to a lesser or greater extent. It is an evolutionary process that synthesises ongoing experiences with inherited assumptions and insights. It filters out flaws and integrates more plausible insights. We shall pick up these aspects in chapter 8.

To sum up

To become effective, science and faith have to become accessible and plausible to the proverbial man and woman in the street. They have to find expressions capable of interacting with the basic assumptions and motivational patterns of contemporary society without abandoning their critical function. They have to dump obsolete patterns of thought without losing their historical awareness or the profundity of their assumptions. They have to be committed to simplicity without becoming simplistic. They have to avoid obscurantism, speculation and fantasy.

While epistemology focuses on the limitations of the observing subject, the pragmatic pursuit of science focuses on the observed object and gets on with the job. It is amazingly successful in doing so. Its criterion is evidence, mathematical stringency and plausible conjecture. If theology wants to be taken seriously by the science community, it must abandon fantasy and superstition, reification of biblical metaphors, Platonic abstractions and inferences from problematic assumptions.

It must refrain from arguing that science is just another kind of religious conviction. It must show how faith assumptions emerge from existential experiences within social and natural contexts. It must demonstrate that notions of the transcendent are inescapable and that different notions of the transcendent have consequences in this world that can be life enhancing or destructive of life.

The mandate and method of science do not cover the whole of experienced reality. Examples are the sense of beauty, satisfaction, fulfilment, meaning, identity, stability, existential significance, historical importance, acceptability and authority. There are instances of 'immanent transcendence', such as the past, the future, the whole of space, the whole of energy, and so on. Faith, in contrast, depends on a notion of the transcendent Source and Destiny of reality as such and a whole. This notion is part of immanent reality, while the transcendent as such is not.

Both science and faith are embedded in history and communicated as story. However, science is geared to a historical story while faith is geared to a significance story. The two are inextricably linked to each other in various ways.

6

Elementary insights of the natural sciences

Experiential realism caters for all facets of reality investigated in academia. Modern cosmology traces the evolution of reality back to the 'big bang' and forward either to the total dispersion of energy or a 'big crunch'. The evolutionary process must be seen in the context of the entropic process, which tends towards the dissolution of all structured and potent energy conglomerations. The evolution of reality proceeds through emergent levels of complexity and volatility—waves, particles, atoms, molecules, amino acids, cells, organs, organisms, neurological functions, instincts, and environmental conditioning. Each level of emergence depends on all previous levels, yet constitutes a new level of reality with its own characteristics and regularities. Evolution led to differentiations within each level of emergence, such as elements at the atomic level and species at the biological level. The emergence of consciousness presupposes the biological organ of the brain, but it marks the threshold between the physical and the spiritual (and social) dimensions of reality, the topic of chapter 7.

The task of this chapter

The aim of Part II is to locate the phenomenon of faith within the context of immanent reality as explored by the sciences. Chapter 5 dealt with the approach of experiential realism as applied by the sciences. This chapter deals with substantive insights we owe to the natural sciences. It moves up the ladder of emergences to the point of transition between the physical and the cognitive levels of reality, which marks the boundary between the natural and the human and social sciences. Chapter 7 will take off from where the present chapter has brought us and move up to conviction as a phenomenon encountered within immanent reality.

I will not attempt to provide a detailed account of any of the positive sciences, let alone a combination of them. There are excellent and accessible overviews on the market that fulfil this task.[166] My aim in these two chapters is to investigate in which sense a theological discourse falls within a scientific frame of reference based on the principle of experiential realism. Though I will concentrate on the natural sciences, experiential realism is an approach that allows for a comprehensive view of reality as actually experienced by human beings and that is wide enough to include all academic disciplines currently practiced at institutes of higher learning.

Natural science is part of the modern project. Its method is based on evidence, mathematical stringency and plausible conjecture. It seeks to explain and predict. It dismantles unwarranted assumptions, spurious conclusions and superstitious hopes and fears. It transforms the way humans experience, interpret and deal with reality.

It has become the authority in which countless modern people place their trust.[167] It fulfils one of the roles that religion plays in pre-scientific cultures, namely that of making sense of how reality comes into being and how it continues to function.[168]

This chapter offers a whirlwind tour through some major insights of the natural sciences. It concentrates on big bang cosmology, entropy, emergence and evolution. Is this how far we can get before hitting the limits of experiential realism? Not quite. Evolution has produced a creature that is faced with options. Options have consequences. Consequences call for responsibility. Responsibility presupposes a transcendent frame of reference that includes meaning, criteria of acceptability and authority. These are unavoidable and indispensable aspects of the way humans experience reality. But they will be discussed in the next chapter.

Section I
Fundamentals of unfolding reality

Big bang cosmology

Let us look back as far as we can through the spectacles, microscopes and telescopes provided by modern science. How did we get to where we are? According to the overwhelming consensus among contemporary scientists, it all began with the 'big bang'.[169] Roughly 13.8 billion years ago (some estimates go up to 15 billion years) an infinite concentration of energy 'exploded', causing energy to 'fly in all directions'.

Obviously, this is just a primitive way of imagining what happened. The actual process is much more complicated. The 'inflationary model' of the big bang, for instance, suggests that due to circumstances in the very early phases of the universe, gravity first exercised a repulsive rather than an attractive force. It is through inflation, rather than initial concentration, that the masses of energy dispersed in the universe today came into being.[170]

Driven by force fields (the electromagnetic field, the gravitational field, the strong and the weak nuclear force), tiny irregularities triggered the emergence of structured energy conglomerations. This led first to tiny ripples in an otherwise homogeneous expansion, ultimately to concentrations of energy in the form of objects, planets, stars and galaxies, seemingly leaving huge empty spaces in between.

As a result, the ensuing process was not uniform but scattered.[171] The attraction between these energy concentrations (gravity) led to the compaction of energy into waves or particles. Various kinds of particles combined to form greater and more complex forms, notably atoms. All the entities found in the cosmos today, from subatomic particles to galaxies and human bodies, are the result of this process.

There are billions of galaxies in the universe (10^{11}), each consisting of billions of suns (10^{11}) and other objects. Our **solar system** is part of the galaxy called the Milky Way. It consists of a star (the sun) and various planets, of which the earth is one. **The earth** has a core, a mantle, a thin crust, most of which is submerged under seawater, and a thin atmosphere. The crust and the atmosphere of the earth may form the only space within the entire universe in which life can develop and survive.[172] But there may be many others of which we do not know and may never know.

Entropy—the supply of energy that 'works'

Big bang cosmology provides the background to the 'law of entropy'.[173] There are various ways of expressing this law. Basically, it says that, whenever energy gets a chance to do so, it will move from compaction to dispersion, from order to disorder, from disequilibrium to equilibrium, from high potency to low potency, from energy that has the potential to 'do work', to energy whose potential to 'do work' is exhausted. Normally, this happens only in closed systems because open systems interact with other such systems from which unspent energy can be imported.

In scientific parlance, the cosmic process is driven by the transformation of 'low entropy energy' (structured and 'potent' energy) into 'high entropy energy' (unstructured and 'spent' energy). The entropic process provides the 'fuel', as it were, for any process to happen. More powerful processes tend to draw energy out of their respective environments into their vortex, causing the latter to deteriorate and disintegrate. Any construction implies deconstruction elsewhere in the system.

Yet, in time, the dominant concentrations also become victims of the entropic process. Reality is subject to a universal rhythm of emergence, evolution, deterioration and decay. The relative speed with which particular entities form and disintegrate differs. A stone can take billions of years to disintegrate; a living cell can take a day or two. The pathways also vary immensely. They do not only spread in different directions but also form hierarchies of complexity.

Because evolutionary construction draws its energy from entropic deconstruction, the entropic tendency towards disorder is the precondition for order and novelty to emerge.[174] The evolution of order and complexity does not run counter to the entropic process, therefore, but happens within its context.[175] Indeed, some philosophers of science believe that evolution may be the shortest available route to total dispersion.[176] This meandering route is guided by regularities (natural law) that bring about the different phenomena we observe in the world today.

The entire system is in flux. Gravity leads to compaction, entropy to dispersal. So it is the flow of energy that drives the cosmic process. However, differentials in energy concentration tend to balance out. Wherever equilibrium is reached, energy becomes ineffectual. To use a picture, dammed up water can drive powerful turbines or cause devastating floods, but it is powerless when it has reached the tranquil lake below. Only a system that is 'far from equilibrium' has energy at its disposal.

While evolution implies an increase in order, organisation and dynamic process, therefore, the entropic process leads to an increase in dispersion, disorder and stagnation in the cosmos as a whole.[177] Scientists also discuss the possibility that gravity will lead the process backward into a 'big crunch'.[178] Whether entropy leads to ultimate dispersion or whether gravity leads to a giant black hole depends on the amount of matter in the universe. It also depends on whether the expansion of the universe outpaces the entropic process.

The outcome of either way will mark the end of reality as we know it. Where there are no energy differentials, no processes take place, nothing happens, nothing distinguishable exists.[179] This will be the end of the world as we know it. Long before the universe as a whole disintegrates, however, the sun will have burnt out and turned into a 'red giant' and the solar system will have disappeared in a giant fire ball. Ironically, science agrees with the biblical contention that reality has had beginning and will have an end. But science sees no prospect for a new and perfect reality to emerge from the ashes of the old.

What does entropy mean for life? Life on earth can only sustain itself because the earth's system imports low entropy energy from the sun and radiates high entropy energy back into outer space. Within the biosphere, however, all life, barring the most primitive levels,[180] depends on the death of other life and has to face death itself. Death is tantamount to the entropic dissolution of the living organism and each of its cells.

The relation between evolution and entropy can also be shown to be true at other levels of reality, for instance, at the level of social and political structures. It appears even in the spiritual spheres of life.[181] Cognisance of the law of entropy is particularly important for an appropriate assessment of economic reality and its ecological impact.[182] Economic centres grow at the expense of economic peripheries or the natural environment. The first process leads to increasing discrepancies in productivity and life chances, the second to the deterioration of the environment.

The tendency towards entropy is only valid for a closed system, not for an open system. The reason is that an open system can import low entropy energy from its environment to replenish its resources. However, taking a system together with its environment as a whole there is always a preponderance of entropic dissolution over constructive compaction.

At the moment, humanity increasingly uses up more low entropy energy than what the solar system can provide and what lower levels of biological life can process. If there can be no flourishing of life without the sacrifice of other life, our numerical proliferation and our surge towards ever higher living standards can only lead to the eradication of other life and potentially to the destruction of the earth's capacity to regenerate and maintain life as such.

The earth is an open system that can import virtually infinitely large quantities of energy from the sun. The sun is slowly burning out, but it will take an awfully long time to do so, and by that time, we will all be gone. However, if the process of depletion outpaces the process of regeneration—as it presently does—this amounts to a relative closure. It has the same effect as total closure, except that it is drawn out over time. Moreover, life's capacity to regenerate itself slowly succumbs to internal entropic forces.

As I will show in Chapter 11, this insight is of critical importance for the reconceptualisation of the Christian faith. Reality is so precious precisely because it is unique, vulnerable and finite. God, the ultimate Source and Destiny of reality, sacrifices endlessly and abundantly so that we can live. We are invited to participate in God's sacrifice so that other creatures can live. We should not claim more life for ourselves than due to us. We cannot expect a life that never reaches its limits, nor an infinite continuation of the world we know.

Emergence—The staggered unfolding of reality

There are discernible stages in the development of reality marked by different levels of complexity.[183] A higher level of complexity presupposes the relevant lower, less complex level, but constitutes a new and different kind of reality with its own characteristics and regularities. The whole is more than the sum total of its components.

The transition from a lower level of complexity to a higher one is called 'emergence'.[184] Emergence is characterised by an increase of information, organisation, interaction, integration, complexity, flexibility, volatility and transience. Just compare the simplicity and durability of a quartz crystal with the complexity and vulnerability of an earthworm!

Thus atoms are composed of subatomic particles (protons, neutrons and electrons), yet they have characteristics entirely different from those of elementary particles. At the atomic level, hydrogen and oxygen can combine to form water, which is something entirely different from its two gaseous components. Similarly one can assemble the right quantities of all the elements that make up a human body—carbon, phosphorous, oxygen, nitrogen, and so on—and mix them together. But this soup will never make up a living human body.

The most fundamental constituent of reality appears to be energy. Its most basic form seems to be the wave. Waves are organised in fields, such as electromagnetic fields. Superimposed fields lead to energy concentrations. The attraction between energy concentrations due to gravity leads to compaction. Matter is compacted energy, with the result that there is an equivalence between mass and energy.

The most fundamental forms of matter are subatomic particles. They combine to form atoms. Atoms form molecules. Molecules combine to reach ever-higher levels of complexity. Amino acids are sophisticated molecules that combine to form proteins. Proteins are the building blocks of cells, which are the building blocks of organs, which are the building blocks of organisms.

The biological organ of the brain constitutes the infrastructure of the mind; human minds are the building blocks of social structures and processes, which are again part of the ecological system as a whole. We can see, therefore, that the theory of emergence encompasses the whole of reality, from the most basic to the most elevated levels of reality.

Lower levels of organisation are nested in higher levels. The energy trapped in matter, for instance, is enormous. A human body is estimated to harbour energy with the equivalent of seven hydrogen bombs. Energy trapped in matter can be released. As nuclear energy demonstrates, tiny pieces of matter can release enormous amounts of energy.

The lower level always provides a vast pool of possibilities from which the higher level operation selects and combines what is most congenial to satisfy its higher level needs. This process is called "preadaptation". As we move up the hierarchy new "autocatalytic sets will spontaneously emerge".[185] The range of "the flow into the adjacent possible" grows.[186] That means that the future becomes more open.

Kinds of regularity

As can be expected, different levels of emergence are characterised by different kinds of regularity.[187] Crudely speaking, the subatomic (quantum) level is characterised by probability,[188] the physical level by causality,[189] the chemical level by propensity,[190] the biological level by teleology,[191] the instinctual level by flexible responses to environmental challenges, and the personal level by intentionality.

The supra-personal level follows a number of regularities depending on the kinds of collective intentionality that form critical masses within a community or society, the kind of social and natural environment in which it is embedded, the kind of hierarchy according to which a community or society is structured, and the kind of power and determination elites can muster to give direction to a process. Social structures and processes are again embedded in the overarching dynamics of the earth's entire ecosystem.

According to the majority of scientists, time and space are dimensions of an emergent and evolving reality. They do not pre-exist the reality we know in the sense of an absolute time and an absolute space. In the same way, regularities and the underlying patterns of information do not seem to pre-exist but evolve together with the evolving stages of emergence to which they apply.

Vibrant systems of feedback loops operate between the levels. While the lower level constrains the possibilities available to the higher level, the higher level determines how lower level entities and factors operate within the higher level system. There is bottom-up causation: physical laws prevent me from walking through a wall. But there is also downward causation: I can organise a bulldozer and remove the wall.

There are also multiple interchanges at the same level of emergence: a dog can protect me or attack me. Upward and downward causation run through the entire system: when I chase a cat out of the kitchen, all the amino acids, atoms, particles and fields in her body have to move with her. And just imagine the kind of downward causation that happens when humans detonate a nuclear weapon!

Emergent evolution means that there is progression in time, differentiation in space, increasing complexity of relationships, increasing subtlety, increasing vulnerability and shorter life spans. However, we have to distinguish between 'vertical' evolution of reality into a hierarchy of emergences on the one hand and 'horizontal' evolution in various directions at the same level of emergence on the other.[192] The step from inorganic to organic matter is a step upward in the hierarchy of emergence. The simultaneous existence of cats and dogs is a differentiation at the same level of emergence. Let us look at some of these levels in some more detail.

Section II
Levels of emergence

From waves to proteins

Processes at subatomic levels are extremely involved and keep physicists groping for plausible explanations. Expressed in a layman's language the most fundamental 'stuff' of reality seems to be energy. It can take various forms such as low entropy energy that can deliver work, and high entropy energy that cannot; potential and kinetic energy; light, heat, electricity and various forms of radiation. While the total amount of energy appears to remain constant (according to the first law of thermodynamics), different forms of energy can be transformed into other forms, such as light into heat, or heat into light.

Energy is organised in fields, which form waves and particles. To try and make sense of this elusive area as a layman, I imagine that various kinds of particles are formed by the crests of various kinds of waves superimposed upon each other. Throw a number of stones into a pond at various places and observe how such superimpositions lead to all kinds of shapes. If that were a reasonable analogy, subatomic particles of all kinds would form the first level of emergence.

At the next level of emergence, particles—namely protons, neutrons and electrons—combine to form atoms. Atoms come in different sizes, depending on the number of protons in their nucleus, thus forming the 'periodic table' of ninety-two naturally occurring elements. Oxygen, hydrogen, nitrogen, chlorine, copper, silicon, aluminium, carbon, iron, zinc, sodium and potassium are well-known examples.

Atoms combine to form molecules. Two hydrogen atoms and one oxygen atom combine to form water. One sodium atom and one chlorine atom combine to form table salt. One hydrogen atom and one chlorine atom combine to form hydrochloric acid. It is amazing how different the characteristics of these three very simple substances are.

However atoms can combine to form molecules of enormous size and complexity, producing all sorts of extraordinary characteristics. On the way to life, the next emergent step is the appearance of amino acids. When assembled in particular orders, amino acids form proteins. To form *colagen*, for instance, you need 1055 amino acids all arranged in the correct order. Possible combinations of such a number of entities number billions, so the very existence of a protein is a miracle in itself.[193]

The evolution of organisms

The next step is the assembly of the right kind of proteins in the right kind of ways to form cells, organs and organisms, thus living beings. Life in all its forms is a novel level of emergence that follows its own regularities.[194] Organisms consist mainly of four elements—carbon, hydrogen, oxygen and nitrogen—plus smaller quantities of a few others such as sulphur, phosphorus, calcium and iron. Dawkins notes: "There is nothing special about the substances from which living things are made. Living things are collections of molecules, like everything else."[195]

Indeed they are. Molecules are collections of atoms, atoms are collections of particles. That is not something special. Nevertheless the quotation reveals a reductionism similar to Dawkins' reduction of biological life to the 'selfish gene'.[196] What is special is the particular arrangement of these components into complex organisations. A new kind of reality appears when a huge number of components, arranged in a particular way, begin to interact with each other and develops a dynamic of its own.

In biological evolution, there is a stream that runs from algae to plants, and another stream that runs through amoebas to fungi and animals.[197] These streams have differentiated into immense numbers of species, which do not necessarily represent higher levels of emergence.

The first vertebrates are believed to have appeared more than 500 million years ago, amphibians some 400 million years ago, and mammals, to which humans belong, about 200 million years ago. The ascendancy of mammals was made possible by the extinction of the dinosaurs that had dominated the scene before.

The first apes appeared about 25 millions years ago, the first 'hominins' about 7 or 6 million years ago, the first species of the genus *Homo* 2.5 million years ago, 'archaic *Homo sapiens*' perhaps 800,000 years ago and modern humans (*Homo sapiens sapiens*) perhaps 200-150 years ago.[198] These figures are educated guesses based on the interpretation of fossils.

The differentiation of living organisms into species is due to changes in the genetic information systems that determine processes within cells and the interaction between them. Such changes are called *mutations*. Mutations have all sorts of causes. The cancer inducing impact of radiation, for instance, is well-known. Species that are inherently more capable of adapting to specific environments have an advantage over less adapted ones and become dominant. Biological evolution is driven, therefore, by 'random' mutations and the survival of the fittest within specific environments.[199]

It would seem that this simple Darwinian explanation is not wrong but insufficient to account for what actually happens. In the first place, the word 'random' needs to be unpacked. Things don't just happen. Chaos theory has provided us with the picture of an extremely fine-tuned situation (such as a pencil standing vertically on its head), where the slightest impulse can direct all further developments in a variety of different directions. Such impulses can come from anywhere in the universe.

Second, differentiation can also happen through 'random genetic drift', that is, certain characteristics tend to spread within a population without any particular reason for doing so. Third, complexity theory has shown that there are self-organising mechanisms that lead almost automatically to certain outcomes. These can be reconstructed, at least in theory, by means of mathematical formulae.[200] Fourth, biological evolution does not blindly follow the instructions coded in its DNA, but evolves within a complex interaction of heredity, environment and choice.[201] Perpetual responses to repeated environmental experiences can be programmed into genes.

Fifth, emergent regularities lead to a degree of convergence between organisms, even where the underlying building blocks are quite different.[202] Higher animals have some kind of alimentary canal. Animals capable of flying, whether birds or mammals, have some kind of wing surface that can utilise air currents to stay aloft. Most higher plants have some kind of root system. Most land animals have a coordinated leg system to move around. Higher forms of life cannot exist without gravity pulling them towards the earth.

In other words, there are parameters of what is possible and congenial to an operation within a particular environment. These parameters co-determine the kind of phenotype that emerges from the interplay of various factors. Evolution therefore seems to be a more complex process than many evolutionists assume.[203]

Evolution must be seen within the context of the entropic process. This basic fact is often not considered to the extent it deserves. The speed of change differs vastly between different levels of emergence, but all material entities emerge, evolve, deteriorate and decay. Death and deconstruction are 'built into the system' because life and construction depend on them.

The most fundamental convergence between all higher life forms is, therefore, that they are all subject to the interplay between the evolutionary and the entropic processes. With the exception of the most primitive forms,[204] all life feeds on the death of other life and ends in death. Death is an inevitable and inescapable prerequisite and fate of all higher forms of life. But what then is life in the first place?

Life in an emergentist perspective

The second creation narrative in the Bible says that God breathed the breath of life into a lump of clay. So there is a material from which it is made. Today we know that this material is constituted by certain elements. This is obvious.

What then is the scientific equivalent of the 'divine breath of life'? This is one of the instances where theology can be illuminated by modern science, in this case by the theory of emergence. Trillions of complex systems function in perfect coordination. But there are hierarchies within this unfathomable network. And when a key function is lost, say the provision of oxygen to the brain, it is all over and all the other functions disintegrate.

Let me illustrate the intricacy of life with a recent experience. A much loved Siamese cat, healthy and bouncy, dies in front of our eyes for no apparent reason. My hands carefully move through the body to find any injury or spasm but to no avail. The skin is as soft as silk, all muscles are totally relaxed, and the eyes seem to look at me in their sparkling blue. Everything that made up this organism seems to be as present and intact as a minute before. But life has gone. It seems as if all systems that pass information through the body have been switched from one to zero. Immediately the order of the body's chemistry begins to disintegrate. The immune system is no longer operational. Decay ensues at the very moment of death.

Seen in cosmic contexts, life is the highest manifestation of the evolutionary mechanism that causes constructs to emerge in the face of the entropic process, yet feeding on the entropic process. Life is close to the peak of a much larger, much more comprehensive hierarchy of emergence whose most basic building blocks are composed of energy, fields, waves and quanta, but which is built up on this foundation in a great number of levels of complexity culminating in human consciousness. But before we come to that, we need to go back to the organism.

The organism seen from the inside

Just imagine that there are about a million different types of protein in the human body, all interacting with each other! They are the building blocks of cells. We have about ten thousand trillion cells of about 100 different kinds in our bodies.[205] Cells are the building blocks of organs. Organs make up organisms. There are trillions of them around, each with its own self-contained organic integrity. What is it that keeps this bunch of components together and functioning in a coordinated way? Part of the answer is the interaction between the nervous and the endocrinal systems. Let us go into some detail.[206]

At the most basic level, the organism functions through a constant transfer of *information* across an immensely intricate network. This network is constituted by two essential channels of communication. There is an *emotive* system that determines how we feel, and a *motor* system that causes different parts of the body to respond to messages and function in a coordinated way. The two are coupled by multiple feedback loops. Against this background, life can be described as an uninterrupted and all-inclusive exchange of information.

The flow of information within the organism involves two kinds of messengers. The first is the *neural* network. It transmits electrical signals of very short duration (a few thousands of a second), which are called *impulses*. Such impulses are transmitted between adjacent cells and target particular regions of the body.

When a neuron receives sufficient impulses from its environment to change the relation between positive and negative charges beyond a particular threshold, the neuron 'fires' an impulse that travels along the axon and its dendrites and triggers the formation of a chemical (called a *neurotransmitter*). It binds on the connected neuron and sets off a corresponding impulse there.[207] The combined interaction between neurons can form unbelievably complex patterns and processes. Think of an irrigation system or the lighting system of a big building. You switch certain clusters of channels on or off. This can become automated by a supervening process.

The second is the *endocrinal* system. It transmits messages through chemical substances called *hormones* that flow through the entire organism and take much more time to do so (minutes or hours). There is an intricate feedback system in which hormones either enhance or curtail their own secretion to stabilise the internal milieu. Communication takes place when a hormonal messenger finds a receptor of opposite but complementary nature and the interaction leads to a new chemical messenger and so on.

Internal stability

The emotive system is based on the interaction between the external milieu of an organism and its internal milieu. The *external milieu*—temperature, light, the presence of oxygen, physical impacts, sensual impressions, etc.—is subject to considerable variation and the organism has to respond by trying to keep its *internal milieu* relatively constant, a condition called *homoeostasis*.

The internal milieu is facilitated by fluids that are distributed in three 'compartments'—the fluid within the cells, the fluid surrounding the cells, and the fluid of blood plasma and lymph that acts as a means of transportation throughout the body. The *internal secretions* of different glands are poured into this common medium in response to the impact of changes in the outer and inner milieu. They are designed to maintain the internal stability of the system.

It is the disturbance of the inner milieu that makes us feel either great or miserable. To return to the desired balance, fluids are secreted that cause a sense of *attraction* for something needed to restore the balance, or *aversion* against something that disturbs the balance. The same external stimulus, for instance the smell and taste of food, can cause attraction or aversion depending on the state of the inner milieu. *Pleasure* and *pain* are, therefore, emotive expressions of the most basic mechanisms that keep the organism intact.

The evolution of the brain

The organism is composed of a plurality of discrete information systems, but it is the immensely complex interplay between all these systems with multiple feedback loops that causes the organism to function as an entity of its own. Add to that the intricate interaction between the internal and the external milieu and you have a vivid example of the phenomenon of emergence. The next step on the hierarchy of emergence is the brain.

Higher organisms are capable of goal-directed movements. This phenomenon necessitates the presence of a command centre and a system of communication. The nervous system and the endocrinal system are critical components of higher forms of animal life because they form the communications network that conveys information from a pivotal centre, the brain, to more peripheral parts of the organism and vice versa. The brain "provides the machinery of information management and control needed by creatures of increasing versatility."[208] Conversely, increasing versatility is made possible by the powers afforded by a sophisticated brain.

Evolutionary epistemologists see the emergence of the brain as a special case of the general process of the adaptation of species to their respective environments, a process which drives the evolutionary process at the biological level.[209] There may be unfathomable intricacies in the process. The point to be made at this juncture, however, is that the appearance of the brain does not represent a break in the evolutionary process but just a higher level of emergence.

In its turn, the brain forms the substratum for the emergence of a new, much more complex, set of patterns and processes that we experience as the personal dimension of reality. Its complexity and versatility are truly mind-boggling. Let us try to explore the link between the two.[210]

The brain as the command centre of the organism

The stability of the brain is of such fundamental importance for the organism that it has its own encased milieu, called the *cerebral milieu*. There is a special cerebral fluid that makes the communication between different parts of the brain possible. The brain also produces its own hormones. It has controlled entry and exit points. Body hormones, fats and other substances are not automatically accepted into the brain. Whatever reaches the interior must pass through the cell membranes themselves. This safety precaution is called the *blood-brain barrier*.

Brain and spinal cord make up the *central nervous system*, as opposed to the *peripheral nervous system* that connects it to the rest of the body. The latter consists of the *autonomous* nervous system, which regulates automated functions like heart beat and breathing, and the *somatic* system that conveys *sensory* messages from the sense organs to the brain and *motor* messages from the brain to the muscles.

The human brain is three times larger in relative terms than that of other higher mammals. The human brain consists of between 50 and 100 billion (10^{11}) *neurons*, which form about 10 per cent of the brain cells. The others are called *glial cells*, which have a number of supportive functions, for instance in the formation of synapses. Each neuron consists of a body (*soma*) and a root or trunk (*axon*). The axon can be as long as a meter and have a great number of branches (*dendrites*). A special kind of structure (*synapsis*) establishes contact between the branches of different neurons. Each neuron is linked with up to 10,000 synaptic connections.

Structure of the brain

The most primitive form of the brain evolved some 600 million years ago in a simple worm. This worm was structured like a tube with a hollow cavity running from mouth to anus and a large nerve ganglion at the front. All its descendants have maintained this basic shape, including the human being. The human brain has various parts that have evolved successively in pre-human evolution. They are fairly complex and we do not have to go into detail.[211] Most important for our deliberations is the evolutionary sequence of three main parts of the brain:

(a) The so-called *reptile brain*. This is where the most basic survival instincts are located, feeding and breeding; fight or flight; hunger and thirst; activity and sleep.[212] Instincts are hereditary arrangements in primitive areas of the brain that proved themselves profitable for the survival of the species and that function 'mechanically', similar to the change of colour in chameleons.[213] Their patterns of behaviour are basically immutable, though they have been qualified and superseded to a great extent by later developments in the brain.

(b) The *limbic system*. It is situated between the reptile brain and the neocortex, thus between 'instinct' and 'intelligence'. This is the seat of emotions and motivations. It mediates the symbolic representations of the internal and the external world. In comparison with the neocortex, it is rather 'irrational'. Its malfunctions, due for instance to an epileptic seizure, can produce all kinds of unpleasant emotions and sensations, absurd convictions and uncoordinated movements.

(c) The *neocortex*. This is the seat of abstract thought, symbolic representation, language, memory, anticipation, evaluation, reasoning, choice, self-control and planning. In the human being, the *frontal lobes* of the neocortex are more highly developed than in other primates, thus facilitating the superior quality of human intelligence. The neocortex is also able to eliminate disruptive 'noise' emanating from the internal or external environment, thus making concentration possible.

Memory—how information is deposited in the brain

When a foetus develops, the impact of its genetic makeup on the one hand and sense impressions derived from the environment in the womb on the other begins to 'hardwire' the brain into particular patterns. That is, they install relatively stable connections between neurons by means of growing clusters of synapses. Synapses can be switched on or off. This process continues throughout the life of the individual.

All impacts of environments on the individual and its responses to these impacts are stored in particular brain patterns in the form of neural connections. These patterns have the property of being symbolic representations of the original environmental impacts.[214] Brain scanning techniques can pick up the electric currents involved and locate these patterns within the geography of the brain.

The formation of such relatively stable patterns is an example of 'downward causation'. As an emergent reality, memory presupposes the neurological architecture of the brain. But the repeated impact of differentiated experiences selects certain options from the total pool of potential brain states that are most congenial to enhance the orientation of the individual within its environment.[215] New experiences continuously amplify and adjust the resultant pattern. As everywhere else in emergent reality, the law of entropy is circumvented through relatively stable, yet dynamically evolving structures, a phenomenon sometimes called 'negative entropy'.[216]

'Memory' has a number of layers. Genetically transmitted information forms the most fundamental layer. This includes the procurement of the prerequisites of healthy survival such as food, sexual attraction, rest, activity, shelter, protection, emergency responses (flight or fight), etc. As mentioned above, these 'instincts' seem to be located in the most primitive layer, the so-called 'reptile' brain. They are very powerful and cannot easily be dislodged. But they can be superseded (controlled or repressed) by patterns and processes at higher levels, such as personal goals, cultural norms, or collective visions. This constitutes the area of ethical reflection and behaviour.

Second, there are relatively programmed responses to environmental influences. The most original environmental impacts form the most deep-seated, endurable and 'stubborn' patterns. The brain is largely 'conditioned' during early childhood. But a synaptic network (colloquially called memory) is not a static and immutable deposit. It can be called up (or activated) when the individual finds itself under the impact of a completely different set of circumstances.

The network will then adapt to the new situation and sink back into the subconscious in this new form. Memory itself is, therefore, malleable and in constant flux. More recent and superficial impressions are stored in short-term memory. In a secondary process, they are relocated into long-term memory. But they can also get lost, impacted, or changed as such. The oldest memories are the most stable and persistent.

The emergence of the human mind

The estimated number of synaptic switches in the brain varies between a hundred thousand billion (100,000,000,000,000) and a thousand trillion. They can be switched either on or off, thus allowing for a virtually infinite number of discreet networks, greater constellations and overall brain states. Genes determine the basic form and functions of the brain, but experiences, decisions and actions mould actual states of the brain represented by particular systems of synaptic connections. This is sometimes called the 'hard wiring' of the brain. This is a misnomer, because brain states are relatively stable, yet in constant flux and infinitely malleable in principle.

Every single switch changes the state of the brain. Mathematical calculations show that "the number of states in which (the brain) can exist greatly exceeds the number of atoms in the universe."[217]

These are unfathomable numbers. To appreciate their significance, I invite the reader to conduct a little experiment. Draw a number of points on a piece of paper and see how many relationships can exist between them. The connection between two points is either 'switched on' or 'switched off', giving you two possible states. If you have three points there are nine possible states. If you have four points there are fifty-two possible states. So the number of possible states grows 'exponentially', that is, it grows faster and faster as you increase the number of points, until they reach super-astronomic proportions.

Now imagine what happens when you have a hundred thousand billion synaptic switches all interacting with each other! Moreover, as the mathematics of complexity has shown, these relationships themselves can develop incredibly complex, unpredictable and practically infinite kinds of patterns and feedback loops. This explains the unbelievable complexity, versatility and vitality that characterises human mental processes.

It is in this whole area of mind-boggling complexity that the phenomenon we call the human mind emerges. Is this a credible proposition? I think it is. Have you ever wondered how a computer can transform a series of zeros and ones (the digits of *on* and *off*) into a faithful reproduction of a Beethoven symphony? Yet this is what happens even with the limited digital computation invented by humans. Nature is characterised by an infinitely more complex system of information, interaction and communication than that.

The human mind is capable of communication by means of highly differentiated and abstract systems of symbols called language. There are rudimentary parallels in higher animals, but human communication is without parallel. This is an inestimable boon not only in terms of communication, but also in terms of the development of higher intelligence. Human comprehension is aware of sequences in time (thus history) and regularities in space (thus abstract thought). It is able to envision a reality larger than its immediate environment (a world).

It intuits overarching contexts and transcendent foundations and combines them into systems of meaning. It tries to define its individual and collective identity within its concentric contexts. It discerns options with potentially beneficial and detrimental consequences. It formulates goals, values, norms and visions. It is aware of the fact that it is derived, embedded, dependent, vulnerable and mortal. It is aware that it is in need of authority, accountable, under scrutiny, guilty and dependent on reconciliation.

The mind includes faculties such as intuition, observation, comprehension, integration of meaning, conviction, identity, value, acceptability, authority, volition, intention, vision, agency, creativity and conscious and purposeful interaction.[218] Again, some of these characteristics also appear in less developed forms among higher animals. However, the human mind is the most complex phenomenon found in reality that we know of.

It is also the most subtle, the most versatile and the most unstable. Its complexity, volatility and unpredictability surpass those of more simple structures and process such as found in the fields of physics, chemistry or biology by vast orders of magnitude. To appreciate that, compare the complexity, versatility and transience of human thoughts with crystals, chemicals, stones, trees and ant colonies.

The evolution of the human mind

I am not in a position to express an opinion on the various theories of how personhood, thus the most distinguishing characteristic of *Homo sapiens*, has emerged and evolved.[219] The field of hominid evolution is still riddled with widely differing theories and conjectures.

On the one hand one has to take the biology of hominid evolution from *australopitheticus* (3.75 million years ago) through *Homo habilis* (2 million years ago), *Homo erectus* (1.7 million years ago), 'archaic *Homo sapiens*' (0.5 million years ago), and *Homo neanderthalensis* (between 220,000 and 45,000 years ago) to *Homo sapiens* or the modern human (200,000 to 120,000 years ago) into account. Again—these figures are conjectures that vary considerably.

Was human evolution gradual or 'punctuated' (based on a unique and sudden mutation)? Did modern humans emerge in various strands or from a single genetic pool? Did various brain capacities first evolve separately and then at some stage or other suddenly click into each other to form human personhood much like the components of the space station? These possibilities are still widely discussed.

On the other hand one has to ask when and how personhood as an emergent reality materialised. Without doubt we have to assume that it happened in various stages. It appears that human consciousness and intelligence "and with it creative, artistic and religious imagination" emerged during the time between 45,000 and 35,000 years ago.[220] This is sometimes referred to as the 'cultural big bang'.

Discovery of the 'cultural big bang' has put the theory that the human brain evolved gradually through the emergence and addition of elements to an expanding and ultimately excellent complex of cognitive resources under strain. This idea is too simple to account for the evidence observed. One view is that "opportunistic evolution has conscripted old parts of the brain to new functions in a rather untidy fashion and new structures have been added and old ones enlarged in a rather haphazard way."[221] This view is consistent with the Darwinian view that evolution occurs routinely through "random emergence and fixation of innovations" that prove to add to the viability of organisms in particular environments.[222]

A qualification of this rather haphazard construction of the human mind is the view that "in both development and evolution the human mind undergoes (or has undergone) a transformation from being constituted by a series of relatively independent cognitive domains to a situation in which ideas, ways of thinking and knowledge now flow freely between such domains."[223] The components could have evolved gradually or haphazardly, but it is their relatively sudden integration that constituted the breakthrough.

In my own words, it is rather like the assembly of different parts produced in different countries to make up a computer. Once they click into each other, they form a larger functional entity that is much more than the sum total of its components, thus a typical case of emergence. One can also think of the globalisation of financial systems or communicative networks that enhances the fluidity within the system as a whole.

However that may be, within an evolutionary frame of reference it seems clear that the emergence of human intelligence or the typically human 'cognitive resources' was a most effective means first of survival, then of development, and finally of dominance because it provided the "ability to cope intelligently with an intelligible world".[224] Moreover, if evolution is understood as the "principle provider of the organisation of living things and their adaptations", thus as "the process by which knowledge is achieved", as evolutionary epistemologists assume, human intelligence can be understood as the (provisional) peak of the evolutionary process as a whole.[225]

Emergent functions within the mind

As mentioned above, the reptile brain is the seat of instincts. Instincts are augmented by mental conditioning, which is formed through repeated environmental experiences. These can be programmed into the genetic code of a species over long periods of time in the form of 'phylogenetic memories'. This happens, again, through the elimination of those entities that do not fit into the pattern. Together instincts and mental conditioning form the subconscious.

At the most basic level, the reptile brain, lies the phenomenon of *plain reactivity*. An environmental impact leads to an automatic reaction. When I inadvertently touch a red hot plate or a live electric wire, the jerky reaction of my arm is violent, immediate and automatic. My system is programmed to avoid danger. Similarly, the immune system will attack an invading virus no matter what. It cannot think and decide; it functions mechanically.

The next step may be a subconscious kind of *responsiveness*, that is, the capacity to assess potential outcomes, but without the subject being conscious of it. Since early childhood the sudden sight of a snake, for instance, has filled me with uncontrollable terror. Yet this force is not an inescapable determinant of my behaviour. Depending on the circumstances, my actual response can vary between freeze, fight, or flight. This may be correlated with the limbic system.

Then there is *consciousness* proper, that is, the awareness of an imagined, real, or potential environmental impact that calls for an appropriate response. Consciousness differs from the subconscious in that it can objectify, reflect and decide. A storm is brewing, so let me get into safety. A financial opportunity presents itself; go for it! I receive a death threat; what now?

The imagined, real, or impending impact can be analysed. The options available can be evaluated in terms of a particular frame of reference, and the most appropriate response can be envisaged and executed. Coming to my senses, I can let the snake go as a creature that also has a right to live. So this level is the seat of comprehension, rationality and intentionality that is associated with the neocortex.

Emergent levels of consciousness

Consciousness differentiates into various forms and levels of complexity, notably self-consciousness and reflective self-awareness.[226] Each conscious individual is more or less aware of the immense network of relationships in which it is embedded and which constantly impacts its world in a variety of ways. *Self-consciousness* is consciousness becoming aware of its own identity within its concentric environments. Paradoxically the functioning of one's own body, including one's own brain, can form part of the observed 'environment' in this case. I can refer to 'my body', or 'my soul', or 'my heart'.

I realise that my body has fever; that my psyche suffers under a depression; that my wife has given birth; that my firm has earned a contract; that the economy is in a recession. New impressions are integrated into the pre-existing hard-wired, yet malleable and dynamic patterns of memory within the brain. Together these patterns form the system of meaning in the context of which the individual operates. This consideration brings us to a new level of emergence that takes us to the topic of the next chapter.

What is the relation between the *subconscious* and consciousness? Psychiatrists tend to attribute neuroses to the suppression of subconscious forces either by conscious decision or cultural convention. Obviously, there is some truth in this stance. However, mastery over the suggestive power and urge of one's 'reptile brain' as well as control of the irrationalities of the limbic brain belong to the singular capacities of the human being. We are not the helpless victims of lower levels of emergence. There is not only bottom-up causation but also top-down causation.

So on the one hand consciousness operates within the context of the subconscious and is, at least to a considerable extent, constrained by it. On the other hand, consciousness can control, utilise, suppress, or supersede the subconscious, for instance, when a married man remains true to his wife although his instincts attract him to another woman.[227] It is at this point that the embeddedness of individual consciousness in the impersonal supra-structures and processes of collective consciousness—including religious, cultural and social values and norms—kick in to guide consciousness in its decisions and behaviour.

The emergence of subjective experiences

When humans experience, comprehend, assess and respond to the objective world, they do so as conscious subjects. Such experiences are called *qualia*. The sciences, assuming that only matter was real, have often discredited what they deemed 'purely subjective' experiences in favour of 'objective reality'. That was a cardinal mistake. For one, this kind of 'objective reality', cleansed from subjectivity, gives us no clues how to integrate experiences, how to react, and which direction to follow. That again has immense consequences in the real world. We shall come back to that in the next chapter.

At present I am concerned with the second unacceptable consequence of this stance: it creates the impression that subjective experiences are unreal, thus irrelevant. Nothing could be further from the truth. As mentioned above, they are based on underlying brain states. The question is, therefore, whether we can explain, in scientific terms, how brain states translate into human experiences.

You are 'overwhelmed by the beauty' of a landscape or a symphony, you are 'struck with awe' when looking at outer space through a telescope, you 'feel the pain' of a divorce or a professional failure, you are 'indignant' when confronted with a case of callousness or injustice, you are 'horrified' by atrocities such as the genocide in Auschwitz or Rwanda, you are dismayed by the bombing of Dresden and Hiroshima, you are 'stunned' by the human and ecological costs of a catastrophe such as a tsunami or an earthquake.

What is the ontological status of such experiences? We have seen above that the internal equilibrium of the system is maintained when it interacts with the outside world. This happens through the hormonal system and the neural system. Any deviation from the norm causes a countervailing reaction. Abundance leads to repulsion, deficiency leads to attraction. As a result, the organism oscillates between the sensations of pleasure and pain.

A number of chemicals have been discovered that cause subjective experiences such as sexual desire, thirst, hunger, joy, or depression. Sexual hormones and the 'fight or flight' hormones are well-known examples. The mechanisms by which sense impressions, such as the smell of food, an aggressive animal, or the presence of a congenial partner, translate into the presence of the respective chemicals in the bloodstream have been explored in appreciable detail, although the field is still wide open for research.[228]

According to neuroscience, then, subjective experiences are based on brain states caused by the interplay between neural patterns and chemicals released through the impact of sense impressions on the system as a whole. Such feelings are not willed or constructed; they 'pop up' on their own. Particular patterns emerge and evolve through the cumulative impact of genetic factors, hard wiring of cultural assumptions, memory of past experiences, and ongoing environmental impressions. The brain is conditioned to differentiate between what is desirable and what is repugnant.

Depending on how existing brain states were constructed, the impact of new information can lead to very different kinds of reactions and subsequent adaptations. Witnessing savage cruelty, for instance, can lead to abhorrence in a subject that is conditioned by a particular culture and profound satisfaction in another.[229] The smell of male sweat can attract or repel a female. The incessant thump of pop music can lead to exhilaration or irritation.

So far so good. But all this belongs to the explanation of objective processes, not to the explanation of subjective experiences. How can an objective brain state lead to a subjective experience, that is, how it actually feels to be depressed or in love? This is often deemed a mystery. Is it really? This 'mystery' may never be resolved because there may be no mystery in the first place. It may be due simply to a change in perspective from object to subject.

You can explore the operation of a functioning brain from outside by means of brain scan techniques, chemical analyses, and so on. Then your own brain is not involved in the functioning of the other brain and you cannot possibly fathom how this other brain 'feels'—except perhaps by referring back to your own subjective experiences of this nature. Conversely, your experience of how your own brain feels under certain circumstances does not give you the slightest clue of how your brain actually functions in neuroscientific terms—just as the sensations associated with driving a car do not give us a clue of how the engine and the gear box actually work.

Consciousness is the functioning brain *experiencing itself from within* its operations. To use an analogy, it is possible to observe a battle from outside the event (say as a journalist or a historian). But this experience is entirely different from finding yourself within the event (say as a combatant or one caught in the crossfire).

It is similarly possible to observe and describe the manifestations of a religious conviction such as Islam from outside this conviction. It is also possible to try and reconstruct its internal rationale. This is what the academic discipline of Phenomenology of Religion does. But you will never feel its compulsive and disciplinary power unless it has imposed its validity on your own consciousness.

Let us summarise

This chapter has taken us through the levels of emergence from the most elementary forms of energy through particles, atoms, molecules, amino acids, organisms, neural systems to the functioning brain. The brain is the seat of consciousness. It marks the threshold between physical, chemical and biological reality as impersonal infrastructure, on the one hand, and the evolution of the personal, spiritual and social dimensions of human reality on the other. This is the topic of the next chapter.

The emergence and evolution of meaning

Human consciousness is the highest level of emergence we know of. It presupposes the biological organ of the brain, as well as the entire non-personal infrastructure of emergences. Individual consciousness is embedded in collective consciousness. Humans are able to transcend their immediate experiences and locate their place and identity within their concentric contexts. Traditions and experiences are consolidated in systems of meaning that define acceptability and allocate authority to act. Intentionality and agency operate in situations that are highly sensitive to initial conditions. At these junctures, 'switches' send the processes in different directions. They have vastly beneficial or detrimental consequences. This fact calls for alertness and accountability. The concept of God refers to the ultimate reference point of a system of meaning. The content of such a concept has far-reaching consequences in all aspects of life. In the biblical tradition, God is the name for the transcendent Source and Destiny of reality as such and as a whole, interpreted as benevolent intentionality. Formally it must embrace the personal level of emergence or it is defunct. Substantively it is geared to a dynamic vision of comprehensive well-being or become counterproductive.

The task of this chapter

In Chapter 5, we followed the hierarchy of emergences from the quantum level upwards to the emergence of human consciousness. Although only a tiny fraction of what is knowable in principle is actually known, it is relatively unproblematic to explain these levels in terms of the natural sciences. We now come to the more intangible facets of reality that are usually dealt with by the human and social sciences. Kindly note, however, that in Part II we consistently remain within the boundaries of immanent reality.

The transcendent as such falls outside this realm; the *notion* (or concept) of the transcendent does not. The aim of this chapter is to indicate where the concept of the transcendent fits into an immanentist frame of reference. I pick up the thread where we left off at the end of the last chapter, discuss the nature of human consciousness and move on to systems of meaning and religious convictions. Then I will ask what precisely the 'reality component' of Christian faith assumptions might be in terms of experiential realism.

Section I
Body and soul

The theory of emergence tells us that consciousness is based upon the functioning brain, which is a biological organ. This flies in the face of what Westerners have assumed for more than two millennia. Can we really be so sure of that? Indeed we can. Brain functions can be manipulated. Moods can be changed through the injection or ingestion of certain chemicals. Drugs can influence cognition. The most potent of them can turn normal people into monsters that murder, rape and smash. Motor functions can be set off by the stimulations of certain neurons.

Encephalography can locate certain intellectual or emotional functions in the topography of the brain. Certain patterns of behaviour can change or lapse due to a stroke or the excision of relevant areas from the brain. Previously ingrained memories can be lost when the brain deteriorates in old age. Injuries to the brain or brain disorders can impair intelligence, disrupt the flow of memory, and change personality. Death is described by biologists as the absence of brain activity—a condition that can be established empirically.

There is no question, therefore, that consciousness is based on brain functions. This insight is of fundamental philosophical and theological importance. There is no 'mind' or 'soul' or 'spirit' that subsists and functions without the biological infrastructure of the brain. This fact renders the traditional metaphysics of idealism and dualism obsolete. There is no immortal soul because there is no immortal body. But it also renders traditional materialism obsolete. The sum total of lower level components does not explain the operation of a higher level of complexity. The only appropriate approach to the difference between spirit and body is the theory of emergence.

The insight that personal consciousness is a higher level of emergence built on the entire impersonal infrastructure of emergence has particularly significant repercussions for anthropology. On the one hand, it is not an unfortunate and redeemable accident that our personhood or 'soul' is located in our body, as Platonic metaphysics believed it was. 'Soul' emerges from bodily reality and continues to depend on bodily reality. It is also embedded in the impersonal supra-structure of collective consciousness with its ideological, social, economic and political facets.[230]

Significantly, the biblical account also knows of no bodiless soul. The Hebrew word for soul (*nephesh*) means a living being, as opposed to a corpse. According to the oldest creation narrative, God breathes life into a clump of clay and that is how the human being becomes a "living soul" (Gen. 2:7). When God takes away the breath of life (not the soul), the organism collapses into the earthly material of which it is composed (Gen. 3:19).

According to the New Testament, there is no immortal soul either; 'resurrection' is the gift of a new life to the entire human being, including body and soul. Even the Platonic dualism was meant to depict the difference between essence and existence, good and evil. The same is true for the (originally) Parsist dualism between the present age and the age to come where evil would be eradicated. It was not meant to suggest an ontological dualism.

On the other hand, the personal level of emergence cannot be reduced to the forces and processes at lower (biological and physical) levels of emergence, nor is it the powerless and helpless victim of higher (social) levels of emergence. Both these stances deny what it means to be human. Humans are able to see themselves as biological bodies embedded in natural environments and social agents embedded in social environments. A person can impact both these environments—lower down and higher up.

The reality content of human consciousness

Consciousness is an exceptionally agile part of immanent reality. It will intuitively seek to find its way through the 'hard-wired' structures within the brain and the whirlpool of fleeting impressions coming from both inside and outside the body. The hard-wired structures are themselves the cumulative and constantly changing result of processed impressions. The most fundamental driving force of consciousness is to survive, followed by the motivation to gain and maintain optimal subjective well-being within particular environments. In fact, all life is programmed to survive and prosper. It is a kind of 'teleology' built into all living organisms by evolution.

But consciousness is sensitive enough to be aware of the demands of wider networks of relationships because these relationships impact its well-being very profoundly. They need to be taken into consideration for a more comprehensive equilibrium of well-being to materialise. In comparison with that of higher animals, human consciousness is much more profound, multidimensional, pliable and capable of developing extensive horizons in terms of time, space and power relations.

In terms of **time**, consciousness is more or less aware of past developments that led to the current situation. This situation is fixed because we cannot go back in time. I call that *factuality*. It is also more or less aware of the potential consequences of possible futures. I call the spectrum of options that present themselves *potentiality*. At any point in time, one of these options is being realised. I call that *actuality*. It is in actuality that processes continue to flow.

In terms of **space**, consciousness is more or less aware of the operation of its immediate internal and external environments. These environments constantly shift. However, human consciousness is also capable of transcending its immediate environments towards wider horizons. Environments form 'concentric circles', as it were, from immediate to remote—psyche, body, community, society, nature, the earth, the solar system, the galaxy and the universe as a whole.

Horizons also move from the actually experienced, the known but not experienced, the knowable but not known to the unknowable. As I will argue in the next chapter, the unknowable still falls within the limits of immanent reality and should be distinguished from transcendence proper. In terms of **energy**, consciousness is more or less aware of shifting balances of power between its own resources and others operating in the networks of power flows within these concentric environments.

Intentionality

Different levels of emergence are characterised by different kinds of regularity.[231] At the personal level, we find intentionality, which is an emergent level other than physical causality and biological teleology. Intentionality presupposes the infinite fluidity of brain states, yet constitutes a higher level of complexity. Organic life is about feeding and breeding, avoiding pain and escaping death. Aspirations to reach higher levels of being only emerge at the personal level.

There are forms of intentionality in higher animals, but nowhere do they reach the scope and sophistication found in humans. Human consciousness is aware of horizons much wider than the sphere of immediate experience. It interprets the past and anticipates the future. It sees itself in wider spatial contexts. It assesses the weight of other forces within the situation. It is confronted with the necessity to choose between available options and garner enough power from its own resources or those of its environment to make an impact.

Intentionality presupposes the experience of a discrepancy between currently experienced reality and a potential, desired and envisaged reality. It is the remarkable trait that does not take reality for granted but attempts to master, change and direct it to suit its purposes.[232] Intentionality translates into agency. In contrast with animals, humans are designed to go beyond the current situation and conquer the future.[233]

Although the human mind is a product of evolution, it does not passively submit to evolutionary processes but aspires to become the driving force of these processes.[234] Agency becomes creativity. The emergence of intentionality is, therefore, a quantum leap within the evolutionary process. Creativity does not suspend the evolutionary process but tries to employ and direct it. The breeding of domestic animals and plants is a case in point.

Science investigates reality with the explicit aim of mastering it. Technology uses this knowledge to harness the evolutionary process for human ends.[235] Science and technology are connected with multiple feedback loops. Technology dismantles reality, uses selected parts to construct something new, and discards the rest. Doing so, technology acts as a catalyst that enhances evolutionary change and channels it in particular directions.

It also becomes a catalyst that accelerates the entropic process. Economic growth is an acceleration of the throughput from resource base, via extraction, processing, distribution, consumption, to waste. The modern economy thus reflects both the enhancement of evolution and the acceleration of the entropic process. The faster the process of evolution, the faster the entropic process! We can see, therefore, that the current ecological problem is rooted very profoundly in the nature of human consciousness.

The complex nature of agency

Because it depends on lower levels of emergence, the human person is also determined by probability, causality, propensity, teleology and instinctual conditioning at the infra-personal levels of emergence. Physical reductionism is premised on this phenomenon. This is called 'upward causation'. However, causation also runs in the opposite direction. Humans have the capacity to deliberately impact the infra-personal levels of emergence. This is called 'downward causation'.[236] There is a complex system of feedback loops between bottom-up causation and top-down causation throughout the entire hierarchy of emergences.

Intentionality can extend the range of options by enlisting supportive powers from its internal and external environments and moving them against obstructive powers that determine a particular situation. In the end, the attainment of freedom and mastery is a power game between human intentionality and the impersonal forces of reality.

The individual person is again embedded in collective consciousness. Social processes can impact personal decisions that can again impact biological functions, chemical reactions, physical, and even subatomic processes—and vice versa. Intentionality thus operates within the constraints imposed by both infra-personal and supra-personal levels of emergence. However, within these parameters, the human being is confronted with a range of options, thus with the gift of (constrained) freedom and the demand of (limited) accountability. The greater the realm of options, the greater the realm of freedom and, by implication, the realm of accountability.

Social structures and processes are again embedded in the earth's ecosystem as a whole, which forms a further level of emergence with its own dynamics following its own regularities. The ecosystem impacts the social system and vice versa. To analyse the functioning of reality at this level of emergence is of paramount importance for dealing with the economic-ecological crisis.[237] Severe imbalances within the ecosystem will be evened out in the long run and this may lead to the extinction of the human species.[238]

In sum, evolution has produced a creature that is not only determined by causality as physical reality is, nor only guided by instincts as biological reality is, but faced with the necessity to choose between options. It has to weigh these options against the criteria of acceptability, utility, or desirability. It has to take action based on decisions.

In this way, it can become creative and bring about situations and entities that did not exist before and that would not have materialised without its interventions. It has to face the consequences of its decisions and account for their impact on reality as a whole. Let us analyse the mechanism of decision making in some greater detail.

Switches

Within the parameters set by the past, the future is open. It is a kind of 'field' of unrealised potentials. The higher the level of complexity, the greater the area of the "adjacent possible".[239] The probability of their realisation is staggered according to the forces that impact on the situation. Left to its own devices the world process will follow the path of least resistance. Expressed in Darwinian terms, options with greater 'fitness' within environmental niches will survive.

Yet intentionality can influence processes to move in desired directions. Intentionality and creativity are geared to desired outcomes; they have an 'arrow', a direction. How does this 'direction' actually materialise? Any conscious effort to bring about changes in direction depends on the capacity of the human subject to bring pressure to bear on the constellation of forces obtaining in a given situation. Left to its own devices water will flow downhill. But human intentionality can bring countervailing forces to bear that make water flow upwards.

As mentioned above, I find it helpful to distinguish between factuality, potentiality and actuality as expressions of past, future and present.[240] **Factuality** is the situation in which we find ourselves at any given point in time. Because it is the outcome of past developments, it cannot be changed. We cannot access the past; in fact, it no longer exists. So there is nothing we can do about factuality. Moreover, it lays down the parameters within which any future can occur.

However, depending on circumstances, these parameters open up a limited but substantial range of possible futures. I call this openness of the future **potentiality**. I am now here and cannot be simultaneously in Amsterdam or Tokyo. I can also not live tomorrow or yesterday. But departing from here I can move horizontally in all kinds of directions, following all kinds of trajectories for considerable distances. Yet this freedom is subject to certain parameters. I cannot move vertically up to the sun or vertically down to the centre of the earth. I also need time for such movements. I also need considerable resources of energy to reach a desired destination.

Actuality is the point at which one of these possibilities is realised, turns into present reality, and subsequently into a fixed part of the past, which then co-determines a new range of possible futures. It leaves behind the whole range of potentials that have not been realised and that will never be realised because they have also become a part of the past. Humans become aware of options as reality moves into the future. They anticipate their possible consequences and try to let beneficial options materialise.

Every single moment is, therefore, the starting point of a process that moves into the future. If a process is amenable to a change of direction in a particular situation, it presents us with a series of options. I call realised options 'switches', a metaphor taken from the railways. A switch thrown in Johannesburg can direct a train either to Durban or to Cape Town. A ridiculously insignificant switch in the life of my mother has led me—quite against her intentions—to become a South African theologian rather than a Namibian farmer as previously envisaged.

Switch theory is capable of resolving one of the tricky controversies between the modern and the 'postmodern' approaches. On the one hand it confirms the hard causal certainties highlighted by modernity. On the other hand it confirms the relative validity of the 'postmodern' insight that the 'text' of a given reality can be 'read' in many different ways, thus exploring 'cracks' and 'crevices' in the system and, following this lead, making reality serve the disadvantaged and the suffering better. The former refers to the past with its immutable parameters, the latter to the multiplicity of options that these parameters open up for the future in any particular situation. If understood in this way, 'postmodernism' loses its character of epistemic opportunism.

Where can switches occur?

We have to account for the statement that the future presents us with a spectrum of options, because a consistently deterministic worldview will not allow for this possibility. Causality rules supreme. This stance is not wrong, but it has absolutised a relative truth. The true part of the deterministic assumption is that situations are not equally amenable to switches.

Some of these starting points are heavily predetermined by the past. Others are characterised by high "sensitivity to initial conditions", as chaos theory puts it. Tiny variations at the beginning of a process can lead to vastly different outcomes.[241] It is at these points that the future opens up a spectrum of options. To deny these points of openness to options is the flawed part of the story.

Available options are scattered in a continuous spectrum from total flexibility to total determination. The further an option is from infinite contingency, or at least from considerable pliability, the more other forces exert themselves and the more energy is needed to realise it. In all practical spheres of life—transport, building, politics, education and human relationships—this is self-evident.

To illustrate, imagine a rock with a sharp edge at one of the peaks of the Andes mountains. A drop of rain is falling precisely in the direction of the edge. The slightest breeze can cause the drop to fall on either side of the edge. The breeze would then initiate a process that would lead the drop either to the Atlantic or the Pacific Ocean. The same kind of mechanism seems to operate at the level of chemical reactions, the formation of complex molecules, mutations in genes, and many others.[242]

Chaos theory has taught us that there are situations that are so finely balanced that the slightest impulse can move the ensuing process into a variety of directions. Take the proverbial pencil that stands precisely balanced on its head. In the absence of any outside influence, it will continue to do so forever. This does not happen because the probability that such a total absence of influences is virtually zero. But that is not the point.

The point is that the slightest impact on the pencil will send it falling in the direction opposite to the impact. The closer a situation is to such a balance, the smaller the impulse can be that will impact the direction of the ensuing process. Conversely, the more stable and static a situation, the greater an impulse must be to make a difference. I cannot uproot a mature oak tree with my bare hands, but I can do so with appropriate machinery.

In highly balanced situations, the decisive impact to make a difference can be miniscule. The mere imagination of an impending threat can cause the 'flight or fight' hormones to flush through my body and cause an action that changes the course of events. The mere thought of an ice cream can make my mouth water. The biological mathematician Stuart Kauffman believes that life itself can only exist at the edge between chaos and stability.[243]

In a much more profound sense, this is also true for human freedom. In a situation of high sensitivity, the balance of power between the impulse of my action, other factors that impact the situation, and the internal resistance of the process to change appears to be decisive. I cannot prevent a great river from flowing downhill with my bare hands, but a company with the means to build a dam could succeed in doing so. The closer I get to a finely balanced situation, the greater my freedom and the more consequential my decision and action.

At points of extreme sensitivity an ardent wish or a fervent prayer in a seemingly hopeless situation may open up my consciousness to options previously hidden to me and trigger the impetus for developments to move in certain directions. Conversely, a spirited initiative, backed up by personal energy and the necessary physical support can push even less sensitive and more resistant situations in the intended directions. This brings us to the problem of free will.

Free will as an emergent phenomenon

The problem of the freedom of the will has plagued Western thought throughout its history. Against the background of switch theory, the denial of human free will is not only counter-intuitive, but also counter-factual. It is based on one or the other kind of reductionism. Today there are at least four versions of this stance.

(a) Determination by an omnipotent and omniscient *divine will.*
(b) Determination by *Newtonian causality or quantum mechanics.*
(c) Determination by *instincts programmed into the human psyche* by biological evolution.
(d) Determination by impersonal *structures and processes of society.*

All kinds of reductionism are based on the assumption that the individual is a senseless, powerless and helpless victim of determining forces, whether divine predestination, physical law, biological evolution, or social dynamics. All these cases are falsified by the theory of emergence and switch theory. In all these cases, the fluidity between bottom-up and top-down causation is not taken into consideration. The rigidity of the infrastructure of a new level of emergence says nothing about the versatility of the latter. When driving a car, I can lose my way and go in the wrong direction, but the engine, the gear box and the wheels continue to do their work very reliably wherever I may be going.

Let us distinguish between physical determinism and theological determinism. In both cases, there is a practical and a theoretical dimension to the question. Let me tackle them one by one.

1. The question has no **practical** significance, whether in physics or in theology.

(a) Concerning physics, let us consider a parallel phenomenon. The fact that at the subatomic level matter consists overwhelmingly of nothing at all (so that some subatomic particles can fly through a tree trunk without crashing into anything substantial) does not mean that at the macro-level of Newtonian physics I can simply walk through a wall. This is not an illusion but a hard fact. What happens at the quantum level has practically no effect at the level of Newtonian physics. In the same way, any physical determinism, should it indeed arise from the quantum level upwards, in no way takes away from the fact that I am confronted with options, have to take decisions and face the consequences.

One just has to look at what actually happens in human life. The irony is that all the philosophers of science who so fervently defend determinism, freely chose their professions, took on specific research projects and decided to write books on their theories of determinism. They are certainly free enough to argue their case against possible alternatives. They presume that their audiences are free enough to be convinced by their argument, otherwise they would not bother. They would never want to part with their freedom, say under the pressure of a totalitarian regime.[244]

(b) In the same way, any theological determinism, deduced as it is from the assumption that God is the transcendent Source and Destiny of reality as such and as a whole, does not take away from the fact that, as a human being, I have been gifted with the capacity to have intentions, to act, and to be responsible for the consequences of my action. If the theological assumption of determinism were valid, the theological assumption of ethical accountability would be without foundation. Taken seriously it would have to lead to fatalism and despondency.

The Bible never came to the conclusion that because Yahweh hardened the heart of Pharaoh, Pharaoh did not harden his heart. Even the most ardent defenders of divine predestination exhorted their flock to change their ways, do their best in all respects and please the Lord. They never refrained from trying to change the structures of both the church and the society within their sphere of influence, sometimes drastically so. On the contrary—just witness what Paul, Luther, Calvin, or Beza have done!

2. Let us now consider the **theoretical** level. Seen against the background of our deliberations so far, the denial of free will appears to be based on a number of false assumptions.

(a) Physical determinism is the consequence of reductionism. Reductionism is not based on scientific evidence, but on deductions from questionable assumptions. Reductionism allows only bottom-up causation and no top-down causation, which is not only counter-intuitive but counter-factual. Because this assumption is unable to do justice to the phenomena, it is increasingly being rejected in favour of emergence theory by perceptive natural scientists such as Stuart Kauffman and George Ellis.[245]

Emergence theory allows not only for bottom-up causation (which is self-evident) but also for top-down causation (which should be equally self-evident). A wheel would not exist if the atoms of which it is composed would not operate at the atomic level, but when it turns, all these atoms have to turn as well. Human consciousness is the highest level of complexity we know of. It is able to impact all lower levels of complexity—for instance by subjecting a human body to an operation, bringing about chemical reactions, hammering an iron rod into shape, or inducing the collision of subatomic particles.

As we all know, top-down causation is not unlimited. As indicated above, chaos theory gives us the pivotal clue in this respect. There are situations within the causal network where sensitivity to initial conditions is so high that the slightest impact from whatever cause can change the direction of the ensuing process. Human intentionality and agency belong to the factors that can impact such a situation. The weight of the impact depends on the amount of energy they can enlist from their own resources or their environment.

In short, human freedom is real but constrained—and that to a variety of degrees. Of course, this is also true to a lesser degree for animals and plants. The roots of a tree cannot penetrate a granite slab, but they can squeeze into tiny crevices in a disintegrating rock and crack it apart in the long run.

(b) The theory of theological determinism is similarly flawed. Divine and human intentionality and agency do not operate at the same level. So they cannot compete or cooperate with one another. The divine acts through the human and (parallel to the Christological doctrine of Chalcedon) the two levels should be neither separated from, nor confused with each other.

The doctrine of predestination is not based on actual experiences of faith (whether reflected in Scripture or in daily life), but on deductions from questionable assumptions. The assumed divine attributes of perfection, omnipotence, omniscience and eternity found in classical theology are idealised abstractions from the actual flow of life. They have a Platonic rather than a biblical origin. They have caused endless confusion.

In biblical terms, omnipotence means that God is the source of all *actual* power flows at work in any given situation. Its rationale is reassurance rather than speculation. It does not mean that, by definition, God operates under no constraints whatsoever. If that were the case, it would be totally incomprehensible why a God with unlimited power and unconditional benevolence did not simply eradicate depravity, suffering and death with a single, eternally valid and universally effective decree. This is the problem of theodicy, which I shall discuss in Chapter 12.

In fact, there are such constraints, notably space, time, the availability of energy and the different regularities that operate at different levels of emergence. These constraints are just as much 'of God' as the instances where believers believe that God had intervened on their behalf or had predetermined an outcome. These constraints explain why amazing things do happen, and yet why a tenacious phenomenon such as biological death cannot be overcome and has never been overcome. Resurrection must obviously have a meaning different from biological resuscitation.

In sum, free will is not an illusion, but a 'natural' consequence of emergent evolution and as such part of the way in which our reality is constructed. Where counter-intuitive theory imposes itself on what appears to be self-evident fact, experiential realism calls both theologians and scientists down to earth.

Section II
The emergence of collective consciousness

So far we could gain the impression that structures of consciousness are located only in the individual brain. This is demonstrably not the case. The link between individual and collective consciousness is the transfer of information through symbolic representations. The entire hierarchy of emergences is determined by information of various kinds at various levels of emergence. At the biochemical level, for instance, information is carried by the structure of certain molecules. Systems of molecules develop autocatalytic functions. They can settle into distinct patterns. They can replicate themselves. They can impact the patterns of other systems.

Such patterns are primitive symbols that communicate information between discreet entities. At higher levels of emergence the complexity of symbol-formation increases exponentially. Complex systems of information with multiple feedback loops can lead to the self-organisation of an entity at a higher level of emergence, for instance an organism. The functioning brain is the most sophisticated example. As we have seen, the 'hardwiring' or 'conditioning' of the brain comes about through genetic pre-adaptations on the one hand and networks of synapses occasioned by environmental impacts and responses to these impacts on the other.

These impacts are largely based on patterns of information. The scent of honey directs the flight of a bee towards a flower. What makes the human being human is the exceptionally advanced degree of symbolic representation and communication. Systems of meaning are stored, as it were, in the brain. Symbols found in language, facial expressions, computer codes and countless other phenomena act as messengers that transfer meaning from person to person. What happens here is that a construct, represented by a set of symbols in one consciousness, is reconstructed in the consciousness of another consciousness.

The transfer of meaning through symbolic representations is continuous, dynamic, situation-specific and interest-related. As the science of semiotics has taught us, symbols are polysemous, that is, they can carry a number of meanings and associations that may differ between the sender and the receiver of a message. The meaning of a symbol depends on the total context of meaning within which it is located in the case of the sender on the one hand and of the receiver on the other.

Because of this embeddedness of symbols in symbol systems, polysemy does not necessarily lead to anomie. The transferred meaning will adapt to the new context when it moves from the sender's hardwiring to the receiver's hardwiring. There is continuity as well as novelty and evolution within the communicative process. Continuity is constituted by the similarity between the systems of meaning of senders and recipients, novelty by their discrepancies, evolution by the process of the mind sorting out these discrepancies.

In this way, snippets of meaning, logical sequences and more comprehensive systems of meaning can be transferred from person to person, community to community, generation to generation. In time, the interaction between all these inputs may crystallise out to form the dynamic conglomerations of meaning that constitutes a particular culture, mentality, or religious set of assumptions.

Packages of meaning that are able to respond best to the pressing needs and desires of the greatest number of people within a given population will have the greatest impact on the provisional system of meaning of the community as a whole.[246] In this sense, the evolutionary principle of the survival of greater fitness in particular environmental niches is applicable.

Collective consciousness is constituted by the sum total of individual consciousnesses, yet the interaction between these components leads to a new level of emergence with its own characteristics and dynamics. In this sense, collective consciousness represents a supra-personal level of emergence that can co-determine the personal level very profoundly. As in the case within the entire hierarchy of emergences, there is bottom-up and downward causation. Individual consciousness has an impact on community and society; community and society have an impact on individual consciousness.

In a more pluralistic situation, masses of such pockets of meaning continuously flow from person to person, from community to community, all ending up in the melting pot of communal and societal mindsets. If there is an oversupply of alternative and partial packages of meaning, 'turbulence' may be the result. It can take the form of fragmentation, cognitive dissonance, uprootedness, anomie and total disorientation. Some of these phenomena can be observed today in popular 'postmodern' cultures.

Systems of meaning

Human beings are capable of transcending their immediate experience to a much greater degree than higher animals. They transcend their personal recollections into the more distant past, their personal apprehension into the more distant future, their immediate environments into their concentric contexts—body, community, society, nature, cosmos—and their immediately available personal resources towards greater powers. In short, they always envisage some kind of greater whole in terms of time, space and power and try to fathom its structures and regularities. This is the existential root of a system of meaning.

A system of meaning defines our identity, authenticity and authority within the greater whole. How do we arrive at such systems of meaning? Where they are artificially constructed (think of the Hegelian or Marxian dialectic), they tend not to survive very long. More enduring are traditions that emerge and evolve in history. They incorporate successive and diversified collective experiences and are constantly being corrected by history itself. Personal or communal experiences and reflections again augment, correct and transform inherited and previously held patterns. Where this is not the case, a system of meaning morphs into a totalitarian ideology.

A system of meaning is the crystallisation of outcomes of 'trial and error' in response to concrete experiences over long periods of time. Assumptions and patterns of behaviour either worked or they didn't. They have been hard-wired into the collective convictions and consciences of communities. They are informed as much by experiences as by intuitions. They can hail all the way from prehistoric times right down to learning curves that individuals and communities undergo in their daily lives.[247]

A system of meaning integrates the experiences and intuitions of past generations with the apparent requirements of newly experienced situations. This is one of the roots of religious intuitions, myths and rituals.[248] Initially such a system may be diverse and multifocal, but there is a tendency to integrate subsystems under a pivotal centre. Such a consolidated system of meaning represents, therefore, a new level of emergence.

Appropriate religious faith represents the widest attainable horizons in particular cultural contexts. Ideally it will be based on a notion of the transcendent Source and Destiny of reality as a whole. Believers look at reality 'from above' as it were, 'with the eyes of God', discerning their own limited locations, statuses and roles within their concentric contexts. Faith does not have to be ecstatic, irrational or superstitious. It can be down to earth, sober and factual. But in its mature form, it will be geared towards the vision of comprehensive optimal well-being.

The ontological status of a system of meaning

What is the experiential status of a system of meaning? It consists of basic assumptions. It defines acceptability in the form of values, norms, goals and visions. It grants belonging to the acceptable, and disciplines or excludes the unacceptable. It allocates authority to act in a certain way through mandates, statuses and roles. It forms what colloquial language calls 'conscience', what psychologists call a 'super-ego', and what sociologists call a 'sacred canopy' or 'civil religion'.

Materialist reductionism considered such phenomena 'merely subjective', thus not objectively real. So we have to turn our attention to the 'objective ontology of subjectivity'. Brain states are the infrastructure underlying the emergence and evolution of structures of consciousness. They have a built-in propensity to find some sort of coherence.

Ideally, a fully fledged system of meaning will cover all inherited traditions, collective memories and personal experiences made so far. The brain organises the impact of new impressions into the patterns established by previous experiences. So systems of meaning are not wilfully constructed. They impose themselves on our consciousness. It is only when existing and new patterns 'fall into place' that the entire system seems to 'make sense'.

On its own, this mechanism cannot distinguish between patterns based on fallacy, fantasy, fiction, or hearsay, on the one hand, and objective fact on the other. It can only organise and reorganise the existing patterns under the impact of new information. This is of immense importance for our topic because, paradoxically, it provides the reality component of a notion of the transcendent or a system of meaning. The pattern located in our brains is objectively real as such.

On the basis of this objective structure, people can live subjectively in a world populated by angels, demons, saints, witches, or ancestral spirits and experience them as real, communicate with them, battle with them, exorcise them, or be oppressed by them. But just as a scientific theory can be replaced by another on the basis of new empirical evidence or a new mathematical procedure, such systems of meaning can be analysed, critiqued, transformed, even reconstituted through the communication of new content that seems to make sense. They can also change under the impact of environmental experiences.

Not every new information or impression will be accepted as valid. That is why systems of meaning tend to be rigid. There are criteria, built into the system of meaning itself, that filter new impressions. New criteria must be 'hard-wired' in the brain as modifications of the system of meaning before they can take effect. Such criteria are not add-ons. They must be integrated in the very structure of the system of meaning concerned. This structure is robust enough to resist changes that are not in line with its past development.

Yet our brain is continuously bombarded with new information. This may be in line or at variance with what was previously 'taken for granted'. There is, therefore, a constant tug of war between old certainties and new challenges. This 'inertia' assures emotional stability and cognitive coherence. Yet systems of meaning do have the capacity to change. They are fairly tenacious but not fixed. Both 'memory' and the shape of new impressions on the mind are malleable.

If the deviation is large and persuasive, or even fundamental, a state of 'cognitive dissonance' and 'anomie' may occur. Such a state of mind is unpleasant for the mind itself. The brain is flooded with antagonistic chemicals that cause a kind of physical discomfort. The brain will try to sort itself out either by changing the old to accommodate the new (learning), or by transforming the new to fit into the old (rationalisation), or by rejecting the new outright (conservatism), or by abandoning the old in favour of the new (conversion).

In modern times, the volume of new information tends to explode the capacity of the brain to integrate it. 'Making sense' is an onerous chore that makes incessant demands. However, an integrated structure is being sought at all times. It is this kind of consistency or coherence that constitutes the subjective experience of 'truth'.

The sciences hold that, where the right methodological procedures are followed and safeguards against error are heeded, the subjective experience of truth is an always provisional and problematic, yet reasonably reliable reflection of objective truth. Once again, truth is not constructed, but found. It imposes its validity on one's consciousness as the brain sorts out the discrepancies. Often a 'solution' is found by the subconscious overnight or over longer periods of time, when the disparate elements finally fall into place.

As the work of Thomas Kuhn on scientific paradigm shifts has shown, this goes for scientific insight as much as for religious conviction. A scientific system of meaning only imposes different criteria for the kind of information it allows to impact its existing structure than a religious system of meaning. Such scientific criteria are 'empirical evidence', 'mathematical stringency', or 'plausible conjecture'. They are all based on experiential realism.

Religious convictions venture further into the unknown and unknowable. They are based on apperceptions of what ought to be, in contrast to experiences of what ought not to be. They are geared to a notion of the transcendent. Such a notion goes beyond empirical evidence but, to be appropriate, it must incorporate best scientific insight about the immanent, or lose its credibility.

The ethical dimension

Humans are the only creatures that can focus their intentionality and agency upon a greater whole, whether to abuse, exploit and destroy it, or to contribute to its overall well-being. Being aware of wider contexts, responsible human beings could be expected to develop a vision of comprehensive optimal well-being. They could then be expected to tackle any deficiency in well-being in any dimension of life as far as their competence and influence goes.

This expectation would be entirely in line with the evolution of life because there is no life that does not possess an inherent thrust towards survival and well-being, which again depends on the health of its environment. Countless synergies in nature as well as the sparing attitude of many carnivores show that the evolution of such behaviour is entirely 'natural' in terms of the evolutionary process.

Humanity, more than any other species, depends very fundamentally on the well-being of its social and natural environment, the more so the more it increases in numbers and lifestyle demands. The fact that the growth of this responsibility lags so miserably behind the newly acquired scientific and technological powers of the modern human being should be considered a suicidal anomaly in terms of the evolutionary process. I will come back to that phenomenon below.

For now we have to ask how the emergence of such a vision, if it happened at all, could be explained. The constraints of time sequences, environments and balances of power are staggered in terms of their relevance for human life. Depending on the horizons of its awareness, the conscious mind will seek to maintain continuity with the networks of past events that led to the present and weigh the future consequences of possible responses to the current situation.

It will then opt for what it perceives to be the most favourable outcome for the individual, the community in which it is embedded, and the environment on which it depends. Ideally, that is how human history should unfold in time. A science that does not take the character of these historical developments into account is a deficient science. In fact, the pursuit of science itself depends on human history with its endless sequences of reality-transforming switches.

Reality is in the process of becoming. When taking conscious or unconscious decisions, humans impact the direction of the world process as a whole. Minute changes in direction can lead to exponentially deviating and accelerating processes. As a result, the consequences of human decisions can be horrific.

An example may illustrate this fact. In 1914, the German emperor decided to respond to a murder in Sarajevo with military rather than diplomatic means. This set an incredible avalanche in motion: World War I, the defeat of Germany and its allies, the loss of territory, the humiliating Peace of Versailles, the misery of the Great Depression, the abortive Weimar Republic, the rise of Hitler, World War II, the holocaust, the development and deployment of nuclear weapons, another devastating defeat, destruction and division of Germany, the demise of the great European Empires, the independence of former colonies, the rise of the United States and the Soviet Union to superpower status and the Cold War. Without this single cabinet decision in 1914, our world would have been an entirely different place today.

Because the consequences of even minor decisions can be substantial, there should be no exercise of freedom without taking responsibility for the consequences of one's attitudes and actions. However, consequences cannot always be foreseen, not even in the short term and within the range of the interests of a particular group, let alone in the long term and for humanity and the natural world as a whole.

On its own, individual human consciousness would be totally overtaxed by the burden of responsibility. Since even the minute decisions of everyday life have never-ending consequences, we need particular directives that (we hope) will lead processes in acceptable directions. They are based on the collective experiences of countless generations and their sometimes very inappropriate interpretations. They constitute the fundamental ethical imperatives of human life.

Because it is not self-evident what is to be deemed acceptable, ethical imperatives are, whether explicitly or implicitly, embedded in particular systems of meaning. Such 'symbolic universes' acquire some kind of provisional validity or 'sacredness'.[249] They hold the acting subject accountable. They are not based on science. In fact, they normally contain a great number of unsubstantiated and often quite spurious assumptions. That is why they should be required to integrate the insights of 'best science'. They should also be geared to the vision of comprehensive optimal well-being.

Science is an indispensable tool to observe, comprehend and predict reality and its potentials, but its method and mandate do not cover the thrust of human existence towards what ought to be. Science can never be more than a tool. As a tool it will be used by whatever system of meaning avails itself of the powers provided by science and technology. Hitler's extermination camps and the atomic bomb that devastated Hiroshima displayed, after all, great scientific prowess and technological efficiency, without therefore manifesting beneficial human behaviour.

The danger is, therefore, that the exponential acceleration of the processes set in motion by modernity leads to a dearth of appropriate criteria and overshoots the human capacity to control the consequences. It calls for a highly alert and critical mindset, as well as immediate and decisive action. This is not something unattainable. The experience of what reality ought **not** to have become is pervasive and inescapable. This experience should evoke a dynamic and adaptive vision of what reality as a whole **ought** to become, as well as a powerful thrust of human intentionality and agency in the direction of this vision.

However, what ought to be cannot simply be derived from immediate human needs, wants, whims and desires, whether individual or collective. Humans are part of a comprehensive network of relationships, and they are capable of being conscious of this fact. It is this whole, in which humans are embedded, whose comprehensive optimal well-being matters. Humans try to comprehend the particular 'whole' that they deem relevant for their individual and collective lives. The result is the evolution of a system of meaning.

The emergence of 'spirit'

We have seen that any package of meaning can be cast into the form of a symbolic system, communicated to others, and appropriated by these others. It can be shared by a great number of people. It can also be critiqued and transformed. Usually, this happens within communal contexts. A system of meaning forms a tradition that evolves over time and unites a faith community. This is the level of reality that is conventionally called 'spirit', or with a more contemporary term, 'mindset'.

I introduce the concept of 'spirit' here to establish a link between a scientific and a theological kind of discourse. This is important because body-spirit relations have been one of the trickiest problems in the debate between science and religion. What then is spirit in immanentist terms?

Mind or consciousness is not the same as spirit, as sometimes assumed. Spirit is an emergent *structure and orientation* of consciousness.[250] Though based on neural and hormonal mechanisms, mind can be structured in an infinite number of ways. It is also highly dynamic and adaptable. It is like music, which is infinitely variable although it is based on a limited number of pitches, intervals and rhythms. It also shows vast differences in quality. It can also evolve to reach ever higher levels of complexity, excellence and validity. Spiritual growth has been defined by a psychiatrist as "the growth or evolution of consciousness".[251]

Experiential realism will acknowledge that the structures within the conscious mind are part of immanent reality. They can be appropriate or inappropriate reflections of internal and external reality. They can be beneficial or detrimental in terms of the vision of comprehensive optimal well-being. These structures can be described, analysed, scrutinised, transformed, or abandoned. If that were not the case, philosophy, theology and any other quest for ultimate truth—would be pointless.

Characteristics of 'spirit'

Spirit seems to follow at least some regularities. First, spirit is always in evolutionary flux. There is no timeless spirit. Spiritual formations undergo their own historical evolutions. This is not always recognised by religious convictions. It is remarkable that the biblical faith takes its own historicity for granted. Divine self-disclosure happens in human history as a dynamic ongoing process. It manifests itself through the evolving, ever provisional, partial, weak and questionable insight of human beings. As mentioned above, a *notion* of the transcendent is a fallible part of immanent reality that can be and that must be subjected to critique.[252]

Second, spirit is transsubjective in the sense that mental structures are transferred from person to person and from generation to generation through evolving and differentiating traditions. They are constantly impacted and transformed by collective experiences and interests. As a result great numbers of individual minds can be structured in discernible transsubjective patterns. We can speak of the spirit of a constitution, the spirit of an ideology, the spirit of a religion, the spirit of a phase in cultural history (*Zeitgeist*), or the 'ethos' of a commercial enterprise.[253]

Third, there is a constant tug of war between convictions and interests. Convictions claim universal validity and demand the subservience of the individual or group under a greater whole, while interests demand the prioritisation of the needs and desires of an individual or a collective. I have analysed this phenomenon elsewhere and do not have to repeat that here.[254]

Fourth, there seems to be a constant interplay between individual consciousness and what can perhaps be called the 'collective subconscious'. This is the bedrock of meaning that is not consciously held and articulated by a community, but simply taken for granted.

Various factors seem to come into play. At the most self-evident level, a particular spiritual content may be transmitted from generation to generation in ways that go beyond simple verbal communication. One could think of the kind of cultural and religious 'atmosphere' to which the infant, perhaps even the foetus, is exposed and which its subconscious internalises. This may include body language, gestures, facial expressions, garments, domestic and public spaces, foods, kinds of music, furniture, pictures on walls and in books, routines, habits, and so forth.

Psychologists like Carl Gustav Jung and his school go further. They have come up with the idea that myths, fantasies, dreams, deliriums and delusions betray recurring formal structures. Jung calls them 'archetypes'. They can be filled with any number of different contents. Jung believes them to be similar to the geometrical structures of crystals which do not exist until they appear in a real crystal.[255] That analogy probably takes the argument too far. They may rather be similar to what we called regularities at different levels of emergence. One cannot help but think of the phenomenon of convergence in the form of the wings of flying animals that have quite different anatomic derivations.[256] The best 'archetypes' are those that have proved themselves again and again over long periods of time and in various contexts.

Parapsychologists go beyond even that, claiming that some spiritual content can be communicated across time and space and without the involvement of physical energy—from the distant past, from the future, from elsewhere in the universe, and from beyond death to the here and now. There is a tendency to link up this claim with the assumptions of relativity theory in physics on the one hand and postmodern relativity on the other. I cannot help but have my doubts in this regard. Maybe I harbour an obsolete scientific mindset, and I better plead ignorance in this respect. Subatomic physics has indeed come up with some very queer theories about what seems to transpire across time and space.

Fifth, evolving in time, spiritual traditions differentiate into sub-currents in space. This is why we find as great a variety of systems of meaning as we find species at the biological level. One could argue, therefore, that they all have an equal right to exist. Well, they do indeed exist and cannot simply be wished away. We cannot argue with the same force of plausibility, however, that they all have the same quality, profundity and validity, just as we cannot argue that amoebas, lizards and humans have reached the same level of complexity, flexibility and performance.

Finally, when spirit is based on a specific notion of the transcendent Source and Destiny of reality as such and as a whole, it tends to claim ultimate validity. That is to be expected. In this case, a particular spirit is deemed a reflection, representation or 'icon' of the transcendent as such. The assumption of the believers in question is that the transcendent has disclosed or manifested its true nature, or rather its ultimate intentionality, in this particular spiritual formation and orientation. It is at this juncture that faith and naturalism part ways.

When religious convictions claim historical finality and eternal validity for their assumptions, they misunderstand themselves. Notions of the transcendent do emerge and evolve in time. They can also deteriorate and decay. They are not absolute. They respond to changing worldviews and situations of need. Where they don't, they let their believers down. Because everything emerges and evolves in time, this does not disqualify their validity. It just calls for sensitivity, flexibility and humility. The quest for ultimate validity is itself a historical process that can lead to more profound insight and that will never come to a close.

What kind of spirit?

'Spirit' can turn out to be thoroughly counterproductive in terms of the vision of comprehensive optimal well-being, even in terms of the prerequisites of the lower lying level of biological life. It can also degenerate and decay. There is something like spiritual entropy. For the sake of harmony and tolerance, this fact is often overlooked in interfaith relationships. The danger is that we translate the valid assumption that human beings are of equal dignity into the spurious assumption that all the kinds of spiritual structure these people may have internalised are of equal quality and validity.

Hitler's racist and violent convictions, for instance, cannot claim to be on par with the peaceful spirit represented by Ghandi, or the self-effacing love of Mother Theresa. There are criteria and these criteria are located in particular systems of meaning. Ideally, they should reflect a vision of comprehensive optimal well-being. What matters for any kind of philosophy or theology is not the apparatus of the mind as such, but the specific content of 'spirit'.

The biblical faith, for one, is highly sensitive to the ambivalent and problematic character of the human spirit. In fact, the urge to overcome a counterproductive spirit is a potent driving force of faith. We are warned in 1 John 4:1 not to follow every next best kind of spirit, but to test them whether they are 'of God'. The text presupposes a particular definition of 'God', namely 'love'. The assumption is, then, that there is a authentic kind of spirit to which humans should aspire, or in which they can participate, namely the 'divine' spirit manifesting itself in a new human spirit.

Spirit denotes a system of meaning that defines identity, acceptability and authority. While the pursuit of well-being may be fundamental to all living creatures, including the human being, human consciousness can develop widening horizons in concentric contexts. It can also contract to such an extent that it encompasses nothing but the needs and desires of the individual. This 'narcissistic' atrophy has become the overriding tendency in modernity.

An appropriate system of meaning, in contrast, will embed the interests of the individual in the context of the interests of the community, society, nature and cosmic reality. It will also encompass different dimensions of well-being, such as material, biological, psychological, spiritual, communal and cultural needs, in a balanced way.

'God' is a name for the pivotal centre of a system of meaning that covers the whole of reality as far as it is reflected in human consciousness. At least potentially, therefore, a vivid God-consciousness is capable of providing the most inclusive horizons available to humans. More often than not, however, the concept of God has been abused to legitimate the pursuit of individual and collective self-interest at the expense of the interests of others, the community, society and nature. Therefore the quality of a particular God-consciousness must be subjected to critique and reconstruction at all times.

Let us summarise

In the last two chapters, we have used the theory of emergence to integrate all levels of experienced reality. At least in principle, it includes the subject matter of all social, historical and human sciences into a differentiated but consistent whole. The upshot is that reductionism—whether physical, biological, psychological, or sociological—cannot do justice to the complexity of reality that humans actually experience.

The theory of emergence has also led us to the conclusion that any dualism between spirit and matter can no longer be upheld. Religion is part and parcel of a single infinitely differentiated yet highly integrated immanent process. It offers a system of meaning, based on a particular concept of the transcendent, that provides criteria of acceptability and the authority to act.

Following the method of experiential realism, it is not possible to answer the question whether and to which extent this immanent phenomenon actually reflects the transcendent as such. For faith, however, this question presents no problem. The transcendent is, after all, deemed the Source and Destiny of immanent reality and can manifest itself in and through this reality. The critical question is, rather, whether the content of the concept of the transcendent is geared to the vision of comprehensive optimal well-being or not.

Part III

Beyond science —
The murky issue of
transcendence

8 The open universe of faith

The relation between science and faith hinges upon the concept of transcendence. Without transcendence faith would be void of substance. We have to distinguish between the unknown and unknowable within reality on the one hand (immanent transcendence) and the foundational on the other (transcendence proper). The latter is based on the human experience of derivation, dependence, vulnerability, mortality, and accountability. It provides for meaning, criteria of acceptability and authority. Faith does not clash with scientific observations, explanations, or predictions. Rather, it attributes reality as such and as a whole to God, its ultimate Source and Destiny, subjects it to the scrutiny of God and draws its motivation and strength from trust in God. The transcendent does not cooperate or compete with immanent factors but lies at the roots of all of reality. Conceived as a person, God's intentionality and agency do not exclude immanent processes and human agencies but initiates, sustains and empowers them.

The task of this chapter

In Part II, I have analysed the nature of faith as a distinct structure and orientation of consciousness, thus as an 'empirical' phenomenon within the reality we experience. This is as far as the scientific mandate and method allow us to go. However, it says nothing whatsoever about the 'truth' of the assumptions of this faith. What does faith represent—obsolete science, childhood conditioning, projection of wishes, legitimation of elite interests, poetry, hallucination, delusion, superstition, plain deception?

The relation between faith and science as serious approaches to reality would collapse if it could be shown that faith was without substance. Then the only scientifically rational and legitimate stance would be to dismiss faith altogether. This is the tacit assumption of modernity and the overt stance taken by the 'new atheists'.

In Part III, I try to show that the Christian faith does not, of necessity, contradict the insights of modern science. Conversely, modern scientific insights do not automatically lead to naturalist and atheist assumptions. In Chapter 8, I deal with the meaning of transcendence as presupposed by faith, in Chapter 9 with its naturalist alternative.

Transcendence as the pivotal issue between science and faith

Transcendence is the pivot around which the interaction between science and faith revolves. If there were no transcendence, faith would have no referent and science would have no counterpart. Science would have to assume the functions of faith to provide meaning, acceptability and authority. At best, religion would function as a humanly constructed depository of meaning, value and vision. Where this happens, science breaks out of the strict confines of its method and mandate. This is what actually happens in modernity. Its most common form among scientists is naturalism.

Conversely, if faith tried to assume the status and function of science, it would have to provide observations, analyses, explanations and predictions of immanent reality. With that it would deny the transcendent character of its primary referent and go beyond its mandate and competence. This is what happens in Christian fundamentalism and creationism, which assume that 'revelation' gives the believer privileged access to particular information concerning experienced reality that is not available to science.

Both views are flawed. The mandate and method of science as such do not cover transcendence, and faith as such is not equipped to explore immanent reality.[257] Faith only attributes experienced reality to God, its transcendent Source and Destiny, however it may be observed and explained. If transcendence is a valid assumption, however, science and faith each have an indispensable function. They are each released from a chore they are not designed to tackle and free to complement each other.

But then we must also refrain from taking the two apart. Faith is concerned with the transcendence of *immanent* reality as such and as a whole in all its concrete aspects and processes, rather than with some fantasy, speculation, or revealed information about the world. If we first remove God from the world process and then try to smuggle God back into immanent reality as a quasi-immanent causative agent, we make a fool of ourselves. Removed from reality, God is redundant. As a factor among others within reality, God is pure fantasy.

'Immanent transcendence' and transcendence proper

When dealing with the concept of the transcendent, we seem to have reached the utmost limits of an immanentist approach. However, this boundary is an exceptionally murky field. It is important for the interaction between faith and science that its meaning be clear to all participants in the debate.

The first question is what precisely we mean when we use the word 'transcendence'. Is it the experienced but inexplicable, the explicable but not immediately experienced, the knowable but not yet known, the immanent but unknowable, the permanent mystery within immanent reality? All these aspects of reality point to what has been called 'immanent transcendence', that is, aspects of the assumed 'one and only' reality that are not readily accessible to human observation and interpretation.

Or do we mean the *foundational*, that is, something beyond experienced reality that underlies the assumed one reality as such and as a whole, that allows us an imagined perspective on reality from outside reality, whatever that may mean? This is what the concept of the transcendent means in faith and theology.

No amount of appreciation for the importance and excellence of human perceptions of meaning, value and vision can make up for the lost assumption that, in whatever provisional, partial and problematic way, the notion of the transcendent reflects a self-disclosure of the transcendent itself. This is where faith and naturalism part ways.

When I speak of 'transcendence proper', therefore, I refer to the unsurpassable boundary of human observation, comprehension and control. The intended 'object' of faith, the transcendent as such, is located beyond these boundaries, but the human intuition, perception, or notion of the transcendent is located within these boundaries and that is what matters from a scientific point of view.

The second question is what difference a change in perspective from within immanent reality to one from outside this reality would make. Do we need it at all? The third question is what consequences alternative substantive contents of the concept of the transcendent would have in the real world. Is it immaterial whether we take Western liberal assumptions, Marxist-Leninist doctrines, or Muslim fundamentalist convictions for granted? The fourth question is where our notion of the transcendent comes from. What are the experiential roots of an awareness of the transcendent?

Section I
Immanent transcendence

So what do we mean when we speak about 'transcendence'? The basic image is that of "going beyond" certain limits or boundaries. There are such boundaries within immanent reality. Not everything we know or could know is immediately experienced, yet it can become the object of scientific investigation. This phenomenon is called 'immanent transcendence'. Let us consider a few examples.

In terms of time, the past is transcendent because we cannot access something that is no more. The future is transcendent because we cannot access something that is not yet. In terms of space, any situation beyond my immediate environment is transcendent because it is real for me only in my recollection or imagination. In terms of energy, any latent force is transcendent because it has no actual effect on experienced reality.

Any potential that has not been realised in the present, though it could have, is transcendent. Any alternative to the existing social order is transcendent because it might have existed but does not.[258] Any power to influence things that is not at my disposal is transcendent. In terms of complexity, higher levels of emergence contain elements that defy explanation and prediction.

The subconscious is transcendent. The 'other'—whether one's body, somebody else, community, society, or nature—is transcendent. Any authenticity that I could have achieved, but have not, is transcendent. There is also epistemological transcendence: Kant has taught us that the 'thing in itself' is not accessible to us; what we know only reflects processed sense impressions.

The second kind of transcendence marks the boundaries of a particular approach to reality.[259] Boundaries can exist between academic disciplines and between alternative approaches within disciplines. In subatomic physics, for instance, there are a number of different plausible theories built on different assumptions. Physicists may one day find the best theory, but it will still consist of a system of inferences drawn from basic assumptions.

Moreover, no science is able to cover the whole of the reality that sustains and impacts human life. The hierarchy of emergences and the diverse regularities operative at different levels of emergence led to the differentiation of sciences found in academia today. They range from subatomic physics through Newtonian physics, chemistry, biology, psychology, linguistics, history, religious studies to sociology, economics and political science. Each of these fields is 'transcendent' in terms of every other field. All these sciences are based on assumptions and inferences.

In an age of increasing specialisation, it has become necessary and urgent to develop a multi-pronged approach. Commitment to unearth the truth demands that we cover the entire hierarchy of emergences, including personal experiences, in the sphere of legitimate scientific investigation. It would be arbitrary not to do so. Unkind observers have spoken of "professional idiots", that is, people who know everything that can be known of a tiny fraction of reality and be ignorant of its wider contexts. In other words, we have to require experts to transcend their particular area of expertise.

The third boundary refers to the fact that all sciences are confronted with problems they have not yet been able to resolve. It is quite sobering to discover how little particular sciences know with absolute certainty, how much of what seems to be clear to them depends on initial assumptions, how much of their theory is based on logical or mathematical construction, how much depends on conjecture rather than evidence, how partial and provisional the findings of even the most exact experiments in the natural sciences often prove to be, how narrow the confines are of the part of reality that science will ever be able to access.[260]

Impenetrable mysteries encountered by the sciences are manifold. In *physics*, will we ever be able to explain the origin and nature of energy? Of space? Of time? Of natural laws and regularities? Of the reasons for the 'unreasonable' applicability of mathematical theorems to physical reality? Or is it actually only partially applicable? What do we make of the inaccessibility of the past and the unrealised potentials of the future that call for our decisions in the present?

In *cosmology*, we do not know what 'set off' the big bang. We are not even certain that it was a 'bang' in the first place. Various models lead to contradictory interpretations of its immediate aftermath.[261] Nor is the precise age of the universe all that certain. Just as foggy are the possibilities of its remote future—a big crunch or indefinite expansion? We have no clarity on dark matter and dark energy, which seem to constitute the vast preponderance of mass in the universe.

If the spontaneous emergence of virtual particles in a vacuum through field fluctuations shows that the empty space is not empty, with what is it filled? There are speculations about a 'multiverse' (the simultaneous existence of more than one universe), perhaps due to 'hyper-inflation' that disconnected different regions of the initial universe. So far *string theory* is nothing but extremely complex speculation, but speculation all the same. It is easy to populate a mathematical model with more than three (or four) dimensions, but how can we be certain that they correspond with anything at all in physical reality?

Speculations by some cosmologists about the prospects of human life to subsist as a package of *information* beyond its current biological base, and to migrate into outer space in search of alternative sources of energy after our sun has burnt out,[262] seem to be stark nonsense, and that in spite of their impressive sophistication. The same is true for the hope of some believers that this could have some relevance for Paul's idea that we will rise from the dead into a 'spiritual body', whatever that may mean.

In terms of *probability theory*, the world we inhabit, and of which we are a part, should not have come to exist. In view of the remarkable and unlikely fine tuning of the universe, which makes not only life possible but its entire material infrastructure, we do not know why the physics of the world that has come about "lie in this tiny corner of the possibility space".[263] The assertion that God has willed it to come about in this form gives a name to the mystery, but no explanation.

In spite of its stringent mathematical methodology, *subatomic physics* is far from producing a plausible formula for a possible 'theory of everything'.[264] We "cannot even prove the outcome of quantum theory is random."[265] There is no agreement on the precise nature of the entropic process. We do not know the origin and the scope of the arrow of time. There is no conclusive explanation of gravity. The chasm between quantum theory and relativity has yet to be breeched.

In terms of *complexity theory*, our mathematical tools are not powerful enough to predict the behaviour of non-linear sequences, if they are predictable at all, and the behaviour of highly complex systems such as weather patterns.[266] The capacity of our mathematics does not even reach very far beyond the interaction between three or four variable, let alone billions of them: "Friend, you cannot even predict the motion of three coupled pendula."[267]

We do not know whether the *laws of nature* 'pre-existed' or developed with the evolution of the universe. We do not know the connection between chance and law. We do not know whether there are laws that we have not discovered and will never discover. We cannot measure anything smaller than the Planck measurements of space and time. Physics is, like all sciences, in a situation of constant self-transcendence in that it tries to extend the boundaries of the known into the unknown.[268]

In *biology,* we do not know how the first forms of life came about. There are plenty of unanswered questions in the field of cell physiology and genetics. In *palaeontology*, the reconstruction of the evolution of humankind is built on a jawbone here and a foot bone there with vast historical and geographical gaps between the fossils. There is a lack of consensus concerning the origin of racial distinctions among humans.

In *evolutionary anthropology,* we do not know how brain states could lead to the subconscious, consciousness and self-consciousness. In *psychology,* the subconscious may be one of the most elusive and mysterious dimensions of reality. We do not know precisely when and how *language* as a system of symbolic presentations emerged and evolved. We have no idea when and why *religious intuitions* emerged.

Documents and archaeological remnants of ancient cultures and religious traditions, including the biblical faith, do not always yield plausible pictures of the emergence and evolution of these cultures and religions. When it comes to remote history, we largely depend on conjecture.[269]

The behaviour of the *economic system* in a liberal society continues to baffle the experts. This uncertainty leads to vast predictive failures, economic dysfunctions, social imbalances and ecological disasters.

Some of these problems may be resolved in the future, some may not. As Heisenberg said, "The existing scientific concepts cover always only a very limited part of reality, and the other part that has not yet been understood is infinite."[270] Isaac Newton is reported to have said much earlier, "What we know is a drop; what we do not know is an ocean."[271]

Without any doubt there are aspects of immanent reality that may never become accessible to human knowledge through scientific observation, mathematical constructs and logical inferences. Should the universe expand faster than the speed of light, for instance, information emanating from beyond this threshold cannot ever reach us. It is a simple truth that human means of observation, comprehension and prediction are designed by evolution in such a way that they allow us to survive and prosper, but not to venture much beyond those limits.[272]

Section II
Transcendence proper

The examples mentioned above all belong to 'immanent transcendence', not to the transcendence intended, for instance, by the biblical faith. They show that there is more to immanent reality than we can access by means of scientific exploration, mathematical reconstruction and reliable prediction. However, as I use the term, transcendence does not refer to gaps in human knowledge of immanent reality, however fundamental and intractable they may be. It represents the openness of immanent reality as such and as a whole for something beyond its boundaries.

It is clear that we have arrived at the uttermost edge of the mandate and method of science. Whatever goes beyond this edge is beyond scientific investigation. It is also beyond human observation, interpretation and manipulation. It even goes beyond human imagination. That is why it can only be referred to by means of metaphors taken from the experience of immanent reality. In which sense is it possible and meaningful to transcend 'immanent transcendence' from an experiential point of view? What is the status of the idea of radical transcendence within the scientific frame of reference?

This question leads us into the core of the characteristically human experience of reality, a core to which the natural sciences are not designed to penetrate. The experienced need for meaning, criteria of acceptability and authority explodes the limitations of empirical observation, rational comprehension and deliberate manipulation. It leads to intuitions, visions, traditions and rational constructs that intuit a greater sphere from which immanent reality is derived, in which it is embedded, which permeates this reality in all its manifestations and which gives meaning to reality as a whole.

At this juncture, it is of critical importance to distinguish between the transcendent as such and an intuition, notion, or concept of the transcendent. While the transcendent as such is inaccessible, a concept of the transcendent is part of immanent reality. Therefore, it always and by necessity utilises metaphors taken from the experience of immanent reality. We have no other language. Such concepts can be more or less appropriate. They can be described, critiqued, corrected, transformed, or abandoned.

Experiential roots of a concept of the transcendent

Why develop a concept of the transcendent at all? Can we not do without it? Is this not the 'invisible gardener' that makes no difference and can be ignored?[273] It is clear that it cannot serve as an explanatory tool. Its function is quite different. At its existential core, a concept of the transcendent is rooted in the awareness of one's derivation and dependence, one's identity and belonging, one's role and status, one's potential authenticity and actual accountability, one's sense of failure and one's need for restitution, one's guilt and need for reconciliation, and one's awareness of vulnerability and mortality.

In terms of one's existence and one's life world, the qualitative content of a concept of the transcendent is rooted in the experience of what ought **not** to have become and the vision of what **ought** to become. It provides provisional and partial meaning without which human life is not possible. It produces values, norms, tasks, goals and visions. It allocates authority in the form of statuses and roles. It is a comprehensive sort of awareness that ideally covers one's very existence and one's entire life world.

When all these kinds of aspects solidify into a reasonably coherent but dynamic structure of consciousness that is able to integrate ongoing experiences, a system of meaning has emerged.[274] A system of meaning is not an arbitrary construct that could be replaced or ignored at will. It 'falls into place'. It imposes its validity upon one's consciousness. It provides the individual subject and the community in which it is embedded with a sense of validity and stability.

Systems of meaning can be shallow and fleeting, or profound and enduring. But they are not constructs arbitrarily set up by autonomous subjects. Convictions are embedded in dynamic social systems and evolve in the form of traditions. Subjectively, a conviction comes about when one is confronted with a truth claim that one cannot ignore or escape without losing one's integrity, authenticity and authority. Where one tries to do that, the mechanisms of self-justification, rationalisation and ideology kick in.

Does it make a difference?

The sceptical natural scientist typically asks whether the assumption of a transcendent Source and Destiny of reality makes a difference to the reality we experience. This question is not always given the attention it deserves. The answer is that within the confines of the scientific enterprise of observation, explanation and prediction it demonstrably does **not** make a difference.

Science works under a methodological assumption that deliberately (and rightfully) ignores a potential transcendent Source, Agency and Destiny.[275] This also means that the neurological explanation of religious intuitions and experiences in no way removes or invalidates the notion of transcendence. As the transcendent Source and Destiny of reality, God acts **through** immanent reality and not alongside, in cooperation or in competition with it.

This in no way implies that there is 'no work to do' for such a 'higher agency'.[276] That could only be the case if God was a factor among others within immanent reality. It is precisely the other way round. Once you assume, as the Christian faith does, that reality as such and as a whole is derived, dependent and accountable, it is self-evident that without the 'work' of such an 'agency' reality as such and as a whole would simply not have emerged and evolved. Not a single entity would exist; not a single event would occur. There would be no science and no object for science to investigate.

If God is assumed to be the transcendent Source of reality as such and as a whole in its infinite variety, flux and complexity the question whether God's 'intervention' makes a difference becomes meaningless. Even the question whether such a transcendent God 'exists' becomes meaningless—and that for two reasons. One, if God is the transcendent Source and Destiny of reality as such and as a whole, God cannot 'exist' the way immanent entities exist or act the way immanent agencies function.

Two, at least in the Christian faith, the issue is not the existence of God, but the particular divine intentionality that is deemed to manifest itself in reality. In all its forms, the biblical faith takes it for granted that the reality we experience is not closed in upon itself. It is not self-generated, self-maintaining, self-sufficient, self-responsible and self-destructive. It is also not meaningless. It is open, derived, dependent, accountable and saturated with meaning. The concept of God gives expression to the 'wherefrom' and the 'whereto' of this openness.

The fundamental irrationality of convictions

It can easily be shown that numerous empirical improbabilities and logical irrationalities occur in convictions and theologies of all kinds. In this book I attempt to remove them as far as possible. But the assumption of the transcendent as such cannot be accounted for in empirical or logical terms. If the content of a conviction were entirely clear and self-evident, it would no longer be a conviction. All that we can rightfully expect of a conviction is that it integrates the outcomes of 'best science' rather than perpetuating obsolete knowledge.

This unavoidable irrationality is due to the fact that conviction is derived from a kind of source different from observation and reason, namely an intuition of one's location in the whole of reality on the one hand and the discrepancy between what ought not to have become and what ought to become on the other. In this sense, theology and metaphysics are similar to aesthetics and ethics, both of which are not based on empirical fact or logical argument, but on a qualitatively different kind of human experience.

If that is the case, attempts to find empirical grounds or logical arguments for faith statements appear to be subsequent rationalisations rather than first-order sources. This is true for theism as much as for naturalism. If this book were interpreted as an attempt to 'rescue transcendence' by means of 'apologetics' (as a colleague recently alleged), it would be thoroughly misunderstood. Its analytical methodology moves in exactly the opposite direction.

Observation and reason are simply too restricted in their scope to cover the awareness of the whole in which we are embedded. Yet conceptualisations of the whole are indispensable. We need a coherent 'symbolic universe' that is capable of conveying the content of our intuitions along the pathways of a tradition. The urge to arrive at some sort of coherence within such a symbolic universe is essential because the human psyche is conditioned to discard implausible propositions. It is for this reason that a religious or metaphysical system of meaning must integrate the findings of best science or lose its credibility.

Section III
Functions of a notion of the transcendent

In its most fundamental sense, then, the term 'transcendent Source and Destiny of reality' signifies the assumption or conviction that reality is open for, embedded in, penetrated by, and sustained through something beyond its immanent appearance. At first sight, there is nothing more to it than that. But rightly understood, 'nothing more' means that 'everything' is at stake.[277]

Existential experiences and social-cultural formations have to do with meaning, identity, value, appeal, acceptability, authority, goodness, truth and beauty. Experienced reality is a package. If the scientific pursuit of insight is consistent, it has to recognise the multidimensional character of human experience. Empirical observation and scientific explanation deal with one dimension of experienced reality, the appreciation of beauty with another, the awareness of what ought to be with another, and the awareness of ultimate dependence and responsibility with yet another.[278]

Because it is all about meaning, which falls outside the scope of the scientific enterprise, the methodological asceticism of the sciences does not make the assumption of a transcendent Source and Destiny superfluous—unless, of course, one resigns oneself to the meaninglessness of reality, which many consistent naturalists actually do. This must have the most far-reaching consequences for human existence and human relationships to the rest of reality.

There are three particularly pertinent aspects to the transcendent, namely, the human need for meaning, acceptability and authority. Meaning, acceptability and authority circumscribe the area of human experience where faith is located.

First, meaning

While all living beings are programmed to choose survival, well-being and the avoidance of pain, humans have a heightened sense of apprehension. The sum total of immanent factors that seem to make life, identity, meaning and authenticity possible is diffuse and ambivalent. It does not constitute a consistent system of meaning. It does not yield an ultimate reference point. It gives no direction to life.

The painful awareness of the dependency, vulnerability and mortality of all living beings is exacerbated by the human capacity to see the self in greater contexts, in its long term duration, in one's entire life world, even within reality as a whole. There are aspects and trends in reality that are beyond human control. Where do they ultimately come from? What is their rationale? Where will they lead us? What will be their consequences? How reliable will they be?

These profoundly existential questions have found a variety of answers. They are consolidated in the form of a system of meaning. Systems of meaning are not based on evidence but on intuition and conviction. Intuitions and convictions are not invented. They impress their validity on human consciousness.[279] They emerge and evolve in traditions. They can be informed, extended and corrected by scientific insight, but scientific insight itself is embedded in frames of reference and struggles to find a valid overall system of meaning.[280]

We are confronted with fateful experiences that we cannot escape and that we do not understand. The search for meaning is often expressed in anthropomorphic terms as a quest for the purpose of what has happened. Purpose presupposes intentionality and agency. Humans cannot help but ask, why did this awful accident happen? Why did I develop cancer? Why was a lunatic like Hitler able to mislead a whole nation and cause a global catastrophe? What was it good for? And if it was good for nothing, why was it allowed to take place?

Scientist may say that there are no answers to such questions because they are wrongly stated. Purpose presupposes a personal motivation. It does not apply to impersonal phenomena such as the evolution of species, natural disasters, social dynamics, or the cosmic process as a whole. Meaning is a value judgement, not an inherent quality of reality. Entities do not have meaning in themselves; they simply exist the way they have become and operate according to their inherent regularities. These can be described fairly accurately. Events have no meaning in themselves; they only have causes and consequences. In countless instances, these causes can be established with considerable precision.

However, for humans, this argument begs the question. In the first place, something can be meaningful without having been purposed if it fulfils an indispensable function within a greater context. The heart fulfils a function in a biological body; entropic dissolution fulfils a function in the cosmic process as a whole. The engine, gearbox and steering of a car are pure mechanisms, but they are the preconditions for the capacity of a car to move and the capacity of a driver to reach a destination.

In the second place, if phenomena have no intrinsic meaning, they have to be *given* a meaning by humans because humans have to respond to them in one way or another. I am confronted with a fact—what does it signify, which challenge does it pose, which options does it present? What are the consequences of each option? Which option should I try to realise? For what reason? To whose benefit? On which authority? And why is all that important?

Second, acceptability

The experiences of derivation, dependence, accountability, vulnerability and mortality produce uncertainty, anxiety and fear. Meaninglessness, chaos, fateful experiences, guilt, suffering and death all question one's right of existence at the most fundamental level. Humans need affirmation, forgiveness, acceptance and belonging to overcome anxiety and build up trust.

This kind of exercise produces a set of criteria for what should be deemed appropriate and acceptable for the identity concerned, thus for what can be considered authentic human existence. Because consequences of developments, decisions and actions can be beneficial or detrimental, they demand responsibility.[281] Responsibility must be geared to the prerequisites and the enhancement of the flow of life.

As stated in the previous chapter, this moral imperative is in line with the evolution of life in all its forms. There is no life that does not strive for its own survival and flourishing. But life is always embedded in wider contexts on which it depends and to which it contributes. It is the whole network of relationships within reality that is at stake with every event, decision and act.

Tentative answers are provided by visions, values, norms, goals, social conventions and sanctions. When internalised, such criteria give rise to conscience, solidarity, commitment and dedication. They grant or withhold a sense of acceptability and belonging without which humans cannot live a meaningful life. Values and norms are part of a system of meaning, however rudimentary and implicit it may be.[282]

Natural science is restricted to an analysis of what reality **has** become and **may** become. It has no means to determine what reality **ought** to become. Scientific facts and theories must be taken into account to make a system of meaning plausible, but meaning as such is not amenable to scientific inquiry or verification. Science can explore and explain the evolutionary process and the flow of life, but it cannot say why it should be preserved and enhanced.

Science can describe the content of particular visions, values and norms, but it cannot establish their validity. Visions, values and norms may be attributed to the pre-historical conditioning of the human race, the socialisation of the individual during early childhood, the voice of conscience, or a divine decree. But whatever their source, their practical effect in daily life cannot be disputed. They can be deemed 'subjective', but woe unto a society where they are being scorned or dismantled.[283]

The question is, to quote a few examples, whether homosexuality constitutes acceptable behaviour; whether impoverished peasants must be prioritised in economic policy; whether a species threatened by extinction must be protected, left to die out, or deliberately eradicated; whether patriarchal dominance must be upheld; whether armed conflict is an acceptable form of diplomacy; whether the death penalty can be justified in a liberal society. Such questions cannot be answered by science. Science can only point out what the probable consequences of different options might be. Yet nobody should presume that these questions are irrelevant.

Third, authority

To be confident of their acceptability and belonging, humans must establish their individual, communal, societal and natural identities and functions within the greater whole. Systems of meaning define our individual and collective identities within the concentric contexts of the whole of reality: body, community, society, the earth and the universe. They grant the authority to think and act within their specific parameters of acceptability. They allocate certain statuses and roles. This is often done in the form of rituals.

Without explicit or implicit confirmation of their authority, humans become insecure, anxious, lethargic, or fatalistic. Again science cannot allocate such an authority. Cultures and social structures are equally unable to do so as such and on their own because they are mere social conventions and themselves depend for their validity on a system of meaning, however implicit, pluralistic, or provisional it may be.[284]

These three dimensions of experienced reality culminate in certain ultimate assumptions that are based not on evidence but on conviction. While convictions differ widely, they provide meaning, reassurance, direction and courage when confronted with the vicissitudes of reality.

The notion of God

Humans have a need to anchor validity and stability in something that goes beyond the transience, ambiguity and capriciousness of immanent reality. Therefore a system of meaning may consolidate around an overarching authority or converge upon a transcendent centre. That is where the notion of the 'divine' is located within human experience.

In terms of time, 'God' is deemed the transcendent Source and Destiny of reality as such and as a whole. In terms of space, God may be perceived as the 'grounding and bounding' of reality. In terms of power, God is the Source of all the energy in the universe and the regularities according to which it functions.

These concepts are metaphors and metaphors should not be pressed. 'Source' is not meant to imply that the 'flow' that continues beyond the source can happen without God. 'Destiny' is not meant to imply that the outcome is predestined in theological terms or causally determined in scientific terms. 'Grounding' does not imply a static worldview. Gravity, after all, grants us both stability and mobility on a moving planet. 'Bounding' is not meant to imply that reality is not permeable for God. 'Power' is not meant to imply mere influence but rather, to use another metaphor, 'driving force'. 'Regularity' is not mean to imply the absence of intentionality. On the contrary, regularity is the consequence of God's intentionality.

There is an existential and social need for transcendent affirmation. When offered by individuals, communities and institutions, affirmation, acceptance and belonging often prove to be feeble, ambiguous, unreliable, misleading, hypocritical, or based on ulterior motives. Moreover, they tend to be subject to certain conditions: you either conform to the norm, or you are rejected.

The Christian faith, in contrast, anchors acceptance in the benevolent intentionality of the transcendent Source and Destiny of reality, thus offering a kind of certainty, an embrace, a peace, a joy, a purpose and a task that is not dislodged by the vicissitudes of changing situations and human relationships.

Note that when we make this observation, we have not left the sphere of immanent reality. We are not busy with the transcendent as such, which is inaccessible by definition, but with a *notion,* or apperception, or intuition of the transcendent. A notion of the transcendent is part of immanent reality. It has certain characteristics that can be described, critiqued, augmented, transformed, or abandoned. It falls within the range of academic disciplines such as phenomenology or religion, psychology of religion, philosophy of religion and theology.

It has to be acknowledged, however, that part of the *concept* of God is the assumption that it has come about through some kind of self-disclosure of God's transcendent intentionality *as such.* Where this is not the case, faith has been abandoned in favour of naturalism, even where the importance and excellence of religious systems of meaning, values and visions are acknowledged.[285]

Once reality has acquired a focal point in the consciousness of the believer, the orientation of the self becomes eccentric, that is, it no longer considers itself to be the centre of its life world. The self is drawn, as it were, towards the real centre of the universe by a kind of 'gravity'. Or to use another picture, consciousness acquires a 'compass' that indicates the direction towards a 'north pole', as it were, wherever heavy gales and strong currents sweep the ship.

When an individual or community acknowledges its eccentric position, it recognises that it has an assigned location within a vast network of relationships. Its identity is being defined within the greater whole. It is as if the earth were to discover its place within the solar system, the place of the solar system within the galaxy and the place of the galaxy within the universe. Where this does not happen, the ego, its particular in-group and its life world continue to form the commanding centre of the universe.

Idolatry

Alternatively some *immanent* aspects of reality become the centre of one's life world and acquire the status and power of the absolute. It can be a luxury car, a political ideology, a charismatic leader, or the 'bottom line' of an enterprise. It can also take the form of nature as a whole. In terms of genuine faith, such phenomena represent 'idolatry', that is, the absolutisation of the derived and dependent.

Such an absolutisation is deemed inappropriate in terms of a vision of reality as a whole. It can have immense repercussions for human attitudes and behaviour, social structures and processes as well as the human impact on the natural environment. The notion of a transcendent Source and Destiny of reality, in contrast, de-absolutises all immanent entities and, in this way, liberates the subject from immanent enslavements.

A concept of 'God' can have a great variety of forms and change over time, but it always claims ultimate validity, at least for the time being. That is why it tends to become inflexible and dogmatic. However, a concept of the transcendent that is mistaken to be identical with the transcendent is not an expression of faith but of idolatry. It absolutises the relative. It takes great maturity for faith to recognise that validity cannot remain static. In fact, it must stay ahead of the flow of reality or become obsolete and redundant.

Section IV
Is the content of a conviction irrelevant?

Dynamism and animism provide diffuse and intractable systems of meaning and result in much uncertainty, fear and suspicion.[286] Polytheism too reveals a lack of integration. It reflects the experience of a diffuse reality. The lack of integration has also led to the current split between science and faith and the resultant cognitive dissonance of modern believers. Naturalism attributes ultimacy to the natural world. Its religious form would be pantheism.

When experienced as compelling truth, the concept of God lays claim on the human subject. It expects of her an authentic human existence in terms of the system of meaning as a whole. To be credible, it must cover the existential and social experiences enumerated above; be more or less in line with inherited collective memories contained in relevant traditions and, at the same time, filter and integrate the influx of new information emanating from continuing sense perception and alternative systems of meaning.

But that is only the formal aspect. Fundamental for its convicting power is its substantive *content*. Quite obviously, there are huge qualitative differences between convictions. Some lead to withdrawal from the world and others into involvement with the world. Some enslave and smother initiative; others liberate and motivate. Some stress the interests of the individual, others the community, the nation, all humans, all life on earth, or the universe as a whole.

It makes a difference whether one's ultimate point of reference evokes cynicism and fatalism; legitimates the reckless pursuit of power, wealth and status; demands a cold and merciless sense of justice; suggests social and intergenerational equity, or spreads a warm atmosphere of self-giving, redeeming concern for the other, the community and the earth.[287]

It is here that the most fundamental philosophical and theological discussions occur. Is 'God' a figment of the imagination, a basic principle, an impersonal mechanism, a power, a 'pure' spirit, or a person? If God is a person, what is this person up to? What does God expect of us? Does God demand perfection before he is willing to accept us, or does God accept us, suffering the fact that we are unacceptable, so as to transform us in God's fellowship?

These are anthropomorphic expressions of the deeper questions of meaning, acceptability and authority that seek an answer. That is why the struggle for 'ultimate truth' cannot be suspended or abandoned. What the natural sciences can expect of philosophy and theology is that their proposed systems of meaning cover and integrate best science. This critical and reconstructive process is of highest significance for individual human existence, community, society, humanity and nature. It will determine whether humankind and all life on earth will have a future and what kind of future it will be.

Which version is valid?

We have no way of assessing the validity of the notion of the transcendent in relation to the transcendent as such. But there are two fairly reliable immanent criteria: (a) Does it cover the whole of reality or just the desire of individuals and communities? (b) Does it represent a dynamically unfolding vision of comprehensive optimal well-being? At least in the final stages of biblical history, the transcendent stands for the benevolent intentionality of the Source and Destiny of reality as such and as a whole, not for the interests of a community or the cravings of individuals at the expense of others.

The question is which notion of the transcendent can claim to be the most appropriate approximation and representation of the vision of comprehensive optimal well-being of reality as a whole. This criterion is in line with the inherent thrust of the evolutionary process towards the realisation of potentials at inorganic levels of emergence, the survival and well-being of organisms and the need for fulfilment and authenticity at the level of human consciousness.

One does not need to be a believer to recognise the validity of this criterion. But one can say that it is not met by a religion that binds us to an ancestral past, imprisons us in a pre-defined code of laws, or allows us to escape from reality into an imaginary spiritual world. Rightly understood, the Christian faith could meet this criterion. It can unlock both freedom from reality and responsibility for reality, thus making dynamic redemptive responses to changing situations of need possible. Alas, this potential has rarely if ever been realised.

A vision of the well-being of the whole of reality cannot be taken for granted. Insight into the validity of transcendent assumptions emerges and evolves in historical processes called traditions. A tradition is a set of assumptions that responds creatively and redemptively to changing situations of need and changing views of reality and changes as a result of this response. In this process, the notion of God may gradually acquire more comprehensive horizons and greater depths. This is what happened with the biblical tradition over a millennium of ancient history.

When everything is said and done, therefore, what matters is the formal capacity of a conviction to relate to actual experiences and their interpretations on the one hand and the quality of their content on the other. Human insight travels in socially entrenched traditions that evolve in time and differentiate in space.

That is true for science as much as it is true for religion. The important question is not whether truth is historical or eternal—that question should have been settled in favour of history a long time ago—but which historically conditioned and moulded perception of the truth has the most compelling claim to validity while we journey through time and space.

What about postmodern deconstruction?

The radicalisation of the modern quest for freedom and the heightened awareness of relativity in postmodern thought do not obviate the quest for meaning, reassurance, orientation and authority.[288] On the contrary—it exacerbates the need for grounding and direction. The devastating symptoms of spiritual uprootedness are all around us to see. Even scientists cannot do without convictions because the assurance of faith—whatever its content—gives purpose and meaning to the scientific enterprise as such and underpins the validity of whatever ethical orientation they may have.[289]

Postmodern deconstruction may very well dismantle all spiritual foundations, constructs and constraints by virtue of its own consistent rationality, but it is hard to imagine how that could possibly lead to some sort of validity, stability and commitment—all of which we so desperately need for our survival and prosperity.[290]

One should not presume that this kind of transcendence will leave the physical world intact. In fact, consistent physical reductionism is a parallel to postmodern deconstruction—bodies are, after all, composed of atoms, which consist mainly of vast open spaces and tiny energy conglomerations, which are organised in fields that cannot be attributed any substantive ontology. So do we exist at all? Maybe not, but then it is strange that I seem to be sitting here and writing this essay.

One could argue that we will attain the most radical form of freedom when our own bodily structure has dissolved into nothingness—thus after our own deaths. In Platonic thought, the spirit, thus liberated from all bodily constraints, may be thought of as pure intentionality and agency, thus total freedom. But modern science exposed this as a pipe dream. The theory of emergence tells us that the spirit cannot survive without its physical infrastructure. Total freedom is, at the same time, non-existence, thus powerlessness and meaninglessness, thus also the final suspension of all responsibility. Is that a desirable outcome?

The Christian concept of the transcendent goes in precisely the opposite direction. Its constitutive 'elements' are creative power and redeeming love, thus constructive energy and benevolent intentionality. Again no substantive ontology can be attributed to these elements, so transcendence is maintained, but they are deemed the preconditions of the existence of the world we know and the life we live. They are also the criteria of an authentic human life.

The idea that we can have freedom without a structural framework is a dangerous illusion. It is dynamic structure, after all, that characterises reality as opposed to entropic dissolution. The same is true for the concept of the transcendent. Admittedly it will be a deeply inadequate and provisional reflection of the transcendent, assuming that it is a reflection of anything at all. Yet it provides our consciousness with a dynamic structure and orientation. And it is structure and orientation that make freedom and responsibility possible.

Let me use a picture taken from cosmology to make my point. Our planet may indeed be a little speck of dust moving about in the larger universe among trillions and trillions of others. Yet without the feeble power of its gravitational pull, no creature could live. We would all be like astronauts detached from their spaceship, drifting aimlessly, helplessly and meaninglessly into the vast voids of cosmic space.[291] Apart from the obvious fact that biological life could not exist under such circumstances, there would also be no possibility to discern a privileged direction, to formulate a goal, to move anywhere, to accomplish anything, and to make any impact whatsoever.

To extend the picture to our current topic, it is not just any planet that can provide us with these indispensable prerequisites of wholesome human existence. It is a very specific and utterly unique planet. Not every planet can support life. In the same way, religious systems of meaning are unique. They display immense qualitative differences. They are also deeply ingrained in structured individual and communal consciousness. They cannot simply be exchanged at will without causing extensive worldview disruption, cognitive dissonance and anomie.

We need both, first, a sense of radical transcendence that keeps us from absolutising ourselves, our world and our worldviews, and second, a notion of the transcendent whose content is the creative power and benevolent intentionality of the ultimate Source and Destiny of reality. Such a notion can do justice to the flux of experienced reality; cut us to size; liberate and empower us; orient our consciousness towards a dynamic vision of comprehensive optimal well-being and thus alert us to any deficiency in well-being in any dimension of life.

If that is the case, once again, the question is not whether 'God exists', but what *kind* of 'ultimate' is believed to operate at the natural, personal, communal and social levels. The question whether God exists in this sense is as futile a question as whether Mozart's Requiem or Beethoven's ninth symphony exist. A complete record of the sound waves that make up such a composition does not constitute the composition. A symphony is an emergent reality whose flowing and pleasing shape cannot be explained in terms of its physical infrastructure. The question is not whether a symphony exists but what kind of music it represents, just as the theological question is not whether God exists, but to what kind of divinity we entrust ourselves.

Section V
The relation between the transcendent and the immanent

The relation between transcendence and immanence has been an enduring problem in theology and philosophy of religion. In fact, it is quite simple to solve. The decisive clue is the distinction between the transcendent and the supernatural. The supernatural is believed to be a factor within immanent reality that is not subject to the regularities that govern this reality. Interpreted as a supernatural agent, an omnipotent God must be able to suspend or override natural law. This naive assumption of the believing community has caused endless confusion in the interaction between science and faith.

Transcendence is something quite different. 'God' is the name for the assumed ultimate Source and Destiny of reality. As such, 'God' is not deemed outside reality; 'God' is not deemed inside reality either. The creative power of God is deemed the precondition of the existence of reality as such and as a whole in all its concrete manifestations. God's benevolent intentionality is deemed the meaning that gives orientation to reality as a whole, again in all its concrete manifestations.

If that is the case, God's intentionality and agency do not **compete** with, **supplement,** or **cooperate** with immanent causation or human intentionality and agency. Theologically speaking, it is never right to say that **either** God acts **or** humans act so that God does not act when humans act and vice versa. If that were the case, God would **compete** with the human being. It is also never right to say that God **cooperates** with humans so that God is doing God's part and humans are doing their part.[292] God's action also does not **supplement** human action. God also never acts in cooperation with, in addition to, or in competition with the impersonal regularities and processes of reality.

In all three these cases, God would be conceptualised as a causative factor operating at the same ontological and causative level as experienced reality. God's action and human action together would constitute a zero-sum game. Either one of them would do the work at the exclusion of the other, or they would divide it between themselves.

Because God is not a factor within immanent reality among others, but its transcendent Source and Destiny, God always acts **in** and **through** immanent events and human actions.[293] In biblical terms, a God that is part of immanent reality is an idol. The transcendence of God does not imply, therefore, that God cannot manifest God's intentionality and agency in and through the immanent reality of which it is the transcendent Source and Destiny.

I always had great fun with my students when I asked them what made the noise when a trumpet was being played—the trumpet or the trumpeter. It could not be the trumpet alone, nor the trumpeter alone, because without either, there would be no music. Trumpet and trumpeter could also not each do their part, say 50 per cent each, because if we take away the trumpet or the trumpeter, we are not left with half a music, but with no music at all. So both produce the music—and each produces the full 100 per cent of it, because the trumpeter makes the music by means of the trumpet.

The way faith conceptualises God, everything is from God through God and to God. "From him and through him and to him are all things" (Rom. 11: 36). God's origination of reality includes even faith itself. It is considered a gift of God, yet something that humans do experience and enact. "Work out your salvation with fear and trembling, **because** it is God who is at work in you enabling you both to will and to act in line with God's good purposes" (Phil. 2: 12 f).

If God did not act, we could not act in the first place. God's initiative does not make human initiative superfluous but evokes it. God's intentionality does not cancel out human intentionality but involves it. God's freedom and mastery do not enslave humans but liberate and empower them. God's agency empowers human agency.

According to Ludwig **Feuerbach**, God is the projection of unrealistic dreams (immortality, universality, eternity) and idealised human potentials (power, wisdom, creativity, justice, love, and so on) into an imagined divine being. The legitimate aspects of this projection, Feuerbach argued, must be retrieved and re-appropriated by the human race, while the idealised projection itself must be dismantled. Building on Feuerbach's observations, Karl **Marx** described religion as the 'opium' of an oppressed social class, meant to keep it subservient and to lessen its suffering. With the emergence of the classless society, it will die a natural death.

According to Sigmund **Freud**, God is an internalised father figure or superego that watches over our behaviour and represses our healthy instincts. This father figure must be overcome if humans are to realise their innate potential. **Darwinian** evolutionists explain the emergence and evolution of religious and ethical concepts in terms of the phylogenetic history of humankind.

Which of these theories come closest to the truth is not a theological, but a scientific question. Certainly there are serious flaws in most of them. However, the critique of these contentions must emanate from within the scientific discourse, rather than from 'biblical revelation'. Biblical revelation is about the intentionality of God that manifests itself in experienced reality, however it comes about. There is no reason why faith cannot accept, or even engage in, 'best science' in this regard.

Whatever explanation of immanent reality we might settle for, including the immanent phenomenon of faith, it has no impact on the question whether 'there is a God', because scientific explanations cannot, and do not, cover the transcendent Source and Destiny of the reality that is being explained. They only cover the way human notions of the transcendent emerge and function. If faith understands itself, therefore, it should not want to quarrel with science on this score. It cannot know better.

Section VI
Transcendence proper precludes the 'God of the gaps'

So the transcendent God and immanent reality do not operate at the same level. God's action does not supplement, compete, or cooperate with immanent processes. Whenever one looks for a space within immanent causal networks, where God's spontaneous and specific action could be deemed possible, God's agency is taken to be a factor among other such factors within immanent reality, rather than the transcendent Source and Destiny of reality as a whole.[294] There are at least three versions of this approach.

Cluster Publications

A Non-Profit Project
Pietermaritzburg Cluster
of Theological Institutions

Box 11980
Dorpspruit
3206

Complementary copy from author – with gratitude
and appreciation for your positive assessment of the
manuscript.

Cluster Publications • Box 11980 • Dorpspruit • 3206
Tel: 033 846 8602 • E-Mail: cluster@clusterpublications.co.za

Errata

The author and the publishers wish to apologise for the editorial omission of the word **Scriptures** in the following instances

page	par	line
32	2	2
85	5	4
205	7	2
212	2	6
217	6	4
218	2	4
225	4	2
225	4	6
225	4	6
225	5	1
225	5	7
225	6	5
231	3	5
231	5	1
232	2	1
232	7	1
233	2	2
233	3	4

Klaus Nürnberger

One, the naive notion of God's intervention in reality

Nobody doubts that there are serious gaps in our understanding of reality. Examples are, for instance, what triggered the big bang, what led to the first forms of life, what brought about human consciousness, how language originated, how subjective experiences come about, or how humans acquired a sense of transcendence. Assuming that causality obviates divine action, faith easily falls into the trap of assigning God's activity to such gaps in scientific knowledge. It is not always realised that by the same token one is *constraining* it to such gaps.

God then functions as a working hypothesis or 'black box' meant to explain unresolved mysteries in our otherwise closed understanding of causality. But what about the rest of reality? Is causality itself not of God? What kind of Creator is that whose 'action' is limited to particular 'open' situations within reality? Traditional theological concepts of 'creation out of nothing' and 'continuous creation' cover *all aspects of reality*.

They presuppose the notion of a sovereign and truly transcendent Source and Destiny of reality as a whole, including all phenomena, regularities, entities and processes at all levels of emergence.[295] I maintain that this is not just a wild guess or one metaphysical proposition among others. It is essential for spiritual freedom, responsibility and commitment to materialise.

Once we conceptualise God as the transcendent Source and Destiny of reality as such and as a whole, we realise that God's action is identical with the very reality that we experience. For the biblical faith, this is self-evident. Just look at Psalm 104, Job 12: 13-25, or Job 38-41. God "changes times and seasons; deposes kings and sets up kings" (Dan. 2: 21). That experienced reality as such is the manifestation of the action of God is never questioned.

Such a comprehensive divine agency may seem easier to assume for a pre-scientific worldview with a rudimentary and deficient perception of causality than for us today. However, even in biblical times, one was aware of regularities in nature and society, such as the movements of the stars, the rhythm of the seasons, social contracts and institutions, moral norms and the freedom and efficacy of the human will. Yet one was able to ascribe precisely these regularities to God's intentionality and agency.

Today we know better than ancient cultures that the precise way experienced reality functions can only be unlocked by careful observation and reconstruction of reality. The sciences enhance human observation and comprehension of reality. How reality comes into being and how it functions is established by science, not by faith. Liver functions, the behaviour of light waves, the formation of stars, and the neurological mechanisms that form the infrastructure of the phenomenon of faith itself are examples.

If there are gaps in our knowledge, they must be closed *by science* or remain open. Theology cannot close them and does not need to try. In terms of faith, all the immanent forces and regularities that determine these phenomena are, as such, manifestations of divine 'action'. The behaviour of causal networks, random, chance and autocatalytic, self-generated, self-contained and autonomous agencies are included in divine action. The question of faith is how to interpret this functioning whole as an action of God.

It is of critical importance for the interface between science and faith to make this distinction. Time and again, the God of the gaps has exposed faith to ridicule because, as these 'missing links' are being closed by plausible scientific theories or empirically established findings, God becomes progressively redundant as a quasi-immanent agent. A concept of God's action reduced to gaps in the causal chain also fails to cater for the spiritual needs of believers who put their trust in the unlimited sovereignty and mastery of God. The concept of God's 'omnipotence' is crippled from the outset and further eroded by advances in scientific theory.

Two, God leaves space for reality to operate

Another rather popular idea says that God withdraws sufficiently from reality to give it the space needed to evolve and function. If God allowed his absolute power to exert itself to its full extent, it would crush any 'autonomous' process within reality. Natural law would not be able to operate. Human decision making and action would be impossible. It would be a case of divine determinism, akin to the idea of predestination.

This is an example for the way deduction from questionable assumptions can entangle theology in logical impasses. On the one hand, the argument presupposes an abstract understanding of God's omnipotence as unconstrained power and control. On the other hand, it assumes that 'autonomous' processes can happen without God. In fact, the argument is a form of Deism—the idea that God set the world process in motion and then allowed it to function on its own—which means that God has become redundant.

The presupposed notion of omnipotence is an idealised abstraction from experienced reality. It has Platonic rather than biblical origins. In the Bible, references to God's overarching power are meant to reassure afflicted believers that God is still in charge, rather than satisfy metaphysical curiosity. An experiential approach to divine omnipotence would say that all power operating within reality is the power of God. This power obviously operates according to certain regularities. There is no paradox here.

Three, 'non-interventionist divine action'

The God of the gaps has been discredited a long time ago and should have been abandoned by now. The same is true for the Deist idea that God has withdrawn from reality so as to allow it to operate. But the assumption that God and the world compete, cooperate, or supplement each other lingers on even in the otherwise much more sophisticated current science-religion debate, notably the endeavour to find some space for "non-interventionist objective divine action" in the immanent processes of reality.[296]

Here theologians and philosophers refer to the seemingly impenetrable mysteries of field theory, indeterminacy, quantum physics, complexity theory, chaos theory or emergence theory and interpret them as potential access points of divine agency into an otherwise closed causal network. Doing so they are clearly busy with a God of the gaps, not with the transcendent Source and Destiny of reality as such and as a whole—including known regularities, indeterminacies and mysteries.[297] In my view, this is a fundamental and irredeemable flaw.[298]

I am aware of the fact that those who are looking for specific divine interventions in reality presuppose that God is also the origin of the whole of reality. But the latter, more comprehensive view of origination remains quite abstract and unmediated with the specific concept of divine intervention. In fact, it is a parallel to the Deist notion. The distinction between the 'ontological' dependence of the whole of reality on God, on the one hand, and specific divine actions within reality, on the other, as proposed by some Catholic theologians,[299] offers no solution. Conceptual 'ontology' is an abstraction from history and has no ontological status of its own. The world can only be deemed ontologically dependent on God if it is historically dependent on God.

The God of the gaps does not do justice to the biblical conviction that, being the transcendent Source of reality, a sovereign God acts directly (not only in an abstract, general, 'ontological' sense, or as the provider of the infrastructure within which reality operates, but very specifically) in and through **all** phenomena, entities, regularities and processes, and therefore does not need any 'space' to act within the regularities of immanent reality in the first place.

The scientific theory of an immanent world process that functions without sporadic divine interventions—including contingency, probability, random, chance and sensitivity to initial conditions—must be matched with a theological theory that attributes *this entire process* to God's creative and redemptive 'action' and does not concentrate on bits and pieces of it. Any kind of compromise between the assumption of an omnipotent Creator and the assumption of an autonomous creation leads into insoluble logical impasses. It also fails to satisfy the needs of a popular spirituality that believes in the unconstrained mastery of God over reality.

The proponents of a zero-sum game in the distribution of divine agency, on the one hand, and human agency or natural causality, on the other, also do not address the biblical notion of God as the transcendent *Destiny* of reality. As the Destiny of reality, God sets criteria of acceptability for reality as a whole and extrapolates a vision of comprehensive optimal well-being for this reality. It alerts us to any deficiency in well-being in any aspect of life.

It appears, therefore, that the zero-sum proponents concentrate on a fairly small area within a much greater complex of problems raised by the encounter between the Christian faith in a Creator and the scientific worldview. They also assume that the insights of science into the regularities found in reality ought to be given priority over divine agency. Faith in the power of God to 'intervene' must then find spaces within the otherwise deterministic and tightly organised scientific paradigm.

Such a procedure bypasses the real problem in the encounter between a pre-scientific (personal) and a scientific (largely mechanical) set of assumptions. The God of the gaps is a theological imposition on the reality that the sciences explore. Scientific findings as such simply do not suggest the presence of a divine agent anywhere within the system, whether there are epistemological or ontological gaps or not.

The procedure regularly backfires when the gaps are closed by the progression of scientific insight or happily left open in the form of random or chance. In both cases, God becomes redundant as part of the explanation.[300] One can only avoid this impasse if one accepts that God's assumed gift of life, protection, provision, guidance, healing, or response to prayer work in and through both regularities and possible indeterminacies operative in the cosmic process. That is the direction I pursue in this treatise.

Four, the good God

The most popular version of attributing only part of experienced reality to God is to consider God the source of all beneficial or desirable circumstances and occurrences but cringe at the idea that God could also be the source of detrimental or undesirable conditions and events. Sin and death, it would seem, cannot possibly emanate from a powerful and benevolent God. There must be some other source of evil.

If that were the case God could no longer be deemed the transcendent Source and Destiny of the whole of reality. The restriction of the agency of God to the desirable and profitable inevitably creates a metaphysical vacuum that has to be filled somehow. Moreover, in an evolutionary universe, it is simply impossible to separate 'desirable' and 'undesirable' processes, because whatever may be conducive or detrimental to survival and prosperity is due to the dialectical interaction between entropy and gravity.

Perhaps the oldest way of separating good and evil found in the Bible is to assign all good to the righteous God and all evil to the sinful human being, thus inadvertently elevating humanity to a metaphysical status on par with God. That was the position of deuteronomic orthodoxy. Natural catastrophes like earthquakes and droughts were understood as punishment for human sins, or alternatively, as a cosmic order out of kilter because of human sin.

Another option was to assume an evil counter-god. The classical example is found in ancient Parsism, which interpreted world history as the struggle between a good god (*Ahuramazda*) and an evil counterpart (*Angra Mainyu*).[301] Most of the time, the Jewish appropriation of Parsist patterns of thought filtered out this metaphysical dualism.

One way of doing so was to integrate the evil force as a necessary element in God's mastery over the universe. *Satan* (the Hebrew word for accuser) became the public prosecutor in God's government whose function it was to test the integrity of believers (Job 1-2). In the New Testament version of this myth, *Satan*, the accuser, is demoted and replaced with Christ, the advocate (Rev. 12: 7-12).

But in the Book of Jubilees, *Mastema* plays the role of a proper devil. Apocalyptic theology also moved closer to Parsism by assuming that the current world was beyond repair and would be replaced in the 'age to come' with a new creation in which there would be no evil anymore. Some texts in the New Testament pick up this dualism, even if only metaphorically. I will come back to this topic in Chapter 12.

A zero-sum theology is not in a position to tackle the problem posed by ambiguities experienced in reality, including meaninglessness, futility, injustices, suffering, death and entropy. This is the old problem of *theodicy*—the most formidable challenge to the biblical faith ever to have arisen. They also do not provide the motivation to overcome deficiencies as far as they can be overcome.

This is serious! Especially after World War II, including Auschwitz, Hiroshima and Dresden, the faith of countless people in Western countries came to grief because the assumption of a powerful and loving God no longer made sense. If faith is to regain its credibility, it must address the problem of theodicy and do so urgently. It must do so by integrating rather than bypassing or ignoring scientific insight. I will come to that in Part IV.

To sum up

The issue of transcendence is pivotal for the relation between science and faith. True science knows of the limits of its mandate and method. Whether reality depends on a transcendent Source and Destiny, or whether it is closed in upon itself, thus self-contained and self-sufficient, is a question that science cannot answer. Nobody can. We can and do know, however, that human beings cannot live authentic lives without a transcendent frame of reference. Being descriptive and predictive, science cannot provide existential and social grounding and orientation.

Conversely, faith does not provide explanations and predictions. It is geared to a system of meaning that defines identity, determines criteria of acceptability, and grants authority to act. Such a system of meaning is rooted in transcendent assumptions, however rudimentary, superficial and provisional they may be. The theological debate focuses not on the existence of the transcendent, but on the content and quality of the *concept* of the transcendent.

We have to distinguish between immanent transcendence—which includes the unknown and the unknowable within reality—and transcendence proper which goes beyond immanent reality as such and as a whole. The intuition or notion of the transcendent is based on the experience of derivation, dependence, vulnerability, accountability, failure, guilt, suffering and mortality. The immanent networks that provide for these needs are ambiguous and unpredictable and are transcended towards a superior authority.

A concept of the transcendent provides meaning, criteria of acceptability and authority. Its content is of immense importance for individual life, society and nature. Decisive is the question to which degree it is informed by a vision of comprehensive optimal well-being. Where God, understood to be the ultimate Source and Destiny of reality, is a person, God's intentions and actions do not supplement, compete, or cooperate with immanent processes and agencies, but use God's creation for God's purposes.

Awareness of transcendence precludes the claim of modern humanity to ultimate autonomy, ownership and mastery of reality. It precludes the attempt to find a space for God's action in an otherwise-closed causal network. It precludes the naturalist assumption that immanent reality is all there is, thus closed in upon itself. However, these preclusions are based on existential, communal and ecological requirements, rather than on speculative metaphysics.

The closed universe of naturalism

Naturalism assumes that the universe is closed in upon itself, while faith presupposes that the universe is open to a transcendent Source and Destiny of reality. While there is an affinity between the methodological restriction of the sciences to immanent reality and naturalism as a metaphysical assumption, a naturalistic metaphysic does not follow from empirical science. It is based, rather, on the modern quest for autonomy and mastery. Its concern for nature may be based on the Western humanist tradition with its classical and Christian roots. A theoretical comparison between two metaphysical stances cannot lead to a conclusive verdict concerning their validity. It is their potential consequence for human existence, society and nature that matters. Awareness of a transcendent Source and Destiny of reality has potential advantages over naturalism, but they are not necessarily realised. It facilitates a personal attitude towards the world in which we are embedded. It sees reality 'from above' and as a whole. It acknowledges its eccentric position within its concentric contexts. It can lead to freedom from reality (including self) and responsibility for reality (including self).

The task of this chapter

In Chapter 8, I tried to show that the concept of the transcendent is rooted in the existential, social and ecological experience of being human. I also argued that, at least from the perspective of faith, such a concept is indispensable. This assumption is not only questioned, but rejected by naturalism. Naturalism is the most formidable modern alternative to the concept of the transcendent. It is the stance most commonly found among scientists.

Throughout this book, I try to show that science has good reasons to dismiss obsolete, counterfactual, counterproductive and inappropriate assumptions found in traditional faith communities. However, this in no way implies that naturalism is a viable alternative to an appropriately reconceptualised faith. This stance needs to be substantiated, and I intend to do so by way of a comparison between the two alternatives.

Before doing so, let me emphasise that conventional faith is badly in need of the incisive challenge posed by naturalism as an alternative metaphysic. Conventional faith has made itself comfortable in the quaint and cosy cocoon of its age old traditions, shielding itself from the harsh realities of the universe in which it lives and which the sciences inexorably reveal. As such, faith has become quite incapable of being what it is meant to be, namely an instrument of the creative and redemptive intentionality of God. By challenging the very foundations of faith as such, naturalism is capable of shaking believers awake. 'Pure' science alone is too diffuse and too neutral to do the trick.

Average believers have reduced their horizons to their tiny life worlds and their brief life times; their universe is anthropocentric; their faith is geared to the individual human being; it encompasses only the spiritual dimension of life; its certainties are largely built on untenable assumptions; it indulges in wishful thinking. A serious encounter with naturalism is capable of exploding these constraints and exposing faith to the chilly atmosphere of cosmic space, cosmic time and cosmic regularities and inevitabilities.

And yet even the wider horizons of naturalism are too restricted to allow for a concept of the transcendent Source and Destiny of reality. That is what this chapter is all about. I remind the reader, however, that this book is on science and faith, rather than on faith and naturalism. This chapter only serves to sharpen the contours of the notion of a transcendent Source and Destiny of reality against the background of its naturalist alternative.[302].

Section I
The basic assumption of naturalism

Science concentrates on whatever falls within the limits of human interest, observation, comprehension, manipulation and competence. Whatever may lie beyond these boundaries is not be part of the agenda. The question of transcendence is simply left open. This is not a metaphysical, nor a religious, but a methodological assumption. I have dealt with that in Chapter 5.

The scene immediately changes, however, when methodological agnosticism morphs into an atheist worldview assumption.[303] Differences in worldview assumptions have existential, communal, social and ecological consequences and must be taken into account as part of experienced reality.

To understand the fundamental difference between the biblical faith and naturalism, let us first try to imagine the whole of the reality we experience—all of space, all of time, all of energy, all levels of emergence and all the regularities operational at these levels. Both naturalism and the biblical faith have such comprehensive horizons.

Naturalism perceives reality to be a limited amount of organised energy contained in cosmic space moving through cosmic time as an entity that has clear-cut spatial and temporal boundaries, but that is, nonetheless, entirely self-generated, self-contained, self-directed, self-propelled and self-destructive. The twin principles of gravity and entropy rule supreme. They operate according to definite regularities and thus sustain the process of evolution.

So far an enlightened faith would be able to agree. Now let us ask ourselves whether it is plausible to assume that this reality is derived from, embedded in, dependent upon, and under scrutiny of something beyond itself, or whether it is all there is, thus entirely self-generated and self-sufficient. This is where the two approaches part ways.

Naturalism assumes that there is nothing beyond this cosmic whole—no cause, no rationale, no meaning, no destiny, no criterion of authenticity, no qualitatively different future. The world we experience is all there is and ever will be. There is no transcendence. Reality is closed in upon itself.[304]

This is not a scientific theory but a metaphysical assumption.[305] Science cannot prove or disprove it. As we have seen, however, the methodological restriction of science to what is, at least ideally, accessible to empirical observation, explanation and prediction, suggests by mere default that whatever falls outside this realm is irrelevant.

When this suggestion is accepted as self-evident, all assumed transcendent foundations, systems of meaning, values, norms, goals and visions are considered purely 'subjective'. The concentration of science on the observed object rather than the observing subject again suggests that anything subjective is unreal. And what is unreal is ultimately irrelevant.[306]

Alternatively, one can be persuaded that reality is not closed in upon itself. It is embedded in a greater whole or, to be more precise, open for a transcendent Source and Destiny. It is derived, dependent, empowered and accountable to the criterion of comprehensive optimal well-being. Being derived, dependent, subject to criteria and accountable, it is saturated with meaning. That is the assumption of the biblical faith.

This assumption would be purely theoretical and existentially irrelevant if it were not for the fact that humans require a sense of validity, reassurance and direction. The way we perceive the transcendent Source and Destiny of reality determines our system of meaning, our identity, our criteria of acceptability, our authority to act in certain ways, and thus our social system and our ethical orientation.

Radical naturalists seem to have lost the antenna for this kind of question. For them 'ultimate questions' are about observation and explanation, rather than grounding and orientation. And it is science and only science that is deemed competent to answer such questions.[307] The fundamental difference is that faith does not absolutise immanent reality, while naturalism does. With that, it forfeits the transcendent basis for a purposeful life.

The 'beefed-up baboon'

This deficiency manifests itself, I think, in the cynicism of naturalists such as Francois Durand.[308] His stance is much more elementary than the aggressive atheism of Dawkins, or the circumspect concern of Kauffman and Swimme. Durand persuasively exposes the irrational and atrocious behaviour found in male dominated religions of all kinds as a typical manifestation of primate evolutionary history.

In primates the alpha male has extraordinary powers and privileges. He subjugates the females and reduces the gamma males of the troop to emasculated serfs. They have no chance to copulate and procreate. Subordinates are conditioned to need decisive and powerful leadership. Deviants are ostracised and outcasts are murdered. A fierce and merciless ingroup-outgroup hostility is maintained.

According to Durand, male dominated religion is rooted in precisely this kind of pre-historic conditioning. God is the prototypical alpha male. He is jealous and power hungry. He oppresses his subordinates. He denies them the freedom to unfold their natural potentials, engage in promiscuous sex and enjoy the pleasures of life. He demands humility, subservience and asceticism, demanding all power and glory for himself.

All this was ingrained in the limbic system by the pre-historic evolution of the human being. The fatalism concerning the future of humankind that Durand displays is, therefore, hardly surprising and difficult to fault. Assuming that Durand's analysis is correct, I found the confrontation with our subconscious motivations exceptionally revealing. There is no way we can avoid the subterranean bedrock of our evolutionary history and theologians certainly have to update their concepts of original (and actual) sin.

It is remarkable, however, that this 'beefed-up baboon' is capable of producing Handel's Messiah and Goethe's Faust; land a space craft at a precise spot on Mars; determine the genome of its pre-historic ancestors and prove the validity of the theory of relativity. Is it completely unreasonable to assume, then, that it could also aspire to a slightly more comprehensive, attainable and sustainable vision of creaturely well-being? Could a slightly more appropriate awareness of an all encompassing 'Other' be superior to the self-claimed autonomy, mastery and ownership of the human being?

What if evolution had endowed humans with the potential to rise above these primate instincts, acquire more inclusive horizons, generate a sense of responsibility and develop a vision of comprehensive optimal well-being? Could it be that scientists, who lost their capacity to transcend, become the helpless captives of their own limbic systems? The actual stances of sensitive and responsible naturalists belie this impression. There is no scientific reason why our limbic system should be allowed to tyrannise us.[309]

Naturalism as an atheist position

The question of meaning and purpose does not automatically answer the question of ontology. Is there a God? If there is no transcendence, everything we speak about must be located within the realm of experienced reality, be subject to its causal sequences, and constitute a valid object of scientific research. If not, it is nothing but fantasy.

So to exist, God as a 'supernatural intentionality and agency' would have to be part of nature and subject to scientific investigation.[310] Richard Dawkins correctly argues that there is no evidence for such a being.[311] I would have thought that a 'supernatural' entity cannot be part of nature anyway and the search for evidence is a wild goose chase. Faith has always maintained that God is the Creator of reality rather than a part of creation.

So does God exist? From a faith point of view, this question can be settled fairly easily. As the transcendent Source and Destiny of reality, God quite definitely does **not** exist—at least not in the way that anything in immanent reality exists. So it is a wrong question to ask. However, **notions** of the transcendent do exist. They can be located in human consciousness, described, critiqued, transformed, replaced, or abandoned. Dawkins is in the business of doing just that.

Such notions make an incredible difference to human life. They have immense existential, social and ecological consequences. So notions of the transcendent are 'real' in experiential realist terms. That is as far as we can go. But does the mere assumption of the openness of reality to something beyond not imply some sort of ontology for what lies beyond?

Indeed it does. Faith is not afraid to affirm that there is a transcendent Source and Destiny of reality. But an implication is a deduction made by humans from a human notion that belongs to immanent reality. It says nothing about the existence and character of the transcendent as such. Moreover, when pressed, naturalism too reverts to a deductive train of thought and ends up in a similar impasse. This is most apparent in the use of mathematics by subatomic physics. With all its rigour, mathematics has not been able to produce an ultimate subatomic formula.[312]

Obviously faith does not leave the question open. It presupposes that there is a transcendent Source and Destiny of reality. But naturalism does not leave the question open either. It rejects the notion of transcendence as such and in principle. Nature is a self-generating, self-sufficient, self-sustaining, self-propelled, self-directed, self-defining and self-destructive whole. There is nothing beyond immanent reality. Non-existence has no qualities that can be described. Life is too short, Dawkins says, to spend time on a debate about what does not exist anyway.

This also means that nature itself has assumed the qualities and functions of ultimate reality. Naturalists such as Dawkins, Kauffman and Swimme are struck with awe when they contemplate the mysteries, beauties and majesties found in nature—and rightly so. At least in this respect, theism and naturalism agree. It may even be that by virtue of their intimate dealings with nature, naturalists have more profound sentiments in this respect than believers whose faith is geared to an abstract 'creator'.

Is naturalism a self-evident proposition?

Is it really thinkable that the universe just popped up out of nothing for no reason whatsoever, generated its own rules of operation and will disappear again into nothing by virtue of its inherent dynamics? May be! We do not know and cannot know. Observations of the processes that actually occur within the reality we experience do not lead to conclusive clues. Both the affirmation and the denial of transcendence are 'irrational' in the sense that they lack demonstrable evidence. The point is, however, that humans cannot do without a valid system of meaning.

Let me pick up Hawking and Mlodinow's 'model-dependent realism' at this juncture.[313] We see reality through particular filters that highlight certain features and block out others without thereby denying the greater complexity of objective reality as such. I take it that such a model can either highlight transcendence, so that the boundaries of immanent reality seem permeable, or filter it out, so that these boundaries seem solid. Faith locates the 'mainspring' of reality beyond these boundaries, while for naturalism it must be situated within its boundaries.

For naturalism this mainspring must be its own inherent dynamic. Forgetting the model-dependence of their assumptions, Hawking and Mlodinow opt for the theory that immanent reality generates and propels itself.[314] Then the assumption of a transcendent Source is redundant. Typical for scientists, this 'solution' concentrates on explanation and prediction, rather than grounding and orientation, which is the main spring of faith. I have dealt with that before.

The question at this juncture is, however, whether transcendence is a plausible proposition or not. The answer is that the naturalist 'solution' is just as incapable as the Christian faith to say why there is something rather than nothing—which they set out to explain in the first place.[315] Referring to the 'Game of Life' theory, the authors say: "That's all there is to it: given any initial condition, these laws generate generation after generation".[316] But the authors fail to explain why there should be such initial conditions and such laws.

For Brian Swimme, also a naturalist, there is no doubt that the source of all existence is not only *invisible*, but *nonvisible* and *nonvisualizable*.[317] Is it therefore transcendent in the strong sense of the word? Not necessarily. It just means that the most profound layer of nature remains a mystery to the human mind. But at least the question is kept open.

The formal concept of 'God' as the transcendent Source and Destiny of reality is simply a name for the openness of reality. So let me bring Nancy's postmodern idea of 'dis-enclosure' into the equation. The concern here is the radical openness of reality. Christianity must be deconstructed precisely because, contrary to its original intentions, it 'encloses' reality. To confine the 'subject' of creativity (or the dynamics of the world process as such) to nature would again enclose it—much more so than the theological tradition of a 'Creator' does.

Once 'dis-enclosed', however, this subject (or dynamic) must of necessity go 'beyond' nature. It must be thought of as 'above' nature, 'before' nature, 'after' nature and at 'the root' of nature.[318] We would then end up with a (non-theistic) model akin to panentheism.

This model is clearly unacceptable to the naturalist because it still posits some reality beyond nature. It would probably not satisfy the postmodernist either and that for exactly the opposite reason, namely, that it suggests the enclosure of reality by an assumed greater whole rather than allowing reality to be completely void of boundaries.

Thus 'disenclosed', however, the 'mainspring' of reality would still have to manifest itself in the probability observed by subatomic physics, the causality observed by Newtonian physics and chemistry, the unpredictability observed by complexity theory, the teleology observed by biology, the intentionality and agency observed by the human sciences, the dynamics observed by the social sciences, the sequences of events observed by the historical sciences, as well as the human quest for ultimate meaning, acceptability and authority observed by philosophy and theology.

In its boundless quest for freedom, postmodernism may want to get rid of all these regularities. But that is not how the world in which we are embedded seems to functions. Such regularities are inescapable and indispensable. That much is conceded by naturalism as well as the Christian faith.

The real issue between faith and naturalism then is, once again, whether nature is closed in upon itself, self-generated, self-propelled and self-accountable (or non-accountable) in all respects, or whether it is driven by an origin and power beyond its boundaries. Having no evidence for their assumption, naturalism is no less 'irrational' than theism is.

Why these alternatives are important for the topic of this book

It is clear that humans must operate within a frame of reference that is at least in consonance with the general thrust of the evolutionary dynamic. To me, that thrust clearly points towards the realisation of potentials, the flourishing of life and fulfilment in terms of human consciousness. We can generalise this thrust as an arrow pointing towards comprehensive optimal well-being.

To gain criteria for human attitudes and behaviour, we can then explore the consequences of different approaches to reality and measure them against the goal of comprehensive optimal well-being. The potential consequences of the two metaphysical alternatives on offer for human life, social relationships and the ecological future of the planet are immense. And that is the only meaningful level at which the merits of purely theoretical metaphysical alternatives can be assessed.

So let us look at the potential differences between the Christian faith and naturalism in this light. I have to emphasise that *potential* characteristics and differences are not identical with *actual* characteristics and differences. The callousness of believers and the sensitivity of naturalists are not necessarily true reflections of their underlying assumptions. Naturalism is the metaphysic that underlies modernity. What is modernity actually doing to the world that is accessible to us?

Once again, I am not a modernity-basher. Having spent a considerable time of my life in a pre-modern social context, I highly appreciate the achievements of modernity in fields such as science, technology, communication, health, social organisation and human rights. If anything, I am an unrepentant modernist. However, a diagnosis is not supposed to highlight the healthy body functions of a patient, but to expose the one dysfunction that can cause the system to collapse. The sticky point is the arrogance, self-centredness and self-indulgence of modern humanity. And this characteristic is not an aberration of the original rationale, but its explicit and deliberate goal.

Let me recap some observations I made in Part I. While popular postmodernity has taken the modern quest for freedom and mastery to extremes, its horizons have in fact become extremely constricted—much more so than community-centred pre-modern cultures. In spite of unprecedented access to the widest possible information, its awareness of the whole in terms of time, space and power has been lost. The modern mind has become focused on individual desire and collective self-interest at the expense of its social and natural environment.

As a result, 'emancipated' humans have become slaves of their own cravings and the ceaseless 'brainwashing' of the advertising and entertainment industries. Motivated by nothing but the financial 'bottom line' and powered by the most sophisticated strategists, industry and commerce deliberately and systematically dismantle all inhibitions and justify unlimited indulgence.[319] I will argue below that transcendence is the prerequisite of true human freedom from reality and alert human responsibility for reality. It is this particular assumption that has gone missing.

The success of modernity to deliver the goods entices vast masses of people to throw all caution to the wind. The social processes unleashed by modernity are sweeping across humanity. They create institutionalised structures and processes that spiral out of control. Seemingly the massive body of the Titanic cannot swerve in time to avoid the iceberg. In the meantime, the passengers continue to eat, drink, dance, make love, engage in intrigues and plan their business trips, totally oblivious to the fact that the party will soon be over.

The argument is not that atheist naturalism has caused these developments. The argument is, rather, that naturalism is an expression of the underlying mindset and not very likely to undermine its own foundation. At least in as far as it deems the modern economy a normal manifestation of the evolutionary process, naturalism is incapable of subjecting this economy to critique and demanding ground-breaking transformations. Christians have no convincing track record either. So at least we are in the same boat. The commendable efforts to save a few endangered species or reduce carbon footprints may create at least some ecological awareness, but it will not make a difference to the overall process.

There are highly perceptive and sensitive naturalists and ecologists, of course, who have gained comprehensive visions, who have lofty ideals, who simply adore nature and who are competent and entitled to enlighten narrow-minded believers. Again, Kauffman and Swimme are examples. Both in terms of scientific insight and global responsibility, most naturalists may be far ahead of the self-centred pack of believers who try to save their 'immortal souls' in a 'heaven' that does not exist and continue to enjoy, in the meantime, their earthly wealth.

Why opt for transcendence then?

However, the current question is not whether naturalists or Christians are more enlightened, but which metaphysical set of assumption would be capable of bringing about change. The point is not the dismal performance of an aberrant faith but the potential impact of the assumption of a transcendent Source and Destiny of reality that involves humans in its creative and redemptive project. That is what really matters, rather than a purely theoretical pair of metaphysical alternatives.

I try to avoid metaphysical speculations as far as possible. It is the existential, social and ecological consequences of the two positions that should demand centre stage. In my own estimation, the conviction that reality is open to a benevolent transcendent Source and Destiny is superior to the assumption that it is closed in upon itself and meaningless as such. But the superiority certainly does not lie in an alternative metaphysics.

As most perceptive scientists confess, scientific explanations can only go so far. Causal explanations do not provide a sense of meaning, purpose and direction. Why should there such an explosive stuff like energy in the first place? Why should the regularities exist that the sciences so badly try to establish? What lies at the bottom of the mathematical beauty that they represent? Why should the mechanisms that play themselves out in evolution actually function? Why should living beings be enthralled by colour combinations, shapes, harmonies and rhythms?

I do not think Dawkins or Hawking and Mlodinow have an answer.[320] Believers have no answer either. Nobody has. That is the character of transcendence. Believers only give the unknowable and ineffable a name. But the dearth of an answer begs the question why an immanent reality whose derivation cannot be accounted for should then be taken as ultimate. This is no less trivial a question than to ask why we should assign ultimacy to a transcendent Source and Destiny of reality that cannot be shown to exist. To do so is a metaphysical commitment, not a scientific fact.

Naturalists can only say that "this is how it is". Of course, this is how it is. Believers have no reason to disagree. They do not have a greater pool of information. They have no better explanation. The assumption that reality is open towards a transcendent Source and Destiny does not give them privileged access to what cannot be known. It just deepens their awareness of what it means to be human and not divine. Their awareness of being derived, dependent, accountable, vulnerable and mortal beings is more profound—at least in principle, if not in practice.

The perceived superiority of an awareness of transcendence lies elsewhere. The relation of believers to an authority higher than themselves and the whole of their life worlds affords them distance necessary to objectify reality. This distance makes it possible to attain inner freedom from, and responsible for reality as a whole including themselves.

On the basis of their basic assumptions, naturalists tend to be satisfied with swimming in the pool of reality as it is without asking further questions. I suggest that humans must ask such questions and formulate answers, however partial and provisional, or cease to be human—not because they are incurably inquisitive, as scientists normally are, but because they need a system of coordinates to live authentic human lives.

This is the crux of the matter. It feeds into all sorts of derived questions. Are humans indeed sovereign explorers of reality, or is reality being disclosed to them? Are humans indeed masters of reality, or is reality entrusted to them? Is reality—including their own bodies—nothing but a quarry to be mined for the exorbitant and immediate satisfaction of their personal desires and collective interests, or should they be grateful to be alive at all and satisfied with the share they need to survive and prosper?

Section II
The naturalist concern for the natural world

I have no doubt that most naturalists would reject with considerable contempt any insinuation that they propagate or legitimate the ruthless exploitation and senseless destruction of nature that we are witnessing at present. As mentioned above, they may be much more sensitive to the ecological problem than self-centred and spiritualising believers, even if only by virtue of their professional preoccupation with nature.

But on what grounds? As Friedrich Nietzsche's ideal of a 'super human being' with his 'will to power' demonstrates, it is absolutely feasible for atheists to arrive at opposite conclusions. Are responsible naturalists simply 'good at heart'? If so, why did evolution lead to both good and evil hearts—just because altruism beyond the ingroup has greater survival value? And if so, why is such goodness not spread across the entire human race? Or have naturalists matured enough to discern that it is more rational to serve than to dominate—a wisdom that the Christian faith cherished all along? [321]

Or are they perhaps embedded in a cultural tradition, where it is deemed more decent and acceptable to be friendly than pushy? And if so, how did this tradition come about? Do they perhaps unwittingly live off a Christian-humanist tradition that is historically rooted in a profound sense of transcendent intentionality, validity and authority, which calls for responsibility for the whole? Are they in fact humanists rather than naturalists? Let us briefly pursue this question.

Humanist ethical excellence

We have to distinguish between atheist naturalism and humanist ideals. They do not necessarily overlap. For the present argument, it suffices to consider the view that a humanist ethics can dispense with transcendence. It is often said that, being focused on human dignity and well-being, humanists are much more loving, much more involved in voluntary activism, much more sacrificial in their dedication than most Christians.[322] That may unfortunately be true in many, if not most cases. Countless secular NGOs and social movements bear evidence to the fact.

But this argument, I think, misses the point. In the first place, when appreciating the excellence of dedicated humanists we tend to concentrate on the heirs of a long religious, cultural, and social tradition where responsibility, duty and service to the community were deeply ingrained in both individual and collective consciousness, leading in turn to a sensitive conscience, a willingness to get actively involved and to sacrifice one's lifetime, privileges, energy and possessions for a greater cause.

This culturally based bedrock is of inestimable value and should not be deprecated in any way. We have to build the future on responsibility, concern and sacrificial attitudes. When we look at instances, however, where such a bedrock has successfully been dismantled in the West, the picture becomes ugly. Just think of the fallout of Nazism and Stalinism, the current behaviour of the financial sector in the affluent West, the mentality and lifestyle of the superrich, or the hedonistic craze of the common consumer culture!

It is much worse among marginalised and uprooted population groups where the impact of modernity has shattered traditional cultural and religious traditions without providing new foundations; where personal disorientation, anomie and spiritual chaos reign supreme; where the ruthless pursuit of self-interest is the only motivation left; where even human life no longer counts.

The impact of modernity is particularly devastating, where the cultural precursor was a dynamistic-magical worldview and a patriarchal-hierarchical order that had led to a dependent and fatalistic personality structure and subdued all individual responsibility and initiative.[323] Once the emancipatory spirit ignites a powder keg of long suppressed urges, it can explode and tear down whatever comes in its way.[324]

In the second place, autonomy and mastery only indicate a *formal* structure of personal authority that can be filled with different contents. Without doubt, there can be and have been benevolent tyrants, who are humble, dedicated and loving, whose sole ambition is to serve the well-being of their subordinates. The point is that this is entirely up to them. There is no higher authority or compelling purpose around.

There is also no doubt that there also can be, and have been, notions of a cynical and ruthless *divine* tyrant that are meant to legitimate the ruthless oppression and exploitation of 'his' *human* representative on earth. Such a notion is found even in the Bible (Ps. 2). This text reflects a tradition that hails from ancient Egypt and has countless parallels in tyrannies the world over. Its rationale is very obvious.

However, such a notion has been deprived of its legitimacy precisely in the course of the evolutionary trajectory of the biblical tradition and some other religious traditions as well.[325] In the biblical case, it has been replaced with the notion of sacrificial benevolence. The notion of divine intentionality is geared to a vision of comprehensive optimal well-being. There is no way such a God can condone or legitimise ruthless oppression and exploitation.

In the third place, sensitive naturalists may be struck by the majesty, beauty and mystery of the world we know and speak passionately of the respect due to nature. But there is no question that, in their view, it is the human being who is ultimately in charge. As the most advanced creature on earth, the human being is master, owner and beneficiary of the earth, including the human being itself. What humans make of this privileged position is, again, up to them. According to Stuart Kauffman, humans invented and have to reinvent the sacred. Naturalism is a genuine expression of the emancipated and absolutised human subject that characterises modernity.

Conversely, nature is being absolutised. It is a self-contained, self-sufficient whole that encapsulates humans as much as stones and ants. It allows no perspective from beyond. The simultaneous existence of mastery and enslavement seems to be paradoxical, but it is not. One can stand in submissive deference and awe before nature as a tyrant and yet manipulate this mindless authority by utilising its power to one's own advantage. Tyrannies generate devious arguments, tricks and procedures among subordinates. That is what science and technology essentially do with nature. While nature is supreme, it is also the dumb and helpless victim of its human subordinates.

Section III
God as a person

Even more troublesome for naturalists than the issue of transcendence is the concept of God as a person. Notions of the transcendent are usually, though not regularly, expressed in personal terms. The biblical faith takes it for granted that God is a person. For naturalism, this makes no sense because the evolutionary process as such displays no purpose or intention.

Again and again, philosophers and theologians have attempted to rid the concept of God from its personal attributes. The 'anthropomorphism' found in the Bible and in many other religions simply did not seem to be appropriate when considering the overwhelmingly impersonal character of the world process as a whole.

Dawkins argues that personal intentionality only emerged at the tail end of the evolutionary process and cannot, therefore, be predicated as the original source of reality.[326] Indeed it did, but that applies to human personhood; it says nothing about a transcendent Source and Destiny of reality as a whole. Scientists such as Einstein and Hawking loved to refer to 'God' as the ultimate expression of the regularities and mechanisms that determine reality as it unfolds and operates. But that is just a conceptual loan from religion that is not supposed to be taken seriously.

The same is true for the identification of God with the concept of 'creativity' by the theologian Gordon Kaufmann (2004) and the biologist Stuart Kauffman (2008). Creativity is an abstract noun derived from the verb 'to create'. As an anthropomorphic metaphor, it requires a subject. If 'nature' is understood as an impersonal mechanism, it cannot provide such a subject. So the naturalist filter renders the metaphor implausible.

A similar linguistic problem appears with Hawking and Mlodinow's book title *The Grand Design* (2010). A design without a designer? Or does this refer to the design of the M-theory as a mathematical construct so that the authors are themselves the grand designers? Faith, in contrast, can make sense of the metaphor of 'creation' because it is geared to a transcendent Source and Destiny of reality characterised by intentionality and agency. But then God cannot be thought of as the inherent creativity (mechanism or dynamic) of the cosmic process itself.

So why do Christians insist that the transcendent is a person? The reason is, quite simply, that this is how God has disclosed 'himself' in the (human) history of the biblical faith. God has 'spoken' through human speech and 'acted' through human action. The proclaimed and believed self-disclosure of God's intentionality implies that God is a person. Ancient Israel was constituted by the assumption of a covenantal relationship between Yahweh and 'his' people. God 'spoke' through law givers and prophets; the people responded in prayer, praise and obedience.

The New Testament widens this personal relationship between God and his people to include humanity as a whole. According to the first chapter of John, God's creative and redemptive intentionality (the 'logos') manifested itself in human reality ('flesh'), that is, in the ministry of Jesus of Nazareth. After the death of Christ, the Spirit of the 'risen Christ' permeated, liberated, transformed and empowered the 'Body of Christ', the community of believers. In all these cases, we are busy with the personal level of emergence. It is impossible for the biblical faith to perceive God in any other way than as a person.

According to the biblical witness, God became a person to humans *because humans are persons*. The underlying mechanistic assumptions of naturalism may not be wrong as far as they go. But given an emergentist rather than a reductionist approach, this concept of 'God' reflects a truncated view of reality. According to the theory of emergence, intentionality is as valid a dimension of reality as causality is, only at a different level of complexity. A notion of the transcendent Source and Destiny of reality *must include* the personal dimension of human reality, or it is defunct.

Intentionality may indeed only operate at the distinctly personal level, but that does not *confine* God to human personhood. God is the transcendent Source and Destiny of reality as a whole, including the entire impersonal infrastructure of personhood, the entire impersonal supra-structure of collective consciousness, as well as their differentiated regularities and processes. Christian spirituality means living in the constant awareness of the transcendent mainspring of **all** of reality—the nuclear reaction, the virus, the snake, the grimace of an adversary, and the bullet of a criminal.

Humans themselves are much more than persons—and they constitute the experiential basis for our anthropomorphic metaphors for God. It is not my intentionality that makes my heart beat, my digestive enzymes to provide me with energy or my cells to mutate into cancer cells. The idea that God is 'pure spirit', that is, pure intentionality unconstrained by the impersonal infrastructure of personhood, is an idealised abstraction from the way we experience our own intentionality. It looks suspiciously like a projection of our own yearning for a mastery that is unfettered by the regularities at lower levels of emergence, as Ludwig Feuerbach suggested.

The problem of theodicy

The concept of God as a person directly links up with the age-old problem of 'theodicy'. Because there is so much natural evil, so much suffering in biological life, and so much depravity among humans in the world, the assumption of a powerful and benevolent God in charge of reality is counter-intuitive, if not counterfactual. This is a valid challenge that faith must take seriously.

As we shall see in Chapters 11 and 12, theodicy is not foreign to theology, but arises from within faith itself. The problem was that pre-scientific faith assumptions tend to over-personalise reality and underestimate the validity of impersonal causal networks. God's mastery as the Creator seems to imply that God was free to suspend natural regularities and intervene wherever he deems fit. Calamities are taken to be the outcome of a personal decision of God.

Where this is the case, faith misunderstands itself. The reassurance of a benevolent intentionality underlying experienced reality emerged and evolved precisely as a redemptive response to experiences of inexplicable fate, meaninglessness, futility, frustration, depravity, guilt, suffering and death. Faith is based on the defiant assertion that evil has no right to exist. It is a protest of what ought to be in the face of the experience of what ought not to be. It does not succumb to negative experiences but overcomes them as far as possible. It tenaciously clings to the reassurance that God is for us and with us and not against us.

With its distinction between personal and impersonal levels, the theory of emergence can help us make sense of the problem of theodicy. God has become a person to humans because humans are persons, but God is also much more than a person, just as humans are much more than persons. God is the Source and Destiny of the entire hierarchy of emergences.

God's creative and redemptive intentionality can be deemed to 'lie at the root' of the structured behaviour of subatomic energy conglomerations, physical and chemical processes, biological organisms culminating in the brain, the hardwiring of the subconscious, the dynamics of individual and collective consciousness and social structures and processes.

God's creative and redemptive intentionality can be deemed to 'lie at the root' of the emergence and evolution of concepts of the transcendent in the form of historical traditions. It can also be deemed 'to lie at the root' of the emergence and evolution of criteria of appropriateness of such concepts and visions of what ought to be.

Assuming that immanent explanations of reality are reasonably plausible, provisional and subject to refinement and correction, therefore, they do not in any way preclude the assumption of faith that they are derived from a transcendent Source of power, subject to a transcendent criterion and geared to a transcendent vision.

Once one is persuaded that a creative and redemptive intentionality underlies experienced reality, one can pick up prototypical instances where such an intentionality seems to manifest itself. One can become grateful for one's statistically improbable chance to exist and prosper. One can face experiences to the contrary boldly and actively, rather than becoming their victim. One can also begin to appreciate the validity and operation of natural laws and other regularities within reality because without them the world we know would not be able to exist and function.

A personal relationship with reality

It is faith in a personal Source and Destiny of reality that allows us to relate to an impersonal nature in personal terms. And that is of critical importance in the context of the ecological crisis. Having become aware of our derivation and dependence, we would not be able to express our gratitude and humility if there was no personal 'Other'. We would have to assume that we are nothing but a natural outgrowth of the evolutionary process and thus entitled to life and well-being on the mere strength of being alive.

But the naked facts of our evolutionary genesis and the coincidental nature of having come into existence as species and individuals should convince us that we are not entitled to anything at all. We exist 'by grace and by grace alone'. Whose grace? Nature as such is not graceful—quite the contrary! Grace can be predicated only as the gift of a personal God who uses nature to bring us about.

Divine grace is an anthropomorphic metaphor and a metaphor is not an argument for the existence of God. But the inevitable result of a reduction of the concept of God to the impersonal mechanisms in reality is that the personal dimension of human existence is cut out of the equation. Being a person, the human being sticks out like a sore thumb that does not seem to belong to the rest of the body.

In its own self-estimation, humankind is then no longer part of the ordinary world, but its sovereign master, owner, judge and beneficiary.[327] It is the only legitimate 'heir' (according to Ps. 2) after God has 'died'. It is no longer the representative of God and a participant in God's creative and redemptive project. Nietzsche's 'super-human being' and its horrific incarnations in Hitler and Stalin give a vivid expression to the absolutisation of the human being after the perceived 'death of God'.

Assuming that it is desirable to relate to reality in personal terms, does it also make sense in a scientific perspective? That depends on what we mean when we talk of personhood. The urge to survive and prosper is shared by all living things. This kind of 'teleology' utilises situations sensitive to initial conditions to 'achieve its purposes'. It has recently become more obvious that intentionality and agency are not restricted to humans, but shared in various degrees by higher animals. Humans may be the peak of an evolutionary process, but they are not alone.

Moreover, the entire cosmic process is based on information. Information presupposes what could be called 'communicative competence'. If one investigates the spread of communicative competence in reality, there is hardly anything that can be excluded. Seen in this light, human communicative competence is only the most advanced manifestation of a characteristic that permeates the entire system. It would be arbitrary to exclude communicative competence from any notion of the transcendent Source and Destiny of reality.

Section IV
Freedom and responsibility

Once we see ourselves as the representatives of God in the world, we become accountable. The presence of an 'Other' is fundamental for a sense of **responsibility**. Where the human person cannot transcend its own personhood towards an overarching personal Other, it becomes absolute. Being the only persons around, humans rise above nature—including themselves as parts of nature—at least in their own estimation. Inadvertently, the human person itself then assumes the characteristics of a quasi-transcendent reality.[328]

Autonomous humans can lovingly tend and preserve reality, as so many lovers of nature do, or they can treat it as a quarry to be mined for their own use and satisfaction, as most other people do. Being responsible only to themselves, this is entirely up to them. In fact, they are not be responsible at all. Where humans deem themselves responsible to a community, or to society, or to humanity as a whole, they still operate within the collective absolutisation of the human species.

This is not some unwarranted theory or blind speculation. It is typical for the ruthlessly self-seeking and self-serving mood of modernity. As I argued in Part I of this book, the self-absolutisation of the human being lies at the root of the crises humanity is confronted with today—crises that have already created untold misery and that may eventually lead to the demise of humankind. Humans have already become the most deadly and devastating parasites nature has ever produced and endured.

If they cannot see reality from above, as it were, 'with the eyes of God', humans remain stuck in the centrality of their own existence. They focus on the immediate satisfaction of their individual desires and collective self-interest. They will not easily develop a sense of responsibility for a natural world entrusted to their care.

Sadly the Christian faith has hardly risen above this anthropocentricity and ego-centricity either—especially where Christians have been concerned about nothing but their assumed eternal fate in heaven or in hell. But this is an aberration. Genuine faith recognises that humans are superior to nature, but in a completely different way.

According to Genesis 1:27-30 humans are created in the 'image of God', which was a title of the king as representative of God. According to Galatians 3:23-4:7 the status of a 'son' or 'daughter of God' is attained through participation in the new life of Christ. In faith humans participate in God's ownership and mastery of nature. Nature is entrusted to them. They are custodians of nature. They are accountable.

Responsibility demands **freedom**. Only free persons can be responsible persons. Humans cannot attain genuine freedom if they attribute ultimacy to the world in which they are embedded—including the instincts located in their reptile brains, the desires located in their subconscious, the needs of their bodies, the pressures of peer groups, the expectations of their communities, the imperatives of their cultures, the lure of fashions and fads, the inexorable functioning of market forces, the power of political movements, and the mechanical functioning of natural processes. Ironically, it is precisely the ostensibly sovereign masters of reality that fall prey to the dictates of a reality of which they inextricably form a part.

Psalm 82 suggests that the gods have been placed above their respective nations by Yahweh for a purpose, namely, to execute justice. Because they are corrupt, 'the foundations of the world are shaken'. As a result, they are demoted and executed like mortal princes. In Ephesians 1: 20-23, there is a similar picture, only that the underlying transcendent intentionality that is placed over the 'spiritual powers' is now represented by Christ—the instrument of God's suffering acceptance of the unacceptable (see Eph. 2).

Just substitute 'gods' and 'spiritual powers' with modern equivalents—nationalism, fascism, liberal capitalism, imperialism and Marxism-Leninism! Would the devastations these ideologies have inflicted upon the continent of Europe over the last two centuries have occurred if these ideologies had been transcended by, and held accountable to, a higher authority that stood for justice and suffering, transforming acceptance of the unacceptable?

Such 'gods' are personified collective mindsets! But we are humans, and humans are persons. It helps us to realise that these 'apocalyptic monsters' have no right to overpower and enslave us, to name them and shame them, to deny them loyalty and subservience, and to declare into their face that they are accountable to a higher authority characterised by social justice, redeeming concern and respect for the sanctity of life. If that had happened in Europe and the Far East, the nightmare of World War II would never have materialised.

Today there are no boundaries within the European Union; a spiritual power has taken control that is more amenable to tolerance and cooperation. So all the devastations of the twentieth century could have been avoided, if another 'god' had been in charge! Now we are saddled with the *Zeitgeist* of the modern consumer culture—with crazes and irrationalities that are no less detrimental in economic and destructive in ecological terms. Is it immaterial whether we transcend this collective spirit or not?

From a faith point of view, neither an absolutised nature nor an absolutised humanity can do justice to what it means to be authentically human. Participation in the freedom and mastery of God infinitely surpasses the self-appropriated freedom and mastery of the human being. Commitment to the creative and redemptive project of God infinitely surpasses an existence dedicated to the pursuit of private or collective interests. The dignity of being an authorised representative of God infinitely surpasses the status of a self-made man or woman.

I am not God—thanks be to God! Humanity is not God—thanks be to God! The currents and cross-currents of the *Zeitgeist* are not God—thanks be to God! The world is not God—thanks be to God! There is a higher criterion and thus a higher authority—an authority to whom we can relate in personal terms and in whose creative and redemptive project we are invited to participate. The moment we scrap transcendence we attribute ultimacy to ourselves and our life worlds and become its victims as a result.

Are humans not mature enough to be free and responsible?

Can we not assert our dignity as adult human beings?[329] Can we not be responsible to community, society and its legitimate institutions? Can we not follow the naturalist approach whose admiration, respect and awe are focused on nature, the reality that we actually experience, rather than on a transcendent Source and Destiny of this reality, an entity that cannot be shown to exist?

Indeed we can, and to a certain extent we may succeed in doing so. But the counter-evidence is massive. What if peer pressure misleads teenagers into premature sex or drug abuse? What if a misguided primal community expects your loyalty or casts you out? What if society is corrupt, unjust and intolerant of deviant opinion? What if the masses become like beach sand swept away by ideologies and demagogues? What if these spiritual powers demand you to toe the line or be thrown to the wolves? What if the church becomes idolatrous and authoritarian? What if nature destroys life with diseases, earthquakes, droughts and floods?

What makes us free to resist, to overcome, to transform in all these cases? What commits us to stake our lives in such a precarious and sacrificial behaviour? Faith in God as the transcendent Source and Destiny of reality prevents humans from appropriating ultimacy both to themselves and to their life worlds. At the same time it empowers humans to resist and move in the opposite direction. Faith allows humans to participate in a higher freedom from reality and a higher responsibility for reality.

It has been said that if there was no God, we would have to invent one. Feuerbach, Marx, Dawkins, Kauffman and their fellow atheists maintain that we have indeed invented one. There is no way we can refute or substantiate this claim. But we can deem faith in God appropriate or inappropriate. To make up our minds, we can look at the consequences of either position. One of the most pertinent instances in this regard is the economic-ecological crisis mentioned in Part I of this book.

Section V
The sacrificial character of reality

Linking up the assumption of our personal accountability with the economic-ecological problem depicted in Part I of this book, we must go one decisive step further. It is rather self-evident that humans are derived and dependent beings. Human life is a gift of nature, not a human achievement. The cost of our existence has been staggering. Speaking in terms of the anthropomorphic metaphors of faith, entropy is a sacrificial process. The sun, our star, makes life possible by slowly disintegrating. All higher forms of life depend on the sacrifice of other life to survive and prosper.

Human life is so precious, not least because it is so costly. The brunt of the burden of our existence is born by plants and animals, but it also includes human effort, investment, suffering, vulnerability and mortality. Within humanity it is the lower classes, the marginalised and the excluded upon whose shoulders a disproportionate share of the sacrifice rests, but also countless previous generations who have suffered, laboured, struggled and died, otherwise we would not exist, as well as future generations that have to pay a heavy price for our extravagance.

Perceptive humans cannot help but have a sense of indebtedness, humility and gratitude. How can it be expressed? To whom can it be expressed? How can it begin to determine one's attitude to reality? How can it lead to contentment and moderation? Subhuman and supra-human reality is impersonal. Without the notion of a personal Other at the root of their indebtedness, humans can very easily presume that, similar to princes in a feudal system, they are entitled to exact unlimited sacrifices from the non-human world on which they depend without giving anything in return.

Believers, in contrast, sense that they owe their lives to a transcendent Source and Destiny. In pre-modern cultures, the acknowledgement of dependence and indebtedness is expressed in the form of sacrifices to the divine, however conceived. Here humans 'give back' parts of their lives to the deity to represent their lives as a whole (*pars pro toto*).

The acknowledgement of dependence becomes more urgent when humans run into a behaviour that is destructive of life, curtails the life chances of others and the community, or militates against justice and loving concern in general. In these cases, humans feel the need to go through a process of restitution, reconciliation and reintegration. The damage has to be repaired and somebody has to bear the cost.

The sacrifice of God to humans

According to the Christian faith the sacrifice of humans to God has turned into the sacrifice of God to humans.[330] Forgiveness means that the victim suffers the pain that the perpetrator should have suffered. If it is God who forgives, it is God who suffers the pain. Christ proclaimed and enacted the redeeming love of God. When the religious and political authorities of his time got rid of him, God's redeeming love exposed God to human depravity in Christ, his earthly representative.

The cross of Christ is, therefore, deemed the prototypical manifestation of God's suffering and redeeming acceptance of the unacceptable. Seen in modern cosmic contexts, this event was only the culmination of a divine sacrifice that encompasses the entire process of emergence, evolution and sustenance of reality. It is God who sacrifices immense chunks of God's creation so that we can live and prosper.

God does not demand sacrifices for 'himself', but invites us to share God's sacrificial, creative and redemptive action in the world. This invitation prevents us from buying off God with token sacrifices. The prophets in the Old Testament and Jesus of Nazareth castigated a kind of piety that allows us to continue with our rapacious behaviour at the expense of other creatures, including humans.

God is also not the idealised abstraction of a perfect being that is removed from the actual suffering of nature and humanity as Platonic idealism might suggest. God expects the dedication of our life as a whole. However, a sacrificial life cannot be popular in a culture that is focused so predominantly on mastery, exploitation, utility and pleasure. But what else can prevent us from sliding ever closer to a point of no return!

In sum, there are ample reasons to believe that a personal relationship with the Source and Destiny of reality is appropriate, if not essential, for authentic human life, social involvement and ecological health. In view of the immense power that genetic programming, environmental conditioning, peer pressure and mass movements exercise over individuals, it is only a commensurate superior spiritual power that could be capable of calling humanity into its own overarching freedom and responsibility.

Must believers be superstitious?

In closing this chapter, let us look at the basic alternatives of atheism and theism from a scientific point of view. The relation between immanence and transcendence throws light on the roots and the stringency of the atheist position. For atheist naturalism, there is no transcendence. An atheism that bases itself on science presupposes that the evolution and operation of reality in all its dimensions—subatomic, physical, biological, mental and spiritual—can be adequately explained without the hypothesis of God.

Indeed they can—at least up to the limits described above. Increasing insight into statistical probability, physical causality, chemical propensity, biological evolution, self-organising systems, psychological processes and social dynamics have progressively closed the gaps previously filled with the assumption of a 'supernatural' agent. A faith that believes it has a better explanation of how reality originated and how it continues to function than 'best science', is superstitious. Faith does not need to be superstitious, because it is not about explanation but about meaning, grounding, orientation and trust.

Theology must concede that pre-Enlightenment religion was a package of assumptions that seemingly included the explanation and manipulation of reality. As science took over the explanation of reality and technology took over the manipulation of reality during the last few centuries of Western history, the atheist position grew exponentially in strength. That can easily be understood. But atheism does not follow from the scientific explanation of reality. There is something much more fundamental at stake, namely, the 'take-over' of the functions of religion by naturalism, which is not always accounted for.

Must scientists be atheists?

The possibility—at least in principle—of explaining immanent reality in all its aspects without recourse to the intervention of a superior being seems to be a necessary condition for the plausibility of atheism. If faith were indeed about the explanation of reality, atheism would seem to have won the argument. If everything within reality had been explained, or could in principle be explained, a 'God hypothesis' would be redundant for explanatory purposes. If that is what faith was all about, faith would have no rationale, validity and justification.

However, as mentioned above, religious conviction is not about explanation. It is about grounding, orientation and vision. It is about meaning, acceptability and authority. It is about self-entrustment to the proclaimed benevolent intentionality underlying everything that exists and happens within reality, whether it can be explained or not. Believers may share in the scientific quest for explanatory lucidity, if they are thus inclined, but such a quest does not constitute their faith.

Conversely, there is also an element of self-entrustment in atheist naturalism. It is not its explanatory power that lies at the root of the naturalist assertion that there is no God, but the postulate that experienced reality is all there is. This is a metaphysical commitment that emerged and evolved in history as much as faith in God did. It has antecedents in antiquity. It has been a constant, though often subdued companion of faith throughout its history. The explanatory agenda of science does not lead to atheism. There is a different kind of agenda at work.

The fundamental ingredient of modern atheism appears to be the emancipatory thrust of modernity. It led to the appropriation by humanity of what was previously perceived to be divine authority, mastery and ownership. With its closure of immanent reality, naturalism has made it possible for the human being, as the only product of nature endowed with higher intelligence and intentionality, to assume entitlement to supreme mastery and ownership of reality. In pre-modern cultures this assumption was considered *hubris*—a kind of arrogance that was deemed destructive of human nature. In modernity it is taken to be self-evident.

Faith in God denies this closure of reality. It assumes that the universe, including the human being, is derived from, dependent on and accountable to its transcendent Source and Destiny. The human being is not autonomous and cannot be. Because humans experience themselves as derived, dependent, embedded, determined, accountable, vulnerable and mortal beings, the existential plausibility of this assumption surpasses that of naturalism. In fact, looking at the evidence, the assumption of human autonomy is arbitrary and irrational.

Once God is no longer deemed an agent or a factor within reality among others, but the transcendent Source and Destiny of reality as such and as a whole, the atheist assertion loses its stringency. A serious scientist does not have to be an atheist. On the contrary, the burden of proof lies on those who assume that the universe is closed in upon itself and that humanity is served best by assuming the role of a sovereign owner and master of reality. Conversely, atheists do not have to be scientists to prove their point, because their prime motivation is not based on scientific insight, but on emancipation, ownership and mastery.

Let us summarise

Naturalism is the most formidable alternative to a concept of the transcendent Source and Destiny of reality. Faith in God implies an open universe, while naturalism assumes a universe closed in upon itself. It is akin to the methodological restriction of science to immanent reality, but it cannot be derived from the latter. It is a metaphysical position with its own agenda.

Its most basic motivation seems to stem from the typically modern quest for mastery and ownership of reality. Such an anthropocentric view of reality stands in marked contrast to the eccentric position of the human being in the biblical faith. In view of the unanswered and unanswerable questions encountered in science, naturalism is no more plausible than the assumption of a transcendent Source and Destiny of reality.

The real issue must be the potential consequences of the two positions for human existence, society and nature. The track record of modernity, of which naturalism is a typical outgrowth, is not very reassuring. Nor is the track record of a narrow-minded, anthropocentric and spiritualising Christianity. Awareness of God's creative and redemptive intentionality leads to a dynamic vision of what reality ought to become.

Potentially at least, the notion of transcendence allows for an objectified view of reality as a whole, thus for freedom from reality and responsibility for reality. Responsibility demands the personal Other. Freedom is based on a link with the transcendent Other. In the presence of a personal God, humans can have a personal relationship with reality. God's relation to reality is essentially sacrificial.

This is of fundamental importance for our dealings with the economic-ecological crisis. The aversion of modern humanity to participate in God's sacrificial benevolence on behalf of creation as a whole has devastating repercussions for a global society seriously out of balance and an overextended natural environment.

By virtue of their preoccupation with nature, naturalists may have considerably wider horizons and a stronger commitment to the well-being of the natural world than the majority of believers. This excellence does not necessarily follow from their naturalist metaphysics. It may be derived, rather, from the heritage of the Western humanist tradition with its roots in classical antiquity and Christianity. However that may be, in view of the present crisis, humanism must be considered a teacher and ally of faith, rather than its adversary.

Part IV

Faith —
the perspective from beyond

10 The approach of faith — integrative transcendence

The reality that we experience can be seen either from an immanent perspective or from a transcendent perspective. Science views reality from within that reality, exploring it as far as it is accessible, at least in principle, to human observation, comprehension and control. Faith looks at the same reality 'from above', as it were, 'with the eyes of God'. This implies that its view of reality must, of necessity, integrate the insights of best science or be defunct. In terms of experiential realism, faith is a particular structure and orientation of individual and collective consciousness. In theological terms, God becomes a person for humans because humans are persons. God uses the proclamation of a message to disclose God's creative and redemptive intentionality. The message is based on intuitions that emerged and evolved in human history in response to situations of need and travelled through time in the form of traditions. Faith participates in God's freedom from, and God's responsibility for the world. This is critical for the economic-ecological impasses modernity faces at present.

The task of this chapter

In Part II, we dealt with faith seen from the immanent perspective of the sciences. In Part III we reflected on the plausibility of a concept of the transcendent in view of its naturalistic alternative. In Part IV we are raising the necessary questions about the particular content of the Christian concept of the transcendent and its relation to modern scientific insight.

In biblical terms, God is the transcendent Source and Destiny of *immanent* reality, rather than an imaginary reality situated somewhere beyond the reality that we experience and that the sciences explore. Our perspective from beyond immanent reality focuses precisely on the reality that we experience and that the sciences explore. The worldview of faith is, therefore, not based on an ostensibly eternal truth that 'dropped vertically from heaven', as it were, but on the proclamation of God's dynamic response to the constant flow of reality and its human interpretations.

If that is the case, the worldview of faith must, of necessity, integrate the insights of 'best science' into its perspective from above. That is what I mean when I speak of the approach of 'integrative transcendence'. Scientific insights are dynamic and provisional. So the integration of such insights is an ongoing task. The worldview of faith must by necessity be as dynamic and versatile as the worldview of science is. Where this is not the case, it loses touch with the flow of experienced reality and its changing interpretations.

In the present chapter I discuss the method of 'integrative transcendence', which I consider the appropriate approach to Christian theology. In the following two chapters I will suggest how 'best science' can be integrated into a perspective 'from above' without sacrificing the constitutive assumptions of the Christian faith. In Chapter 11, I demonstrate that this method is feasible even within a pre-modern experiential theology. In Chapter 12, I move beyond classical theology to more radical reconceptualisations of traditional theological motifs—again with the aim of 'becoming a scientist to the scientists'.

The approach of integrative transcendence

In Part II, I dealt with experienced reality, including the phenomenon of religious conviction, as seen from an **immanent** perspective, which is the perspective of science. In Part IV I deal with that same reality, including the phenomenon of science, as seen from a **transcendent** perspective, which is the perspective of the Christian faith. Faith and science thus see the same reality from two different vantage points. Science explores it from within. Faith sees it 'from above' as it were, 'with the eyes of God'.

Faith entails both an all-inclusive view of reality and the assumption of its dependence on something greater than itself that determines both the source of its existence and the goal of its operation.[331] As the transcendent *Source* of reality this assumed 'beyond' generates, permeates and sustains reality. As the transcendent *Destiny* of reality, it determines the acceptability of its direction and the meaningfulness of its outcomes.[332]

Expressed in Christian theological terms, God is engaged in a creative and redemptive project that aims at the comprehensive optimal well-being of the whole of reality. Human life, like everything else, emerged from this project and will remerge into it. It is a highly sacrificial process in which humans are invited to participate. That is what gives human life and reality as a whole a meaning, purpose, dignity and value.

The biblical faith is not built on scientific *observation* of immanent reality, but on the *proclamation* of God's creative and redemptive benevolence (Rom 10:14ff). It is persuaded that God, the transcendent Source and Destiny of reality, has disclosed God's creative and redemptive intentionality within experienced reality. This does not imply that faith has privileged access to the way in which reality emerged and evolved. For that it depends on the insight of 'best science'. Its concern is to gain the kind of grounding, structure and orientation of human consciousness that can claim ultimate validity and lead to human authenticity.

An analysis of the history of the biblical faith suggests that this faith does not have a particular worldview of its own. It owes its dynamic to the fact that it responds creatively and redemptively to changing situations of need and their various interpretations. The 'Word of God' picks up different worldviews it finds on its way through human history, as it were, and proclaims the message of God's creative and redemptive intentionality within these changing contexts. By doing so it critically transforms the worldview concerned and gives it a new direction.

What the biblical authors did for their times, we must do for our times. We have to "become all things to all people" as Paul put it in 1 Corinthians 9: 19-23. The doctrinal systems constructed at various times must be seen as provisional conceptualisations of stages reached in this integrative process. This is also true for the classical doctrines of the Trinity, Christ, creation, salvation, eschatology, and so on. They are meant to define the evolving identity of the Christian faith rather than to reflect ultimate truth.

The current marginalisation of the Christian faith is at least partially due to its failure to respond creatively and redemptively to rapidly changing insights and worldview assumptions. It will regain and retain its credibility and relevance in the modern world only if it manages to integrate the (admittedly provisional) insights of 'best science'. The validity of Christian faith assumptions cannot be derived from science or tested by science, but Christian worldview assumptions must take account of valid scientific insights or become delusory and irrelevant.[333]

The experiential roots of a notion of the transcendent

Integrative transcendence is an experiential, not a speculative method. The Christian faith does not claim any direct access to God. Christians believe themselves to be taken into the powerful flow of creative and redemptive benevolence that originates in God, reaches us, flows through us to others and further into the wider world. God is behind us; the needs of the world are before us. The moment we turn around, leave the needs of the world behind us and try to face God-self, we stare into an abyss.

We cannot hope to fathom the transcendent Source of the cosmic process as such and as a whole. Our focus has to fall on God's benevolent intentionality that has manifested itself in the history of Israel, culminating in the Christ-event. "Nobody has ever seen God . . . God is love and those who abide in love abide in God and God abides in them" (1 John 4: 12, 16). If there is any 'revelation' at all in the biblical faith, then it takes the form of an existential and communal involvement in God's creative and redemptive project.

Why assume a transcendent Source and Destiny of reality in the first place? Humans experience themselves as derived from, and dependent on, a network of forces and relationships that they cannot fully fathom, explain and predict. They are vulnerable and mortal. They depend on a system of meaning for a definition of their identity, belonging, acceptability and role within the greater whole. They are accountable to an authority higher than their own. They are haunted by inadequacy, guilt and failure. They need reassurance, forgiveness and reconciliation. They need to be allocated authority in the form of statuses and roles.

All this is fact, not fancy. The question is only what it is that serves to provide humans with this kind of spiritual and emotional integration, stability, structure, orientation and direction. They do not need to be conscious of their ultimate foundations. Such foundations can be profound or shallow. They can have a plethora of existential, social and ecological consequences. But they will always be there.

Humans who do not experience themselves as derived, dependent and accountable, live in a fool's paradise. Human consciousness cannot operate without structure and orientation. If humans try to be completely free from assumptions and constraints, they are vulnerable to nasty surprises. They may end up in dead-end-streets. Where they assume unrestricted autonomy, they deceive themselves. Humans are not self-constituting, self-maintaining, self-defining and self-accountable and they cannot be.

The substantive aspect

Awareness of ultimate dependence is only the *formal* aspect of religious belief. The Christian faith simply takes it for granted. What matters is the *content* of the transcendent meaning of human existence as part of the transcendent meaning of reality as a whole. What makes the Christian faith specifically Christian is the proclamation of the benevolent intentionality of God. But God's intentionality can matter only if this God is deemed the ultimate Source and Destiny of reality, thus the ultimate authority over their lives and their life worlds.

In the Christian faith, the proclamation of this content leads to the assurance of faith that the ultimate Source and Destiny of reality is for us and with us and not against us. As argued above, this reassurance is entirely in line with what happens in reality anyway. There is an evolutionary thrust towards the realisation of potential and the flourishing of life.

This thrust manifests itself most clearly as such in human consciousness. It develops its own evolutionary dynamic towards more inclusive horizons and more profound forms of realisation. It is oriented towards a dynamic vision of comprehensive optimal well-being that translates into concern for any deficiency in well-being in any dimension of life.

The performative character of the Christian proclamation

Following the approach of experiential realism, the point of departure for faith and theology cannot be the transcendent itself, nor the problematic means by which the transcendent is deemed to disclose itself (such as biblical metaphors, doctrinal propositions, or metaphysical speculations). What theology can access, describe, explain, critique, transform and proclaim is the *concept* of the transcendent. A concept of the transcendent is part of immanent reality.

How does it come about? According to the theological tradition it is the proclamation of the 'Word of God' that creates the Christian faith. A proclamation is not a descriptive but a performative statement. It is this character of the Christian message that changes human motivations, relationships and communities. It brings about a particular personal and communal 'spirit', that is, a particular structure and orientation of human consciousness.

The existential and social impact of the message is due to the fact that it is experienced as a reflection or 'icon' of the self-disclosure of the transcendent Source and Destiny of reality as such and as a whole, however provisional, partial and problematic it may be. It defines one's identity within the whole. It defines one's authenticity as a human being. It defines what one's life world ought to be. It cannot be ignored without forfeiting one's integrity. It is like a little barred window that allows the rays of the sun to fall into a prison cell and connect the inmates with the glorious world outside.

It is the impact of this message that forms the particular content of the concept of God in the minds of its recipients. God's proclaimed intentionality moves from God to humans, and through humans to the rest of the world, never the other way round. When proclaimed, the message of divine benevolence either imposes itself on human consciousness and conscience or it makes no impression.

The evolutionary origins of the message

Faith does not suck its concept of the transcendent Source and Destiny of reality out of the thumb. It is based on a message that comes from outside the consciousness of the recipient individual or community. This message crystallised out in a millennium of ancient Israelite history. It represents a particular strand within the wider evolutionary history of religious intuition. It goes back to the first inklings representatives of the species *homo sapiens* had of transcendence. It culminated in the Christ-event. It evolved through a process of collective discernment and worldview formation.[334] Its driving force was the proclamation of God's creative and redemptive response to changing human needs. It invited its recipient into God's creative and redemptive project.

In the Old Testament, God aims at the well-being of socially weak and vulnerable individuals and families in the face of antagonistic social and spiritual forces. Examples are the well-being of the clan of Abraham against infertility and vulnerability; the freedom of Israelite tribes from Egyptian slavery; the need of nomads for land to settle and cultivate; the need of Israel as a nation for protection against the aggressive intentions of other nations and imperial powers; the need for responsible governance against the oppressive and exploitative tendencies of elites such as kings, priests, the rich and the influential; the need arising from the spiritual fragmentation of the people through deified aspects of reality such as the Canaanite fertility cult; the spiritual subjugation of the nation under the gods of imperial overlords. In post-exilic times, the Jewish community was in danger of disintegrating when national rallying points such as kingship, temple and country collapsed. God became identified with symbolic representations and observances such as circumcision, the Torah, the Sabbath and the Passah.

The New Testament holds that God offered a human prototype of his engagement on behalf of enslaved and suffering humanity in the Christ-event, that is, the appearance, ministry and death of Jesus of Nazareth. Jesus assumed messianic authority to proclaim and enact the creative, redemptive and sacrificial intentionality of the God of Israel on behalf of humankind. Christians are persuaded that this assumption of authority was valid, that is, that it represented and represents the self-disclosure of the intentionality of God, the transcendent Source and Destiny of reality as such and as a whole. God grants participation in the new life of Christ through the 'Spirit of Christ'. The experiential core of the Christian faith is, therefore, the manifestation of the Spirit of Christ in the life of the 'Body of Christ', the community of believers.

The biblical faith has had—and still has—its own emergent and evolutionary history, which is part of the evolutionary history of the cosmic process as a whole. To borrow Kauffman's phrase, it is "at home in the universe". This historical process is not likely ever to come to an end. The biblical tradition relegates ultimate insight and renewal to the 'end of times' whatever that may mean. It is through this interminable process that the awareness of derivation and dependence, vulnerability and mortality, acceptability and obligation, guilt and reconciliation, vision and direction develops and finds its ever new and provisional coherence.

Expressed in theological terms, the biblical message is the form that the self-disclosure of the transcendent took and still takes within immanent reality. Theology is the critical and constructive attempt to come as close to an appropriate concept of the transcendent as possible. It is a task that, similar to the task of science, will never come to a close.

Does it make sense to believe that the structure and orientation of a particular human consciousness (that of Israel culminating in Jesus of Nazareth) has such a foundational significance for humanity as a whole? The answer is that not only all insight, but all of reality follows the pattern of emergence and evolution. There is always a specific beginning in space and time. It is characterised by high sensitivity to initial conditions that determines its evolutionary trajectory. This is what happened with the big bang, the emergence of life and the emergence of *homo sapiens*. Other convictions have come into being in similar ways. Evolution always leads to differentiation. This is nothing strange. The real question is whether the content of this particular conviction is valid or not. Christians are persuaded that it is and build their lives on it.

The immanent character of spiritual experiences

As individual and collective mental construct the Christian concept of the transcendent is part of immanent reality. It emerges and evolves. It can also deteriorate and decay. It can be more or less appropriate. It must at all times be subject to critique and reconceptualisation. African traditionalist, Jewish, Christian, Muslim, Hindu, Buddhist, Marxist, fascist, liberal humanist and naturalist worldviews can all be described, evaluated, transformed, or jettisoned. The Christian faith is not exception.

The theory of emergence allows us to locate the phenomenon of faith within the context of the processes of experienced reality. The Christian faith is, like similar phenomena, a particular form of 'spirit', that is, a particular shape of structured and oriented consciousness. There is nothing irrational, uncanny, or otherworldly about the proclamation of the biblical message, as long as one does not absolutise historically conditioned worldview assumptions and metaphors found in the biblical tradition and its subsequent appropriations.

The fact that all this happens within the human brain through neurological and chemical reactions does not, as such, question the validity of the assumption of

God's self-disclosure. All of human consciousness functions this way. According to the biblical faith, God 'acts' through immanent reality. Once one has internalised the proposition that the transcendent Source and Destiny of reality is intrinsically benevolent, one has no problem with seeing the entire cosmic and evolutionary process, including the functioning brain itself, as an outflow of God's creative and redemptive intentionality. And God would certainly utilise whatever processes 'he' had initiated within reality to manifest 'his' intentionality.

To use a picture, the automatic functioning of the engine, the gearbox and the wheels of my car does not question the fact that the car has been constructed to fulfil a particular purpose and that it can be used to drive towards particular destinations. The parts have their rationale within the functioning whole, which can be geared to a particular purpose.

Let us summarise

Part IV changes the perspective of science from within reality to the perspective of faith from beyond reality. However, this does not mean that we leave the world in which we are embedded. Believers cannot fathom God, the transcendent Source and Destiny of reality, as such. Rather, they are struck by the message of the creative and redemptive intentionality of God, which invites them to become part of God's creative and redemptive project.

This message emerged and evolved in the history of ancient Israel as a sequence of responses to human needs and predicaments. As such, it is part of evolving religious consciousness, which is again part of the evolving world process as a whole. The message itself, however, is not descriptive, but performative. That is, it creates faith in God's benevolent intentionality among responsive recipients.

This process can be rendered in scientific terms as the formation and orientation of human consciousness. God uses the human capacities of communication and comprehension to disclose God's intentionality to humans. The biblical faith thus cannot dispense with the assumption that God is a person for humans because humans are persons, although as the transcendent Source and Destiny of reality as a whole, God is much more than a person.

11

Faith and science — a classical model

In search of a conventional theological model that is capable of integrating 'best science', we may consider Martin Luther's experiential approach to theology. For Luther the reality we encounter in daily life reflects the power, but not the intentionality of God. It is explored through observation and reason rather than uncovered by revelation. In contrast, God's creative and redemptive intentionality is not experienced but proclaimed—and that in the face of all experiences to the contrary. The proclamation of God's benevolence is based on its opposite, the crucifixion of Christ. On the cross, God's unconditional benevolence manifested itself through the fate of Christ, God's messianic representative on earth. It reveals God's sacrificial commitment to a humankind engulfed in self-centredness and depravation. While science is given all the space it needs to explore reality, faith subjects all human pursuits to the criterion of God's creative and redemptive intentionality. Its message of suffering, transforming acceptance of the unacceptable leads to a sacrificial attitude towards a reality in travail.

The task of this chapter

Where are we in the argument? In Part II, we looked at reality, including the phenomenon of faith, from the immanent perspective of science. In Part III, we dealt with the tricky question of transcendence in general. In Part IV, we look at immanent reality, including its scientific interpretation, from the transcendent perspective of faith.

In the last chapter, I presented the approach of integrative transcendence, which looks at the very reality that the sciences explore 'with the eyes of God'. In this chapter, I present the theology of Martin Luther as an example of how a classical Christian theology can take scientific insights on board without losing its transcendent foundations.

Luther's theology as a model of faith-science relationships

Theology deals with insights that have emerged and evolved in history and that are communicated through traditions. Traditions can be redirected if they go astray or reconceptualised to respond more appropriately to changing situations and worldviews. To abandon the tradition as such, however, would jeopardise the identity and integrity of the Christian faith as an emergent and evolutionary phenomenon.

In my view, Martin Luther's theology presents a helpful framework for the interaction between science and faith precisely because, among classical theologians, he followed the experiential approach found in the Bible most consistently.[335] Of course, Luther's theology is pre-modern and extra-scientific. But this is true for all classical theologies.

Luther's case is remarkable in that the basic tenets of his theology are more amenable to a dialogue with science than most others of its kind. If we tease out these tenets, we will get a rather 'modern' Luther, a 'streamlined Luther' as it were. Because the aim is not to engage in historical studies, but to find a plausible contemporary approach, we should not hesitate to do exactly that.[336]

I begin with a bird's eye view of Luther's theological approach.[337] Then I offer a more elaborate exploration of the linkages between science and faith on the basis of Luther's approach.[338] Latin terms are added in brackets for the convenience of readers who are familiar with classical Luther research.

Section I
An overview of Luther's theology

God—manifest power and proclaimed intentionality

In previous chapters, I distinguished between the formal aspect of faith—the general assumption of God as the transcendent Source and Destiny of reality—and the substantive aspect of the faith—the typically biblical ascription of a benevolent intentionality to this God. These two steps can be recognised very clearly in Luther's theology.

Luther distinguishes between the *power* of God, which we experience in reality, and the *intention* of God, which is proclaimed on the basis of the biblical witness. It is this proclamation that creates our faith, rather than empirical experience, historical information, or metaphysical speculation. The proclamation is based on the Christ-event as the provisional outcome of the history of Israel.

The distinction within faith between experienced power and proclaimed intentionality is of critical importance for our topic. One, we are struck with awe at the immensity, complexity and regularity of what has come to be our world, including our own lives. Two, we are struck with awe when we realise what our world, including our own lives, ought to become if it were to gain its authenticity and that it will not reach authenticity on its own.

These two aspects can be found in various forms in most profound convictions. The specific contribution of the Christian faith is the way that the two are linked by the message of God's suffering, transforming acceptance of the unacceptable as manifest in the Christ-event. This is the essence of what Christians call the 'good news' or the gospel message.

The power of God

The experience of God's power in reality (*Deus in vita*) is highly ambiguous. Reality can be wonderful. It can also be horrific. It can even become demonic at times. Our experience of reality does not reveal God's redeeming love, but obscures it (*Deus absconditus*). However, humans are able to discern that experienced reality is governed by regularities such as natural law, social law and moral law (*usus civilis legis*).

Regularities found in reality are of God because God is the Creator of this reality. In this realm, we depend not on biblical revelation, but on our God-given gifts of observation and reason. Following these leads, we can attain secular wisdom, entrench it in legal frameworks (*usus civilis legis*), and gain public acceptability (*iustitia civilis*).

The intention of God

While in daily life we experience God's power but not God's intentions, the gospel message reveals God's intentions but not God's power (*Deus revelatus*). The gospel proclaims God's suffering and transforming acceptance of the unacceptable into God's fellowship (*iustificatio sola gratia*).

God's gift of acceptance stands in contradistinction to a demanding, inciting, judging and condemning law. The law makes God's acceptance conditional on our own moral achievement or excellent disposition. It reveals that we are out of step with God's creative and redemptive intentions. It shows us what ought not to be, but it cannot redeem us (*usus elenchticus legis*).

In contrast, the proclamation of the gospel is a promise. A promise refers to the future. The future is in the process of becoming. The content of the message—God's redeeming love—is not something we can observe, comprehend, or possess. It is hidden in the experiences of life but 'real' in our anticipation.

The message is not a descriptive, informative, or explanatory statement, but a performative statement. It becomes effective as it creates our faith. Only as far as we let ourselves in for it can we participate in its dynamic (*sola fide*). Faith is trust. Trust is self-entrustment. Self-entrustment to the promise of God's redemptive intentionality leads to our transformation because it involves us in the new life of Christ.

The proclamation is not based on an unsubstantiated postulate or baseless speculation. It is gleaned from the biblical . Or to be more precise, it is based on the Christ-event, which the New Testament considered the culmination of the relation between Yahweh and Israel. The gospel message is the outcome of a millennium of growing insight into the meaning of God's creative and redemptive intentionality.

For Luther, previous stages of the biblical tradition have to be measured against the criterion of the gospel. Only "what promulgates Christ" is the Word of God for us. Luther rather unconsciously realised that the theological content of the biblical faith was subject to an evolutionary process. There are stages in this process that did not make it in the long term, that were weeded out, as it were, by the continuing history of the Word of God as the dynamic redemptive responses to human predicaments.

Consequences of the message

The message of God's suffering acceptance of the unacceptable into his fellowship creates a new community. It is composed of those who have been accepted in spite of the fact that they are not acceptable. In this way, it creates a new set of relationships that are qualitatively different from the utilitarian and hedonistic relationships prevalent, for instance, in modernity. These are based on the fulfilment of conditions. Here acceptance is achieved, not granted.

It also changes the way we experience our world, namely, as manifestation of the creative and redemptive love of God. The existence of reality is not self-evident but extraordinary, unique and precious. Life is not something to which we are entitled, but a gift of grace. The eyes of faith see 'through' immanent reality and its apparent ambiguity and meaninglessness to discern the creative and redemptive intentionality of its ultimate Source and Destiny.[339]

The gospel message also defines our existential orientation. We do not take what ought not to have become as final. We move towards what ought to become according to God's redemptive intentionality. Faith draws believers into a movement oriented towards God's vision of comprehensive well-being.[340]

God's intention is revealed under the guise of its contrary

Why is the Christ-event deemed the culmination of the history of the relation between Yahweh and Israel? The proclamation of the gospel message is based on a particular interpretation of the birth of Christ in a manger and the humiliating death of Christ on the cross as found in the New Testament. Here the most horrific catastrophe is proclaimed to be the pivotal redemptive act of God. Luther places considerable emphasis on the fact that God's self-disclosure happens under the guise of its contrary (*sub contrario*).

This is important for faith, because it reassures us that the afflictions, predicaments and catastrophes we encounter in our lives do not question God's commitment to our salvation. The fateful occurrences of life are de-absolutised. They cannot destroy divine-human relationships. They do not have the last word. God's unconditionally redemptive intentionality has the last word. Faith overcomes fatalism, empowers bold action and opens up the future.

The message of the cross also implies that God does not identify with representatives of human success and glory (*theologia gloriae*), but with the guilty, the outcasts, the failures, the oppressed, the suffering and the dying (*theologia crucis*). God does so not to condone misery and depravity, but to liberate, redeem, transform and empower.

Faith is always afflicted faith

It is not immediately apparent that fateful occurrences such as the cross are part of God's creative and redemptive purposes. On the contrary, they seem to contradict such purposes. It is precisely this tension between experience and promise that constitutes a living faith. Faith is a stubborn refusal to grant negative experiences ultimate significance in the eyes of God. It is always a struggle against the 'hidden God' in the name of the 'revealed God', who is again hidden in the cross.

Luther believed that a faith that is not afflicted is not a genuine faith.[341] Affliction arises from experiences of meaninglessness, blind fate, human failure, fear of eternal condemnation, the depravities of humankind, suffering and death. Faith is trust in God's grace. Trust based on the message of the cross is an obstinate refusal to take no for an answer.

Faith sticks to the divine promise, come what may. God is for us and with us and not against us—and that precisely in the midst of agonies and calamities. Faith is a constant struggle against hopelessness and despair. Faith is reassurance, never certainty. We never have God's intentionality in our pockets. We constantly have to fall back on the proclamation coming from outside ourselves.

The criterion of truth

Obviously there are other messages around that call for trust. They create convictions similar in form but different in content. This content may be misleading. If you trust in the wrong message, you may be worse off for it. Faith creates both God and idol, Luther said. Therefore we need a criterion of validity.

As mentioned above, for Luther this criterion was God's redeeming love as manifest in Christ. We are invited to participate in God's creative and redemptive project in the world. Our faith is genuine when we 'carry our cross' in the assurance of God's redemptive intentions. This criterion is valid both for our private lives and for the structures and processes of society.[342] Love needs to be hard where human hearts have been hardened.

It is the message of God's creative and redemptive love as manifest in the crucified Christ that constitutes the canonicity of the biblical canon. The biblical tradition is valid for us only because, and in as far as, it promulgates the gospel of Christ (*was Christum treibet*).

Where this does not seem to be the case, it does not place faith under any obligation, whether in terms of worldview or moral standards. This typically Lutheran way of looking at the biblical tradition is of critical importance in all spheres of life, including the relation between faith and science.

So much for the overview. Now let me explore the possible links between Luther's theological approach and the scientific approach to reality in some detail. To do so, I have to interpret Luther's statements in the light of their possible experiential counterparts.

Section II
Luther's concept of the hidden God

Martin Luther was a late medieval theologian for whom the existence of God was self-evident. Today the existence of God can no longer be taken for granted. It should be fairly obvious that the transcendent Source and Destiny of reality does not 'exist' in the way that immanent reality exists. It cannot possibly be part of the reality of which it is a source.

Under modern presuppositions, the question of the existence of God reappears in another form: is the universe closed in upon itself, or is it open to a transcendent Source and Destiny, however defined? Expressed in anthropological terms: Is the human being autonomous or dependent and accountable? These are perhaps the two most fundamental questions raised by the interface between faith and science today. We have dealt with them in Chapters 8 and 9. For Luther, they would have been silly questions to ask. Reality cannot possibly be self-contained and humans cannot possibly be autonomous.

The awareness of derivation, embeddedness, dependence, vulnerability, mortality, meaning, identity, belonging, acceptability, guilt, reconciliation, authority and hope cannot be bypassed very easily by any human being, not even by a scientist. How that awareness is filled with content and in which kind of language it is articulated is a different matter! Obviously there are notions of the transcendent other than Christian ones. The point is that there is nothing irrational about the assumption that reality is a derived and open system.

Luther's concept of creation

Assuming a universe open to a transcendent Source and Destiny, Luther developed a concept of God that was not common in the sixteenth century. It has been called 'dynamic universalism'.[343] Dynamic universalism assumes that God is the Source of experienced reality in its constant flux, its variability and its complexity. Luther writes:

> It is God who creates all things, who brings them about and upholds them by his almighty power and his right hand, as our confessions say . . . But if he is to create and uphold, he must be present . . . in every creature, in its innermost and outermost being, all around, through and through, above and below, behind and in front, so that there can be nothing more present, nothing more intimately connected with every creature than God and his power (Luther 1529 par 98 f).

Whatever exists and happens within us and around us is the result of God's ongoing creative activity (*creatio continua*). God's power does not just initiate reality (*creatio ex nihilo*). It does not just keep reality going. It *constitutes* reality at every moment of its existence and operation.[344] Therefore God's presence is experienced in the very fabric of life. It is inescapable and undeniable (*Deus in vita*). Human experience of reality is the experience of God's creative power.[345]

One only has to read Psalm 104 or Job 36: 24 to 41: 34 to discern the biblical sources of this assumption. Luther was not a systematic theologian, but a biblical scholar. The Bible uses a plethora of metaphors for its witness to God's creative and redemptive activity underlying experienced reality.[346] In all these cases, God acts through his creation, not apart from it.

The scientific equivalent to pervasive creative power is the concept of energy that operates according to certain regularities. The operation of energy and regularity does not *demonstrate* the existence of God, but *defines* what we mean when we use the concept of God. If we think we know who God is and then try to find him in reality we are busy with an imagined God, not with the real God.[347] As far as we know and ever can know, God's creativity is identical with the world process. In this sense, Kauffman's identification of God with the 'creativity' observed in the world process as a whole is dead right.

However, for faith, energy is the 'stuff' of immanent reality, not its transcendent Source. Natural law is the manual of its operation, not the one who set up the manual. If one would identify energy or the regularities according to which it functions with God himself, as Kauffman does, one would end up with a pantheistic worldview. Then the world process itself would be taken to be divine. In contrast, the biblical faith assumes that energy and natural law are derived from the creative power and intentionality of God.

One could similarly argue that energy is an immanent manifestation of God's creative power and natural law is an immanent manifestation of God's intentionality. But again, faith posits a power and an intentionality that is logically 'prior' to its realisation, rather than identical with it. The point is that science and faith look at the same reality, albeit from different perspectives and following different agendas—description, analysis and explanation on the side of science, reassurance, trust and responsibility on the side of faith.

Luther's concept of natural law

Science explores the structured behaviour of energy in time and space and tries to discern the underlying regularities. Luther would have had no problem with that. Although conceptually more inclusive than 'the laws of nature', 'natural law' has always been a concept of classical theology. In line with many Ancient Near Eastern insights, Old Testament traditions assumed a cosmic order instituted by God.[348] This order included all kinds of regularity: laws of nature, social structures and morality.

The scientific equivalent is the operation of different regularities found at various levels of emergence: probability, causality, propensity, teleology, intentionality. social organisation.[349] For Luther, there was no doubt that God's creative power follows certain discernable patterns. Following these leads, Luther believed, we are able to attain a secular kind of wisdom, entrench it in legal frameworks (*usus civilis legis*), and gain public acceptability (*iustitia civilis*) by submitting to it. Although Luther dealt with these issues in the context of public responsibility, his approach can easily be extended to cover experienced reality as a whole.

At the public level, natural law defines acceptable behaviour. But the rationale of natural law is redemptive love. Depending on the circumstances, love can be hard if need be. The task of princes, magistrates, the police, soldiers, judges and hangmen was to curtail evil so that humanity can prosper. By implication, this should be true for science, technology, commerce and the management of the consumer household. How feasible is this idea in terms of modern science?

Science and morality

Science generally believes that 'moral law' falls outside its sphere of competence. Value judgements are deemed subjective. Darwinians assume that morality evolved in response to survival pressures—which says nothing at all about validity, purpose, or vision. Meanwhile modernity at large takes the pursuit of self-interest for granted as an overarching principle of life. Many scientists implicitly agree with that assumption without giving the matter much thought.

The axiomatic character of this hidden value judgement is most pronounced in neoclassical economics. As we have seen in Chapter 4, it defines the human being as the *homo oeconomicus* whose profit and pleasure orientation is deemed 'rational behaviour'.[350] This is a formidable assumption to make. And it has formidable existential, social and ecological consequences. Naturalists tend to view the operation of the liberal economy as a normal phenomenon based on evolutionary principles.

This kind of value judgement also surfaces here and there in the natural sciences, often in the wake of the concept of the selection of the fittest. Richard Dawkins' choice of the metaphor of a "selfish" gene is an example. Here the selfishness of the basic unit of the evolutionary process is simply taken for granted; Dawkins just redefines this basic unit as the gene rather than the organism or the species because it is the gene that outlasts organisms and species.[351]

At this point, theology may want to challenge the sciences to think again. Luther categorically stated that any attitude or behaviour, whether private or public, that is not motivated by divine love stands condemned before God. Judgement presupposes a definite intentionality that operates as a criterion of authenticity. Traditional theology speaks of the 'last judgement'. We can reconceptualise it as an assessment of human behaviour based on the ultimate criterion of comprehensive optimal well-being. Does such a statement resonate with the world explored by science?

I want to argue that all forms of life display a determined thrust towards their own survival and the full realisation of their potential. But such a thrust would be self-defeating if the individual organism or species pursued nothing but its self-interest at the expense of its fellow creatures and its wider contexts. Life cannot exist without sacrifice of other life or life chances. No living being can do without drawing on the sacrifice of other such beings and contributing to the sacrifice to make the survival and well-being of other such beings possible.

Multiple instances of symbiosis, synergy and cooperation have evolved over time precisely to achieve the goal of survival and well-being. Predators kill only what is needed for their healthy survival. Biological reality is an integrated symbiotic network. No part can emerge, evolve and subsist on its own. This is not only true for the ecological system, but also for lower levels of emergence. Complex organisations such as organisms, molecules and atoms depend on the integration of their components in larger functional wholes. The dysfunction of any component within the whole leads to serious disruptions if not the disintegration of the latter.

In modern times, this consideration has become a matter of life and death. As I have argued in Chapter 2, humankind cannot allow ruthless rapaciousness to continue for much longer without destroying the infrastructure necessary for its own survival and prosperity. The vision of comprehensive optimal well-being is anything but irrational or 'subjective'.

Intersubjective verification

Science insists on the intersubjective scrutiny of its findings. Luther would have had no problem with that. The exploration of experienced reality and its regularities, he thought, belongs to the public domain and not to the arcane knowledge of faith. To understand and manage reality, you do not use the Bible as if it were the constitution of a state or a scientific textbook. Reality is accessible to human observation and reason, and there is no need for a 'special revelation' in this regard. Revelation only concerns the presence of God's creative and redemptive intentionality within experienced reality.

Confronted with a problem in the public sphere, Luther believed, you can draw from the collective experience and wisdom of humankind as a whole. This would include biblical examples, but also the works of classical antiquity. Of course, the pre-scientific insights of antiquity cannot be used as criteria of the validity of modern scientific findings. By implication, therefore, the sciences are free to explore the way reality emerges, evolves and continues to function and ask their peers to check their findings. Luther's approach gives science all the space it needs to pursue its agenda.

Granted, Luther was as critical of Copernicus as almost everybody else was during his time.[352] But initial aversion to shockingly novel ways of looking at reality is not extraordinary. It can also be observed in the history of the sciences.[353] Luther was also not free from biblicist assumptions. The historical-critical interpretation of the Bible as we know it today did not yet exist during his time. However, Luther shared the keen interest of the humanists in the exploration of classical documents. Luther's criterion of canonicity (what promulgates God's redemptive intentionality in Christ) paved the way for biblical research.[354]

Experienced reality hides God's intentions

In theology, it is the intentionality of God that matters. How can we know the intentions of God? We cannot glean them from our daily lives. Luther was deeply aware of the ambiguity of experienced reality. We experience both the abundant flourishing of life and the agony of suffering and death; providential protection and meaningless fate; unbelievable beauty and revolting ugliness; self-sacrificing love and ghastly atrocity. Even the testify to God's fiery wrath and God's redeeming grace. Believers find the gift of faith and gnawing doubt right within themselves. One is thrown to and fro from reassurance and hope on the one hand frustration and hopelessness on the other. This ambiguity throws faith into constant affliction.

Perceptive scientists are also aware of the profound ambiguities of reality, including the scientific pursuit.[355] Those who are not, are overoptimistic even in purely scientific terms. There is no cosmic evolution without entropy, no life without death.

Because these ambiguities expose human limitations and vulnerability, modernity prefers to ignore them as far as possible. Life continues as if there was no death. Economic growth continues as if there was no ecological destruction. But such attitudes of denial have a nasty habit of backfiring either in the short or the long term.

For Luther, these ambiguities were much more painful than for us today because, as a pre-scientific believer and a reader of the Bible, he deemed meaningless and fateful occurrences a manifestation of God's anger. According to Deuteronomy 28 or 30, fateful experiences reflect God's wrath. Suffering and death were deemed the 'wages of sin'.

Yet for Luther, this retributive kind of justice fundamentally contradicted the gospel of God's mercy in Christ. For him, it is precisely this contradiction that characterises the Christian faith. Faith stubbornly clings to the promise of God's forgiveness in the face of the accusations of the law.

The link between transgression and fate became problematic already in Old Testament times. The Book of Job is a prominent example of the agony that it caused in post-exilic Judaism. With Paul, Luther keenly realised that it was impossible to discern God's redemptive intentions in the existence of a law that incites, judges and condemns. By implication, God's redemptive intentions can also not be discerned in the fateful experiences of life. While God's power is manifest for all to see, God's intentions are not.

Making use of an Old Testament metaphor, Luther called the experience of the ambiguity of reality the 'hidden God' (*Deus absconditus*). God is not hidden because God is not present, but because we cannot discern what God is up to. It is very important to make this distinction. If Luther says that God is *hidden*, he still presupposes that God is the Source and Destiny of experienced reality. We just cannot read his mind.

The Old Testament expression says that God "hides his face". A parallel taken from Africa suggests that the awareness of this existential paradox belongs to the most primal layers of religion. Most African traditionalists perceive the Supreme

Being to be remote and inaccessible, but they still believe that all of reality depends on the power of this 'Supreme Being'.[356]

If modern people sense that God was merely *absent*, as some do, they still allow for the possibility that there is some higher causative factor behind experienced reality as such. Deists and agnostics of various persuasions are examples. Atheists, in contrast, have come to the conclusion that there is no God. If that could be shown to be true, the concept of the transcendent Source and Destiny of reality would be without substance. On the whole, this is the position of modernity in practice if not in theory.

For Luther, this was not an option. He simply could not imagine the creation without the Creator, nor the Creator without the creation. When God's power is experienced in ordinary life, however, God's intentions remain hidden. We should never try to detect God's purposes in the experiences of life, he warned. We should leave alone what we cannot know, because speculation may rob us of the assurance of God's benevolence as the foundation of joyful and fruitful life.

The hiddenness of God and the theory of emergence

In a sense, modern science has both demythologised and radicalised Luther's observation. It argues that we cannot discern any meaning in reality because cosmic evolution, the stages of emergence, the evolution of life and the operation of reality in general are not powered by intentionality, purpose, or teleology in the first place. Tsunamis, car accidents and cancerous growths have causes, but no purpose.[357] Regularities determine how reality operates, but they have no intrinsic meaning.

The theory of emergence gives us some clues in this regard. It is not too difficult to account for the lack of purpose in sub-personal reality. Although *agency* can perhaps be traced far back into the evolutionary trajectory,[358] and *purpose* certainly characterises the functions of components within wholes,[359] *intentionality* in the most profound sense of the word occurs at the personal level of emergence and only there. It presupposes the entire impersonal infrastructure and is embedded in the impersonal superstructure of social reality.

This means, in effect, that there can be no 'pure' intentionality or 'spirit' that is not subject to the constraints found in the material, biological and social world. Conversely it means that if the transcendent Source and Destiny of reality has no personal attributes at all, reality as such and as a whole is meaningless and without purpose whatsoever.

This has deep-going theological repercussions. The concept of God as the transcendent Destiny of reality corresponds with the concept of the ideal human being situated in an ideal world. Is the authentic human being pure intentionality, not constrained by time, space and energy? Obviously not! Why then should we define God as pure intentionality and activity? The concept of God as pure Spirit corresponds with a concept of the human being as a bodiless soul. The soul may be in charge of, or imprisoned in, a mortal body, but it is not constituted by this body. Behind this concept lies the hankering for a human freedom that is unconstrained by its non-personal infrastructure.

This is a Feuerbachian projection of unrealistic wishes into a non-existent heaven. The theory of emergence refutes this possibility: the mind depends on the biological infrastructure of the brain. It is also out of touch with the biblical tradition, which is unable to contemplate a soul without a body.[360] It is Platonism that believes in the liberation of the immortal soul from the mortal body, while biblical faith believes in a restored relation of the whole human being with God, its creator and judge. The concept of 'resurrection' implies a new kind of body, not a bodiless spirit.[361] According to Paul the *Spirit* of Christ permeates, liberates, renews and empowers the *Body of Christ*, the community of believers. It has found a new biological infrastructure and a new collective consciousness.

The link between the Platonic ideal of a bodiless soul and the modern ideal of a sovereign individual is subtle but clear. In a Platonic-Gnostic worldview, God was deemed the epitome of perfection. In Nominalism, this perfect and exclusively spiritual God morphed into pure intentionality, endowed with unlimited power and elevated far above the impersonal structures and processes of reality.[362] Modernity then appropriated divine sovereignty for the human being. Theology should abandon this idea not only on scientific, but also on theological grounds.

Divine predestination and scientific reductionism

Part of God's hiddenness, according to Luther, is divine predestination, the religious equivalent to scientific determinism. Obviously Luther could not have known the modern concept of causality in physics or the concept of evolution in biology and their respective reductionist derivations. His was a personalised kind of determinism that he found in the Bible, and that centred on the sovereignty of God. For Luther, the conviction that God was in charge of all dimensions of reality, including our personal and communal faith, was as fundamental as for Jesus, Paul, or the Old Testament.

Luther keenly observed that humans are not free to think or do as they pleased. They are ridden like donkeys by forces greater than their own.[363] For Luther, the rider is either Christ or the devil. Modern social psychology shows that this idea is not quite as removed from reality as one might suspect at first sight. Just think of peer pressure, the weight of ancestral traditions, or the readiness to conform in totalitarian systems. One should also mention the instincts located in the reptile brain, infant conditioning and various forms of addiction. Consciousness can be structured in beneficial or detrimental ways.

And yet according to Luther, humans are responsible. Responsibility presupposes freedom. Divine responsibility does not exclude human responsibility. Divine agency and human agency do not operate at the same level, whether in competition or cooperation with each other. Divine agency initiates and empowers human agency. In terms of emergence theory, human intentionality and agency can then subdue and override or reconstruct lower levels of emergence by way of 'top-down causation'.

Luther goes one step further than theological determinists such as Calvin and Theodore Beza, Calvin's successor in Geneva. Because original sin and predestination contradict the gospel of God's unconditional, suffering acceptance of the unacceptable, it is a manifestation of the *hidden* God. Believers should never build their faith, their interpretation of God's will, or their practical lives on such doctrines. If they do, Luther warns, they cannot help but lose their spiritual foundations. We should never allow ourselves to drift into the uncharted territory of speculation. What has not been revealed cannot be known and is also not necessary for our salvation.

The doctrine of predestination can be understood as a particular form of reductionism. One derives what exists and happens from an idealised abstraction. In this case, it is not the regularities that govern reality, as in the natural sciences, but the concept of divine omnipotence. Under the impact of Platonic philosophy, the idea of God's creative power has morphed from a pastoral reassurance to an absolutised, timeless and universal principle. With that, it has left its experiential base and hovered off into the realm of metaphysical speculations.

Scientific reductionism is also based on an abstraction from the dynamic vibrancy of experienced reality. Today the sciences struggle to come to an understanding of determinism that does not exclude human freedom. There is not only bottom-up causation, but also top-down causation. I have dealt with this problem in Chapter 6. Reductionist determinists in the sciences have to admit that it is they who freely decided on the initiation, the area and the procedure of their scientific pursuits. That the future presents us with options and that we are constantly engaged in decision making is a simple fact of life.

Similarly humans decide to respond positively or negatively to the message of the gospel, even though life histories, religious traditions and social environments seem to have an overwhelming impact on their decisions. God is not part of reality but its transcendent Source. God's agency works through creaturely agency, not alongside or in competition with it.

This contradiction thus needs a resolution both in science and theology. In Chapter 6, I analysed the difference between factuality, potentiality and actuality in historical processes. We are not free to change the present, which is the outcome of the past, because the past no longer exists. It is no longer accessible and cannot be changed. But depending on the degree of sensitivity to initial conditions, the parameters set by the past open up a wide range of possible futures. It is in this constrained, yet substantial window of opportunity that existing options can be selected and realised.

In sum, it is not just our moral transgressions that throw a dark cloud over God's intentions. Experiences that undermine faith in God's benevolence include pangs of fate, meaninglessness, oppression, injustice, frustration, suffering and death. This is the tricky question of providence. They include the gift of faith and the inability to believe. This is the tricky question of predestination. They include the question of how a powerful and loving God can allow earthquakes, wars and atrocities to happen. This is the tricky problem of theodicy.

It would seem, therefore, that there is a wide area of potential consonance between Luther's concept of the hidden God and the scientific explanation of reality. There are no serious clashes between the two. As a pre-scientific theologian, Luther had no interest in scientific analyses and explanations. He just attributed experienced reality—the reality explored by science—to God, its ultimate Source and Destiny. He did so for a theological reason. Faith cannot do without this assumption of the mastery of God because faith loses its existential rationale, its universal validity and ethical function if it does. It is the model he picked up from the biblical witness.

Section III
Luther's concept of the revealed God

God's intentions are proclaimed in the gospel

I distinguished between the formal assumption that the world is open towards a transcendent Source and Destiny on the one hand and the ascription of a benevolent intentionality to this Source and Destiny. So far, we have discussed Luther's version of the formal aspect, and now for the second. In daily life, we experience God's power but not God's intentions (the hidden God). In contrast, the gospel reveals God's intentions but not God's power (the revealed God).

This is the heart of Luther's theology. Following Paul, Luther formulated the gospel as *justification by grace accepted in faith, rather than by human achievement, excellence, or disposition*. Virtually all Protestant reformers followed his lead in this respect. Recent Lutheran-Catholic discussions have produced at least some congruency concerning this central issue.[364]

The doctrine of justification by grace accepted in faith was a particular contextualisation of the gospel that responded to the Jewish insistence on the validity of the Mosaic Law as means of salvation. During the Reformation, it was meant to refute the legalism of the Catholic Church of the time. But theology is not bound to the use of this legal terminology. In fact, the latter can easily lead to serious misconceptions.[365]

Following the parable of the prodigal son, I translate the gospel message from judicial into communal terms as *God's suffering, transforming acceptance of the unacceptable into God's fellowship.*[366] God, the ultimate source and master of reality, is unconditionally for us and with us and not against us. God accepts us into God's fellowship and invites us to become part of God's creative and redemptive project in the world. That is the gist of a Christian theology.

Can science recognise divine benevolence in reality?

At this juncture, it would seem, faith fundamentally contradicts the scientific view of reality. There is indeed not the slightest evidence of divine benevolence anywhere in the reality that science has so far investigated. It can be found neither in animals, nor in humans. According to Darwinism, altruism in nature and society occurs only because reciprocity and cooperation have survival value. Our common experience seems to affirm the idea that people, in fact all living creatures, are motivated by selfishness.

As his concept of the hidden God shows, Luther agrees that there is no such evidence. In science, a single deviation from an assumed natural law can falsify this law or an entire theory. In contrast, the assumption of an almighty and loving God at the helm of reality is 'falsified' a thousand times in every human life. So where precisely is the 'reality value' of faith in a powerful and benevolent God located, if it exists at all?

For Luther, it is precisely the perceived *absence* of God that demands faith in God.[367] If God's benevolent presence was there for all to see, faith would be redundant. The gospel is a promise, not a statement of fact. Its agenda is reassurance, not explanation. It refers to the future, not to the present. It is not an axiomatic, nor a descriptive, but a performative statement. It creates faith in the validity of what ought to be, thus the illegitimacy and transience of what ought not to be.

God's benevolence is proclaimed, therefore, not on the basis of experienced reality, but in the face of experienced reality. The proclamation is not an analysis or legitimation of existing structures and processes. It is a call to become involved in God's creative, redemptive and transformative project. It does not take reality for granted, but challenges us to change it. The observations of science stand—in their naked brutality. But as far as it is not deemed authentic, existing reality is not accepted as ultimate or legitimate. Reality is confronted with a vision of comprehensive optimal well-being and drawn into a process of transformation.[368]

That is why I argued in Chapter 9 that 'pure' science without a transcending vision is fatalistic. The proclamation of God's unconditional benevolence is accepted by believers, not because of empirical evidence or logical inference, but as an act of entrustment and involvement. Only as far as we let ourselves in for it can we participate in its dynamic (*sola fide*).

Faith is not certainty, nor belief, nor assent to a set of propositions. Faith is **trust**.[369] More precisely, the believer entrusts herself to a movement that goes in a particular direction. Only through participation in God's redemptive project does the promise of redemption gain its plausibility, its reality, and its concrete impact on reality.

The grounds for the proclamation of God's benevolence

The truth of the Christian proclamation cannot be established by exploring how reality came into being and how it continues to function, but only by allowing one's life to be motivated by God's creative and redemptive intentionality. God accepts an unacceptable reality so as to transform it from within. And yet for Luther, God's benevolence is not an arbitrary assumption or an unrealistic fantasy. You can easily be fooled by trusting in somebody or something that does not deserve trust.

According to Luther, the gospel message does not arise in human hearts, nor does it drop into our souls straight from heaven. It is rooted in the collective experience of a community of believers over a millennium of ancient history, as documented in the biblical . Luther emphasised that the Word of God is a word coming from outside our own hearts and our own particular situations. It is an 'external word' (*verbum externum*). It is proclaimed to us by somebody else in the name of the God of Israel as interpreted by Jesus of Nazareth.

In Chapter 6, I have argued that the biblical ascription of a benevolent intentionality to the transcendent Source and Destiny of reality is not the product of human fantasy gone wild. The vision of what ought to be is derived from the experience of what ought not to be. The criterion of what ought not to be is embedded in the evolutionary thrust that aims at life and the realisation of inherent potentials in all forms of life. It develops its own evolutionary dynamic towards more inclusive horizons and more profound forms of realisation.

The biblical tradition came about through a process of cumulative collective discernment and worldview formation.[370] Luther did not interpret Scripture in a fundamentalist way as unmediated and inerrant divine revelation. There is a criterion of truth within the , namely, the Christ-event. The gist of the Christ-event is the disclosure of unconditional divine benevolence.

Texts that do not reflect this truth are not constitutive for the Christian faith, nor do they bind the conscience of believers. There are innumerable texts in the Bible that do not merit the status of "God's Word for us" today. For Luther, it was the message of the suffering, redeeming and transforming love of God, as manifest in Christ that was the Word of God for us.

Expressed in my own terms, the biblical tradition displays a progressive crystallisation and radicalisation of redemptive responses to changing situations of needs and worldviews over a millennium of ancient Israelite-Jewish history. It culminated in the interpretation of the cross of Christ as the prototype of God's redemptive intentionality.

By implication, the continuing power of the gospel lies in its creative and redemptive response to human needs arising from concrete but changing conditions and situations and their changing interpretations. All material, psychological and social needs have a spiritual depth dimension, namely, the need for meaning, acceptability and authority.

The cross of Christ as a paradigmatic event

According to Luther, God's redemptive intentions manifest themselves precisely in humble earthly "masks" such as the temple in Jerusalem, the man Jesus of Nazareth, the church's proclamation and sacraments. The power of God's love manifests itself in what seems to be the exact contrary of redemption (*sub contrario*). The most profound indication of the character of the gospel is the fact that it is the humiliating message of the cross of Christ and not the promise of greatness and glory.

Luther distinguishes authentic theology as a 'theology of the cross' (*theologia crucis*) from the inauthentic 'theology of glory' (*theologia gloriae*). The fact that God is for us and with us and not against us has manifested itself precisely in God's identification with that baby in the manger and with that man on the cross. They are the paradigmatic signs or symbols of God's unconditional acceptance of the unacceptable into his fellowship.

Today we must find a theological interpretation of these historical occurrences that is in line with the gospel message itself. The gospel proclaims God's suffering and transforming acceptance of the unacceptable. The death of Christ was not a sacrifice given by a human being to God, but a sacrifice given by God to humans.

Forgiveness implies that those who forgive have to bear the cost of what they forgive. The parable of the prodigal son is a good example. In the vulnerability of his messianic representative, therefore, God exposed himself to the enmity of humankind.

The death of Jesus on the cross must be taken as the self-giving act of God manifest in the fate of his earthly representative. Identifying with his representative as he died on the cross, God exposed himself to human depravity, waywardness and enmity. And the meaning of the 'resurrection of Jesus' from the dead is that God affirms the validity of the proclamation of the God of Israel as a God of redeeming love by Jesus of Nazareth.

The human corollary to the true God is the true human being. The true human being is a genuine representative of God's intentionality, in contrast with 'normal' humans who assert and defend their individual and collective self-interests. Christ is the human being whose will merges with the creative and redemptive will of God. "Whoever sees me sees the Father" (John 12: 45; 14: 9), which also means that "Nobody has ever seen God . . . those who remain in love remain in God" (1 John 4: 12, 16).

The existential, communal and ecological significance of the concept of the true human being is that believers are invited to break open their narcissistic preoccupation with their self-interests, which Luther calls "being twisted backward into one's self" *(incurvatus in se ipse)*, and participate in the new life of Christ in fellowship with God. The new life of Christ means sharing God's vision of comprehensive well-being and God's concern for any deficiency in well-being in any dimension of life.

Consequences of a theology of the cross

The consequences are far-reaching. To begin with, the fact that God's redemptive intention is proclaimed under the guise of its contrary (*sub contrario*) responds to the experience of an afflicted faith. As mentioned above, affliction arises from a meaningless fate, failure to live up to God's expectations, fear of God's eternal condemnation, the agonies and depravities of humankind in the world, injustice, suffering and death. The message of the cross is that the redeeming love of God is present precisely in situations of human depravity and predicament.

I want to add that the affliction of faith also arises from the fact that science cannot confirm any trace of the redemptive intentions of a loving God in the reality it investigates—which is the only reality we know. So the Christian faith is a counterfactual assumption. It has nothing to show. On the one hand, that makes it vulnerable to an experiential-realist critique. On the other hand, it is precisely this vulnerability that constitutes its inner strength. It is not geared to perfection of any kind. God's redemptive acceptance is proclaimed, rather, as a response to the human need for transformation of the unacceptable.

The Christian faith maintains that the accepting and transforming intentionality of God is present among the illiterate, the failures in business, the rejects of society, the politically powerless, the guilty, the homeless, the jobless, the hopeless, the sick, the addicted, the oppressed, the exploited, the suffering and the dying. God's creative and redemptive intentionality is proclaimed precisely where it is least found—not to legitimate misery and hopelessness but to overcome it.[371] This is what makes faith in Christ indispensable as a corrective and transformative motivation within the imbalances caused by modernity.

In short, faith in the crucified Christ is the most radical protest against what ought not to be that one can think of. The gospel is a rebellious statement, a statement made in the face of the ambiguities of reality, in the face of meaninglessness, injustice, suffering and death. It does not contradict the valid insights of science and technology, but challenges science and technology to come aboard God's creative and redemptive project.

The mediation between power and benevolence

For Luther and the entire Christian tradition, there is only one God who is both powerful and benevolent. In a hidden way, Luther argues, God's redemptive love as manifest in Christ must be as omnipresent as his creative power (*ubiquitas*).[372] Luther based this view not on metaphysical speculations, but on a metaphor found in the New Testament. Christ went to sit "at the right hand of God".

In ancient parlance, the right hand was the seat of the prime minister or executive of a king. It is, therefore, a position of authority and power. For Luther, the right hand of God was identical with God's creative power.[373] God's creative power is, of necessity, present everywhere, otherwise reality would not exist. So if Christ was seated at the right hand of God, this can only mean that the creative and redemptive love of God, as manifest in Christ, is valid and present at all times, in all places, under all circumstances—and that precisely in the context of the operation of God's power in experienced reality.

It is here that Luther's dynamic universalism comes into play. In contrast with Calvin (a Nominalist), Luther believes that God is not in 'heaven', but heaven is where God is. God's creative power is everywhere. So God's benevolence must also be everywhere. Therefore Christ, in whom God's benevolence became manifest, must be everywhere. Universal authority in terms of God's creative activity translates into universal validity of God's redemptive intentionality.

In terms of Luther's theology, therefore, the Trinity can be expressed in a simple, non-speculative way: God, whose **creative power** is *experienced* within the whole of reality, and whose **redeeming love** as manifest in Christ is *proclaimed* in the gospel, is *effective* as God's creative and redemptive **presence** within the community of believers at all times and in all places. All three aspects are part of immanent reality.

The Trinity does not express the eternal harmony of a God somewhere in heaven, or the presence of the divine whole in its earthly parts, or the consummation of the whole of time in God's eternity. These ideas are inferences from speculative assumptions about the nature of God. They can only obscure the existential immediacy and seriousness of what happens here on earth when the gospel is proclaimed and believed.

The Trinity expresses the agonising existential struggle of a faith that clings to God's benevolent intentionality, as manifest in the Christ event, in the face of God's ambivalent power in experienced reality. According to Luther, faith appeals to God against God, struggles with God, and does not let God go until it is blessed. This is an experience that every afflicted believer goes through.[374]

The existential struggle of faith in no way contradicts scientific findings. On the contrary, it is occasioned by the very reality that the sciences explore. Experienced reality is both supportive of life and hostile to life. "Science subjects itself to the facts, faith rebels against them."[375] The promise of the gospel lures the human being to imagine, think and strive beyond the apparent limitations and necessities of reality. It undermines cynicism, fatalism and despondency. It gives human life, including the scientific enterprise, a rationale, a motivation, a direction—not towards mastery and ownership but towards a dynamic vision of comprehensive well-being.

Moreover, for Luther, the proclamation of the gospel—the message that God, the ultimate source and criterion of reality, is for us and with us and not against us—changes the way we see reality.[376] Meaninglessness turns into purpose. Depression turns into joy. Gloom turns into hope. Frustration turns into redeeming love.[377] One begins to see the loving hand of God in all of reality.[378] For Luther, this is not a theory; it is the kind of faith that sustains believers in their daily lives in all its dimensions and whatever the circumstances. Nor is it without an experiential base. Luther explains the first article of the Apostolic Creed as follows:

> I believe that God has created me together with all creatures; that he has given me, and sustains, my body and soul, eyes, ears and all my limbs, my reason and all my senses, together with clothes and shoes, food and drink, house and yard, wife and child, field, livestock and all my property, that he provides me daily and abundantly with all the necessities of this body and life, that he protects me from all danger, and preserves me from all evil. All this he does out of his pure, fatherly and divine goodness and mercy, without any merit or worthiness on my part. For all this I am bound to thank, praise, serve and obey him. This is most certainly true.[379]

Expressed in terms of the relation between faith and science, a believer begins to see that natural law is part of the creative and redemptive intentionality of God. Without natural law, reality could not function and intentionality would remain without effect. Moreover, material reality has a price. The entropic process provides the energy without which nothing could happen. Life presupposes death. Divine love is sacrificial. Authenticity does not depend on perfection, but on participation in the creative, redemptive, sacrificial and transformative project of God. This is a sober, down-to-earth kind of faith.

Section IV
Suffering acceptance and transformation

The proclamation of God's redeeming grace opens up a new life. God's suffering and transforming acceptance of the unacceptable reverses the logical sequence between transformation ('sanctification') and acceptance ('justification'). Contrary to what happens in society, acceptance comes first, transformation follows. The urge for transformation is maintained, but authenticity changes from a *condition* of acceptance to a *consequence* of acceptance (*iustificatio sola gratia*).

The gospel of God's suffering, transforming acceptance of the unacceptable stands in contradistinction to a demanding, inciting, judging and condemning law that makes acceptance conditional on moral achievement or excellent disposition. The law can only reveal our sin (*usus elenchticus legis*). It cannot redeem. Because we are invited to participate in the redemptive action of God, the reversal of the order between transformation and acceptance has immense repercussions in all spheres of life, including the scientific-technological enterprise.[380]

Here are a few examples: Spouses accept each other in spite of disappointed expectations, but strive to find a workable way together. Parents accept their children in spite of their immaturity and waywardness, but lead them in a process of character formation. Citizens accept their governments and administrations in spite of inefficiency and corruption, but insist on improvements. Scientists accept the boundaries of current insight, but attempt to extend them. Technicians work with inefficient gadgets, but upgrade them or design better ones. Employers accept the fallibility of their employees, but give them better training. National economies accept the scarcity of their resources, but try to augment them through trade relations. Believers accept the fact that they and the reality of which they form a part are not perfect, but try to make the best of it.

Acceptance happens in view of transformation. It has an inner rationale. There is a vision—the vision of comprehensive optimal well-being.[381] This vision translates into active concern for any deficiency in well-being in any dimension of life. Acceptance is an invitation to participate in God's creative and redemptive project. It is an expectation that motivates and energises; it does not oppress and humiliate, as the law does.

The insight that the believer is "at the same time justified and a sinner" does not formulate an ontological paradox, but indicates an existential struggle between the power of one's own inauthentic existence and the invitation to participate in the authentic existence of Christ. Luther calls the new life of faith a 'foreign righteousness' (*iustitia aliena*), because it belongs to Christ and we can only access it in faith.

In contrast to Melanchthon and Calvin, Luther does not propagate a "third use of the law" that prescribes a particular human behaviour to be followed after believers have been forgiven their sins. Involved in the new life of Christ and guided by the Spirit of Christ, believers 'know God's mind' and act accordingly. They need no law.[382] But believers also know that, beyond the sphere of participation in God's creative and redemptive love, humankind needs laws, rules and regulations to keep it on track (*civilis usus legis*).

One could argue that moral and statutory laws are the counterparts of natural laws, only at a higher level of emergence. Such provisions express, albeit provisionally, what should be deemed acceptable in private and public life.[383] Humans also need institutions and offices with teeth to enforce them. Such provisions are part of God's creative and redemptive intentionality, and Christians should be actively involved in formulating and enforcing them.

Because participation in the new life of Christ is a constant struggle against our old sinful nature (*simul iustus et peccator*), even believers need such rails to guide them. They can be gleaned from, or informed by, the combined wisdom of humankind, including the Old Testament and classical antiquity. They are not absolute in themselves but derive their validity and motivation from God's creative and redemptive love. However, divine love is given a concrete shape through our God-given observation and reason. This is the sphere of the political, legal, administrative and educational professions. It is the basis of all procedures and rules in all spheres of life.

The meta-criterion: God's vision of comprehensive well-being

As we have seen in Chapter 6, the fact that humans are 'condemned to be free', to take decisions, to account for their motives and live with the consequences of their actions, necessitates the development of criteria. Moral codes and legal systems are meant to set up basic parameters for acceptable behaviour. There are two pitfalls in this regard that Luther's approach is able to avoid.

First, such codes can become inflexible and counterproductive if the historical and situational relativity of their origins is not taken into account. An absolutised law can lose its redemptive and transformative rationale and become an oppressive idol. This has happened often enough in the history of Judaism, Christianity and Islam—and again in feudalism, imperialism, absolutism, fascism and communism. The New Testament is quite outspoken in its rejection of counterproductive traditions and observances (Gal. 5:1; Col. 2:16 ff). As Thomas Kuhn has shown, science also tends to absolutise what it considers to be non-negotiables, though much less so than a faith ostensibly geared to the idea of timeless validity or a sacred past.

Second, apart from their specific contents, all conditions of acceptance tend to enslave. This phenomenon operates in all cultures and in all dimensions of life. One has to toe the line of Marxist-Leninist or fascist ideology, otherwise one ends up in gulags and concentration camps. One is driven by the liberal--capitalist achievement norm, or one drops out of the formal economy. One has to conform to currently fashionable patterns of consumption, or one faces rejection and ridicule. One is terrorised by peer pressure, bound to the example of significant others and reference groups such as film stars and sports heroes. As a scientist, one is not open to the possibility that the underlying assumptions of one's discipline might be flawed.

For Luther, as for the New Testament, there is a singular, flexible and dynamic meta-norm, namely, participation in the creative and redemptive love of God as manifest in Christ. This is not a rule that is defined once and for all, but a new motivation, namely, the Spirit of God's creative and redeeming love in Christ. It is nothing we have to achieve. We are involved in, and empowered by, a dynamic movement other than our own, namely, the new life of Christ (*iustitia aliena*). It is clear, therefore, that Luther's approach is capable of doing justice to a reality in constant and accelerating flux, which is the reality we are faced with today.

Luther applied the basic Christian motivation of redeeming love to all aspects of life, including social relations, politics and the economy. Love is inherently public. According to the New Testament, God has a vision—the vision of comprehensive well-being. This implies that any deficiency in well-being in any dimension of reality is the target of God's concern, thus the concern of the community of believers. This approach has had far-reaching consequences for the way humans perceive and deal with reality, at least in the West.[384]

Suffering acceptance and sacrifice

Acceptance of the unacceptable implies suffering. This is true for both the accepting and the accepted party. God exposes Christ, his representative, to human, social and earthly imperfection and waywardness and suffers the consequences. That is the meaning of the incarnation and the cross of Christ. According to the New Testament, it is no longer humans that give sacrifices to God, but God who gives sacrifices to humans.[385]

When called into God's fellowship, believers expose themselves with Christ to the imperfections of reality. In doing so, they become part of God's suffering. They take the cross upon themselves. The purpose of doing so is not to reach perfection, but to be involved in transformation.

Once we translate the gospel of suffering, transforming acceptance from its narrow moral connotations into the wide spheres of physical, social, earthly and cosmic reality, it leads to surprising insights.[386] For one, the idea of perfection is exposed as what it is, namely, a conceptual fetish that has no reality content whatsoever. Our world is not perfect and never has been. Perfection can also not be achieved. It will never happen. It is a pipe dream.

The gospel is realistic. God accepts an imperfect world and suffers under its imperfection. The law of entropy is the price God pays (and we have to pay) for the possibility of an evolutionary process.[387] Shifts in the earth's crust that lead to earthquakes and tsunamis is the price God pays (and we have to pay) for an earth that allows for the development of life. Death is the price God pays (and we have to pay) for having life in the first place.[388]

Seen in this light, it is God, the Creator, Owner and Master of reality, who sacrifices so that we can live. But God also expects us to be involved in this sacrifice so that others can live. Unnecessary suffering, imposed deficiencies and premature death are void of meaning and must be overcome. Life should flourish.

Judicious frugality by the rich; suffering for others by the healthy; retirement from active duty by the elderly to make way for the next generation, and death at a mature age are part of God's sacrifice for the sake of the flourishing of life. As such, they are immensely meaningful. This is not only true for family, community and society. It is true for the whole of humanity and the rest of the natural world.

Once we enter these larger realms, faith quite naturally links up with science. Mass famine and ecological disaster can only be averted if a critical mass of humankind accepts the necessity of sacrifice and becomes willing to act upon it. This planet will not be saved by a message that legitimates the profit and pleasure seeking of "economic man", but by the sacrificial message of the cross.[389] There is a future to be won and—following the one-world assumption of the sciences—it is to be won right here in this world and not in a heaven of make-believe.

Suffering acceptance and the biblical scriptures

Another surprising consequence of the gospel of God's suffering acceptance of the unacceptable concerns the character of its biblical sources. The biblical message is unique and precious beyond measure, but its worldview assumptions have long become obsolete. Sharing God's suffering acceptance, we have to forgive our ancient ancestors in the faith that their view of reality was not up to the scientific standards and ethical norms applicable today. They could not possibly have known better at the time.

The doctrine of scriptural perfection and inerrancy has led to the standoff between scientists, creationists and fundamentalists. This is tragic. "Only the " (*sola Scriptura*) was, and still is, one of the pillars of Protestant theology. "Back to the sources" (*ad fontes*) was the humanist battle cry that aimed at the restoration of classical excellence in Western culture. The Reformers applied it to the Bible. "Only the " was a statement against the Catholic principle of " and tradition" because a tradition gone astray had come to dominate church and society.

For Luther, the emphasis on the has a decisive, but limited significance. The fundamental criterion of truth is God's suffering and transforming acceptance of the unacceptable in Christ—and that in contradistinction to a demanding, inciting, judging and condemning law. Whatever does not reflect this dialectic is not binding for us. In contrast to Catholicism, Calvin and the Protestant Orthodoxy of the seventeenth century, Luther applies this criterion not only to the ecclesial tradition, but also to the biblical themselves.

Luther does not do so consistently, and there are lots of biblicist arguments in his works. But Luther's approach as such is not based on the irrational assumptions of biblical fundamentalism. It was only during the aftermath of the Reformation that Protestantism developed a rigid concept of revelation. It was based on the assertion of the inerrancy of the as counterpart to the assumption of the infallibility of ecclesial authority in Catholicism. Both approaches are fundamentalist. As can easily be shown, both of them are intrinsically flawed.

They also contradict the biblical witness. According to the Bible, humans are not perfect, infallible, or inerrant. This is demonstrably true also for their assumptions, interpretations and actions, whether they are derived from 'divine revelation' or not.

Revelation, if it takes place at all, can reach humans only through the mediation of human understanding. This realisation, self-evident as it may seem, is of decisive importance for overcoming the fruitless conflict between the 'pseudo-science' of 'creationism' and 'true science'.[390]

The Lutheran approach has no problems with recognising the metaphorical or mythological character of biblical statements concerning the origin, character and destiny of experienced reality. It can also concede without hesitation that Ancient Near Eastern worldviews are pre-scientific and have now become obsolete. The right theological procedure is to deduce the nature of 'creation' from the demonstrable facts of evolution and not the nature of origination from the ostensible facts of creation as depicted in the Bible.

A critical theology will try to unearth the theological intentions behind biblical statements on creation and eschatology and subject them to theological scrutiny. Biblical 'revelation' cannot substitute for scientific research and insight. The biblical tradition itself was flexible enough to constantly reconceptualise inherited patterns of thought in response to changing situations and worldviews.[391]

Suffering acceptance and other convictions

There are other notions of the transcendent and their definitions of what ought to be. More than ever before, humanity now finds itself in a pluralistic situation. Luther keenly observed that such notions are also based on "proclamation", that is, on intuitions, traditions and performative statements, rather than on empirical evidence or logical deduction. They are also accepted in 'faith'. Faith makes both gods and idols, Luther famously remarked. This shows that he did not confuse our concepts of the transcendent with the transcendent as such.

Notions of the divine, or 'ultimate truth', are part of immanent reality. They are subject to scrutiny, critique, correction, transformation, or rejection. If that were not the case, theology would be pointless. Theological and religious conflicts of all times would not have happened. Christians do not agree on what is to be considered divine truth. There are also traditions other than Christianity, each with its non-negotiable assumptions.

Idols are absolutised entities or assumptions. Science follows a particular rationale that can be absolutised. What is variously called 'naturalism', 'scientism', or 'physicalism' is no exception. Darwinism can be absolutised. There are social ideologies that legitimate the pursuit of collective interests. There are personal desires that enslave people spiritually. For Luther, the question is not, therefore, whether God exists. The question is to *what kind of God* we entrust our lives and *what this God does to us*.

What are the consequences of a Marxist-Leninist, a liberal-capitalist, or a fascist worldview? What are the consequences when 'scientism' becomes the religion of a 'secular' West? What are the consequences of an absolutised human being whose "will to power" (Nietzsche) is all that counts? What are the consequences of the definition of the human being as a profit and pleasure maximising 'economic man' for whom society and nature is nothing but a quarry to be mined in the pursuit of profit and gratification?

These examples show that humankind cannot afford to suspend the struggle for the truth. Not all notions of the transcendent are equally valid. History should have taught us by now that this struggle should not take the form of totalitarian impositions of religious or ideological certainties on a population, nor brainwashing techniques, nor demagogic rhetoric, nor fighting it out with modern weaponry. However, we must all expose our convictions to alternative convictions and allow the truth to manifest itself.

Luther was adamant that consciences cannot be forced into a mould or commandeered for alien purposes. That was quite naive. Conscience is not the infallible voice of God in our hearts. It is part of the structure of consciousness, which is determined through initial socialisation and ongoing social dynamics. The normal individual tends to internalise assumptions, values and norms that are currently in vogue. Conscience is also subject to individual rationalisations and ideological manipulations. Persuasion is used regularly and pervasively in politics, marketing and evangelisation—often with sophisticated socio-psychological means. We have to be alert and expose forces that push the population in harmful directions.

God's suffering, transformative acceptance of the unacceptable translates into religious tolerance. However, it must be understood that Christian 'tolerance' is not the same as theological or ethical indifference. In the original sense of the word, 'tolerance' means 'endurance', 'forbearance', or 'carrying a burden'. Tolerance entails suffering. As Paul has it, to 'fulfil the law of Christ, we have to bear each other's burdens' (Gal. 6: 2). We must even bear each other as burdens. One cannot claim that Lutherans have been very good at recognising this fact.[392]

More than that, Christians have no monopoly on valid insight. Throughout its history, the biblical faith absorbed, critiqued, transformed and integrated insights from its spiritual and cultural environments. We can learn from other convictions. They can expose neglected issues and correct untenable assumptions within our own faith. They can make contributions to the insights of others. This includes not only the sciences as such, but also 'naturalism' and 'scientism' as convictions.[393]

But we can, and we must, scrutinise and challenge the convictions of others, as much as our own, against the vision of comprehensive optimal well-being. We must do so in a spirit of respect, yet with the candidness demanded by a serious struggle for the truth. This is not a sideshow or a hobby that we can just as well do without. The future of humanity and the natural world depends on the transcendent assumptions with which we approach reality.

Conclusion

Luther's experiential theology shows that scientific insight can be integrated into the worldview of faith without forfeiting its rationale. His concept of the 'hidden God' caters for the fact that our experiences of reality reveals the operation of energy according to certain regularities that can be explored and utilised by making use of the human gifts of observation and reason. Luther's approach thus allows for all the space the sciences need to conduct their research. It also allows for the responsible utilisation of this research.

Yet the experience of reality does not reflect a creative and redemptive intentionality. On the contrary, reality is experienced as highly ambiguous, even demonic at times. The entropic process in cosmic reality as a whole and the inescapable necessity of death for life are the most disconcerting facts for any faith in the creative and redemptive intentionality of a God deemed to be in charge of this reality.

For Luther, God's creative and redemptive intentionality is proclaimed precisely in the face of the ambiguity and vicissitude of experienced reality. This is done on the basis of the biblical witness. The latter culminated in the exact opposite of comprehensive well-being, namely, the cross of Christ. It is in the manger and on the cross that God's sacrificial and transformative acceptance of the unacceptable manifests itself. It is here that the Christian faith is anchored and nowhere else (1 Cor. 1: 22-2:8).

So the disclosure of God's benevolent intentionality has at least three consequences for our approach to the reality we experience:

1. God's creative and redemptive intentionality encounters reality precisely in the problematic situation in which it finds itself and leads it towards God's vision of comprehensive optimal well-being as far as that is possible under any set of circumstances.
2. The message of God's suffering, transformative acceptance of the unacceptable is based on a tradition that emerged and evolved in history in response to incessant meaninglessness, injustice, suffering and death. As the transcendent Source and Destiny of reality, God suffers under the constraints, sensitivities and necessities imposed by the regularities that make reality possible and the abuse of the freedom without which humanity would not be human.
3. The message of God's suffering, transformative acceptance of the unacceptable leads to a new understanding of reality as a whole. For the believer, reality displays God's benevolent intentionality in the form of an evolutionary thrust towards the realisation of its full potential. This thrust points powerfully towards a vision of comprehensive optimal well-being of reality as a whole.

12 Faith and science — a modern approach

> The perspective 'from above' focuses on the very reality that the sciences explore. Correctly understood, therefore, 'best science' corrects and enriches traditional faith assumptions. (a) The *biblical tradition* does not convey a timeless and inerrant truth, but displays an evolutionary dynamic that responds creatively and redemptively to changing human needs and interpretations. (b) *God* is not a factor among others within immanent reality, but its transcendent Source and Destiny. (c) The biblical projections of *what ought to be* to the 'beginning' of time, the 'end' of time, an alternative space 'above' or a hidden 'essence' should not be confused with empirical descriptions of what was, is or and will be. (d) The concept of *'creation'* refers to the creative power of God. (e) The concept of *'eschatological' fulfilment* refers to the redemptive intentionality of God. (f) A *miracle* is an unexpected, awe-inspiring and redemptive event, but it does not imply the suspension of natural law. (g) The theory of emergence throws light on the problem of *theodicy*: God is a person for humans but he is also the Source of all non-personal levels of emergence. (h) The significance of the *human being* must be seen in relation to its vast earthly and cosmic contexts. (i) The time span of individual human consciousness emerges from a vast causal network and spawns a vast network of consequences. What matters is not its infinite prolongation, but its contribution to God's creative and redemptive project. (j) *'Christ'* is the authentic human being through whom God manifests and enacts God's redemptive intentionality and in whose life we are meant to participate. (k) The substantive content of this new life is God's suffering, transforming acceptance of the unacceptable. (l) The Trinity refers to God's creative and redemptive intentionality that manifested itself in Jesus of Nazareth and continues to be present and effective as the 'Spirit of Christ' that permeates, transforms, motivates and empowers the 'Body of Christ'—the community of believers.

Introduction

The need for a reconceptualisation of the Christian message

In the last chapter, I presented the experiential theology of Martin Luther as a potential framework for the integration of faith and science. It demonstrated that such an integration is possible even on the basis of a traditional Christian theology. However, we no longer live in the times of Luther, Calvin and the Council of Trent. For faith to become plausible and relevant for our scientifically informed contemporaries, we must reconceptualise the Christian message and do so much more boldly.

In view of the predicaments faced by humanity today, such a reconceptualisation has become a necessity, rather than an intellectual pastime. Paul believed that he had to become "all things to all people" if he wanted to be a participant of the gospel (1 Cor. 9: 19-23). God's Word always enters human reality, picks up humans where it finds them—in their worldviews, their commitments and their failures—and leads them a few steps forward towards his vision of comprehensive optimal well-being.

Can believers in general and theologians in particular become 'scientists to scientists'? The Christian faith has been badly intimidated and marginalised by the challenges of scientific insights and the dismissive certainties of secular modernity. It has withdrawn into the comfort sphere of its traditional certainties. It must regain its confidence, boldness and joy. We have a message to convey. That message is of critical importance at all times, and in modern times more than ever before. But we can only do so if we do not cling to propositions that no longer hold water.

I am conscious of the fact that reconceptualisations may rock the spiritual foundations of many a dear and sincere believer. Private individuals and faith communities may want to maintain and cherish their spiritual heritage and nobody wants to prevent them from doing so. But believers and theologians have a task to fulfil in this world.[394] They have to blaze a trail into the future, rather than nursing past certainties.

Science collides head-on with traditional worldview assumptions that have been superseded by modern discoveries and explanations.[395] The earth is not flat. The human male was not created before the animals, and the animals were not created before the human female. A mega-city cannot have the form of a perfect cube and descend as such from heaven.

Astrology changed into astronomy, alchemy changed into chemistry, the sword changed into the intercontinental ballistic missile and the ox wagon was replaced with the jumbo jet. Theology changed into—what?[396] Theology must get its act together or continue to be ridiculed, sidelined and ignored. The integrity of the community of believers and the credibility of its message are at stake. Self-delusion and make-believe are not Christian virtues!

If we insist on a literal-ontological interpretation of the biblical tradition, we will leave scientists worth their professions puzzled and, more likely than not, outside the fold. The vast majority of the population in Western societies have already voted with their feet. This fact alone should cause us sleepless nights.

Science and technology depend for their very existence on conceptual precision, empirical evidence and mathematical stringency.[397] The sciences are capable of *illuminating and enriching our faith*, rendering our perceptions of the world more precise and making our assumptions more credible. As far as the sciences reflect objective truth, they cannot destroy a faith that claims to stand for the truth. The observational and interpretative prowess of the scientists is a gift of God that is meant to be utilised. The world that the sciences describe is the real world, the world that God created.

Reconceptualisation has always been the task of systematic theology as an academic discipline, conducted on behalf of the community of believers. Systematic theology is experimental thinking.[398] One chooses a particular approach and sees how far one gets. Countless forays have been made in the direction of integrating 'best science' into the Christian worldview, some more appropriate than others.

In what follows, I make a few rudimentary, tightly argued suggestions on how I think it could be done. It is a systematic theology in miniature, if you will. I will indicate how my contentions link up with the biblical sources, because Christians of all persuasions read the Bible, and most problems we have today in the science-faith interface stem from an inappropriate reading of the Bible. A discussion of the massive post-biblical and contemporary debate is impossible in such a short chapter.

Section I
The foundational sources of Christianity

The assumptions of the scriptural inerrancy found in certain strands of Protestantism, on the one hand, and doctrinal infallibility found in Catholicism, on the other hand, present insurmountable obstacles for people who have internalised the post-Enlightenment criteria of empirical evidence, historical plausibility and rationality.

Both these assumptions have arisen in the history of the church to establish and legitimate episcopal or pastoral authority. They continue to be popular because the faithful long for stable spiritual foundations. However, both assumptions severely compromise the integrity of the believers themselves and the credibility of their message. What is more, both of them are out of sync with the character of the themselves.[399] The history of the biblical witness reveals a process of constant reconceptualisations in response to changing situations and worldviews.

We must rid ourselves of demonstrably false assumptions and counterproductive precepts. To be a Christian, you do not **have** to believe that the sun stood still for an army to win a battle (Joshua 10:12ff),[400] or that a decaying corpse was resuscitated after three days in the grave (John 11). Such claims cannot survive critical investigation as demanded by science. And you **should** not believe that a father has the right to sell his daughter into slavery (Ex. 21:7f), that a nation—any nation!—has the right to conquer foreign territory and drive out its previous inhabitants (Deut. 7), or that women are not meant to teach and preach in the church (1 Cor 14:33-36). Such precepts are out of step with the love of God as manifest in Christ.

It is typical of the biblical that they convey a story, rather than a collection of dogmatic propositions or a code of legal stipulations. Essentially it is the story of the relation between Yahweh, the God of Israel, and Israel, deemed the chosen people of God, culminating in the Christ-event. This particular story is located in the greater story of the relation between God the Creator and humankind and embedded in the history of the earth and the universe.

This story is composed of different elements, some of which originated as independent traditions and were subsequently combined to form an ongoing sequence of events. Main elements are the promises to the prime ancestors of Israel, the exodus from Egypt, the desert wanderings, the Sinaitic covenant, the conquest of the land, the squabbles with neighbouring tribes, the unification of the nation under the institution of the kingship, the successive destructions of the Northern kingdom by the Assyrians and the Southern kingdom by the Babylonians, the dispersion of Israel and the Babylonian exile of Judah, the re-establishment of the cultic centre in Jerusalem under the priesthood, the Christ-event and the formation of the Christian church. This story is embedded in a mythological prehistory going back to creation and a post-history leading to an eschatological consummation.

While the themselves have emerged and evolved in actual history, they represent a level of discourse that goes beyond a purely historical account. The rationale of the contents of their message is existential and social significance rather than historical fact. Of primary significance in this regard is the unfolding drama of the relation between God and humanity. Everything else is stage management. The relation between God and humanity is characterised by God's creative and redemptive benevolence responding to concrete human needs and interacting with human waywardness and depravity. Throughout, the story has a redemptive agenda.

Because it is existential and social significance that matters, rather than historical fact, the authors use all communicative tools at their disposal—myth, legend, fiction, poetry, parable, metaphor, miracle story, vision, symbolic action and ritual. The depictions of the history of the kings of Israel and Judah, for instance, explicitly refer to the official annals of the kings found in the archives (2 Kings 21:17). Their own agenda is not a historical account, but a theological assessment of a particular reign.

Moreover, because the biblical authors want to communicate a message to their contemporaries, rather than reiterate past events, they do not feel bound to the exact wording of the traditions they inherited but freely change the contents to make the message more persuasive or more applicable to new circumstances. That is why John's gospel, for instance, deviates so drastically from the Synoptic gospels.

They also take motifs gleaned from their cultural and religious environments on board, critically integrating them into their own view of reality. That is why you find two completely different creation narratives in Genesis 1 and 2. In both these respects, they resemble the preachers of all times who interpret biblical texts for their congregations in ever fresh ways.

The biblical Canon

Because it responds to changing human needs and worldviews, the creative and redemptive thrust of the biblical message is subject to an evolutionary process. Formulations of the will of Yahweh, for instance, changed from moral excellence in Deuteronomy, to ritual purity in the Priestly Source, to life skills in Wisdom literature, to human involvement in the love of God in the writings of John. The underlying motive is always the creative and redemptive intentionality and agency of God.

The dynamic thrust of the message implies that the are, essentially, open ended. By their own account, they are heading towards a revelation of God 'from face to face' in the eschatological future. Because the message responds to different situations, it differentiates into separate traditions. This differentiation already began in biblical times, for instance, between the prophetic and the priestly tradition in the Old Testament, or between the approaches of Paul, the Synoptic gospels, John and the Letter to the Hebrews in the New Testament.

To maintain some kind of common basis, contain excessive proliferation and forestall heresy, some considered to be foundational were bundled together in the biblical canon and declared authoritative. But that measure did not arrest the evolutionary thrust. Once contained in a canon, the dynamic of the message explodes into 'polysemy', which is the differentiated interpretation of given texts.

The composition of the canon has become accepted by all of Christianity. It is indispensable as a common frame of reference among Christians. But that does not remove the necessity of critical investigation, constructive interpretation and contemporary application. The biblical act rather like the barrel of the gun that is short, but open in front. It is meant to guide the direction of the bullet (the 'Word of God') on its way to distant destination.

The performative nature of the message

For the relation between faith and science, it is imperative to recognise the difference between the descriptive statements of science and the performative statements of faith. The biblical message does not present us with observations, explanations and predictions, but warns the wayward believer of the consequences of injudicious actions and reassures the afflicted believer of the benevolent inten- tionality of God in the face of experiences to the contrary.

It is the *preached* 'Word of God' that creates faith in the consciousness of the believing community. It calls its hearers to entrust themselves to God and invites them to participate in God's creative and redemptive project. Doing so, it opens up a vision that goes beyond the constraints of experienced reality and overcomes hopelessness, fatalism and despondency.

A more subtle obstacle for scientifically informed people is the assumption of divine perfection in the Christian tradition. Reality, as exposed by the natural sciences, is in evolutionary flux and subject to the law of entropy. It is not perfect and cannot be. Perfection cannot even be imagined, let alone demonstrated or expected to happen.

The idea of perfection has a Greek, rather than a biblical ancestry. Platonic thought abstracted essence from existence, eternity from history, universality from concrete situations in space, and harmony from the power games found in experienced reality. Pre-existent ideas trapped in fleeting matter are due to return to their eternal origins. An immortal soul is imprisoned in a mortal body, only to be released upon the death of the latter. Aristotelian thought replaced the pre-existent 'idea' with a 'form' that strains to acquire perfection, thus overcoming the imperfections of its material encasement.

The biblical tradition, in contrast, is geared to transformation rather than perfection. It emerged and evolved in a millennium of ancient Israelite history. The thrust of the message was constituted by the ongoing creative and redemptive response of the 'Word of God' to changing human needs and predicaments in all dimensions of life. By its very nature, this message presupposes the human experience that earthly reality in general and human life in particular are everything but perfect. They are in need of redemption, transformation and empowerment. A god who would insist on perfection would have to dispose of an imperfect world.

The redemptive content of the message

It is the redemptive thrust of the biblical message that powered the process of individual and collective consciousness formation in biblical times. Insight into the benevolent intentionality of God, the transcendent Source and Destiny of reality, acquired ever wider horizons—from clan, to tribe, to nation, to all nations, and to all cosmic powers.

As it encountered different kinds of needs and predicaments, it became ever more differentiated—from fertility, water and pasture, to social cohesion, national identity, judicial integrity, political stability, military strength, international peace and economic prosperity, individual integrity and existential authenticity.

It also became more profound. The character of the royal representative of God changed from that of a ruthless tyrant (Ps. 2) to that of a servant of his subordinates (Mark 10: 35-45). The sacrifice of the human firstborn to God changed to the sacrifice of the divine firstborn to humanity. The biological ancestor of Israel became the father of the faith of all people.

Most profound, however, was the change from conditional to unconditional acceptance of humans into the fellowship of God. According to deuteronomic law, obedience led to the blessing of Yahweh and disobedience to the curse of Yahweh (Deut. 28 or 30). Divine acceptance presupposed human righteousness. Jesus, in contrast, proclaimed and enacted the suffering, transforming acceptance of the unacceptable into God's fellowship. That was the message developed by Paul and John, the most influential authors of the New Testament.[401]

The cosmic significance of the message

Today we have to release this message from its traditional confinement to the spiritual needs of the individual human being. Forgiveness of guilt indeed implies the sacrifice of what is forgiven by the forgiver. But God's sacrificial acceptance of the unacceptable has comprehensive dimensions.

Evolution presupposes entropy; life implies death; higher organisms can only survive by ingesting other organisms; human life necessitates the death of plant and animal life; the validity of an insight can only be accepted at the expense of rival assumptions; the greater needs and interests of a community or society curtail the needs and interests of the individual, and the health of the natural world can only be maintained if humanity is willing to make definite and substantial sacrifices.

The sciences have exposed the inexorable character of regularities that govern the cosmic process and it is imperative that faith and theology take it on board. If God is understood to be the transcendent Source and Destiny of reality as such and as a whole, God's suffering, transforming acceptance of the unacceptable is the pervasive and non-negotiable precondition for the existence of reality as such.

Human life and everything else emerged from the creative and redemptive project of God and will eventually remerge into it.

The reversal of the order from transformation as *precondition* of acceptance to transformation as *consequence* of acceptance has immense repercussion for the way we deal with others, community, society and nature. The world is not perfect and never has been. It means that the world process must be accepted, as imperfect as it may be, and channelled into the most wholesome directions available at any point in time. Acceptance implies forbearance, patience, responsibility and the willingness to suffer for the sake of a transformation that goes in the direction of comprehensive optimal well-being.

The sacraments—the tangible message

Traditional perceptions of what happens in religious rituals such as the sacraments present particular difficulties to people informed by science. They also puzzle countless contemporary believers. It must be emphasised, therefore, that the sacraments are meant to be enactments of the message of the gospel—a 'visible word' as the Reformers called it—rather than the distribution of mysterious substances.

Let me use a picture. In the case of a bank note, an intangible meaning (say the value of 100 Rand) is identified with a piece of paper. This is done by somebody who has the authority to do so, namely the Governor of the Reserve Bank. The intention is to make the transfer of the intangible value possible. The substance of the paper does not change, nor is another substance added. But if I burn up the paper, I lose the value identified with it.

Similarly, the sacraments are meant to transfer the intangible meaning of the self-giving act of God on the cross of Christ to individuals and communities. The message is identified with water, bread and wine by someone who has the authority to do so. In this form it can be allocated concretely to its recipients. The substances of water, bread and wine remain what they are.[402] No substances are added.[403]

Section II
The transcendent presence of God

It is not by accident that I deal with the concept of God only now. It is the validity of the message that calls for ultimate authority. Ultimate authority can only be located in the transcendent Source and Destiny of reality as such and as a whole. The concept of 'source' refers to the dynamic that powers the world process; the concept of 'destiny' refers to the direction in which the process ought to move. Because I have dealt with these concepts in previous chapters, a short summary suffices at this juncture.

The assumption of a transcendent Source and Destiny marks the point where the biblical faith parts ways with naturalism. Naturalism is an alternative to faith in God. But the awareness of transcendence does not contradict modern scientific insight as such. The reality that the sciences explore is void of any evidence of a supernatural agent or purpose of any kind. Because of its methodological restriction to immanent reality, this cannot be otherwise.

Causal reductionism is being overcome in the natural sciences by various versions of the theory of emergence. There may be occurrences of probability, chance, random, indeterminacy, unpredictability, non-linear processes, sensitivity to initial conditions and underdetermined or undetermined situations. But all this does not imply the possibility or likelihood of a divine intervention that suspends or overrides the regularities that operate at various levels of emergence.

The assumption of a supernatural agency within immanent reality is also inappropriate in terms of the biblical faith. Granted, the pre-scientific worldview of antiquity did not distinguish very clearly between 'natural' and 'supernatural' causation. Naive Christian spirituality does not do so either. In a way, everything is deemed both natural and supernatural. But informed by modern science, we must respect this distinction because it presents a serious obstacle to a fruitful interface between science and faith.

There is a difference between 'immanent transcendence'—which is the presence of the unknown and the unknowable within immanent reality as acknowledged by the sciences—and 'transcendence proper'—which is a name for the boundary of immanent reality as such. More precisely, transcendence proper is the assumed openness of reality as a whole towards a transcendent Source and Destiny of reality.

As the source of reality, it originates, permeates and empowers reality in all its dimensions and at all levels of emergence. Expressed in philosophical terms it is the precondition of the possibility of the existence of anything whatsoever. As the destiny of reality, it is engaged in a creative and redemptive project that aims at the comprehensive optimal well-being of all of reality. It is geared to the optimal realisation of potentiality, the healthy survival of organisms and the fulfilment of human life. It does not contradict the theory of evolution or the law of entropy. But it gives reality a direction, a purpose, a meaning, a point.

Thus understood it is situated neither outside nor inside reality. It cannot be taken out of reality nor brought back into reality. Transcendence proper makes it impossible to collapse big bang cosmology and the theological metaphor of divine creation into one. As we shall see below, these two discourses simply do not refer to the same animal. It also precludes the idea that God has intervened where the sciences have not yet closed the gaps in the understanding of the causal networks that drive the evolutionary and entropic process forward—such as the emergence of energy, the emergence of life and the emergence of consciousness.

It also precludes the restriction of God's agency to underdetermined situations where non-interventionist divine actions might be deemed possible by people sold to the absolute validity of natural law.[404] For faith, divine action is not merely possible but indispensable for the operation of the entire world process. It also precludes the restriction of divine agency to the desirable and life-enhancing aspects of reality and the exclusion of 'natural evil'—entropy, destruction, suffering and death. This is just too simple and too cosy a concept of God. It is also not biblical. Creation and redemption are costly and sacrificial pursuits.

Section III
Science and the four biblical projections of what ought to be

According to the biblical faith the preached 'Word of God' proclaims God's creative and redemptive benevolence. This must be more than an occasional, momentary and situational reassurance. If its validity is to be trusted, it must hold true at all times and in all places. It is like the love and commitment of one's parents that one can depend upon whatever the circumstances.

The Bible contains four typical ways of visualising God's benevolent intentionality: a mythological beginning of time, when everything was very good; an 'eschatological' end of time, when everything will be very good; an alternative space ('heaven above'), where everything is very good, and a hidden 'essence' of reality, which is very good in spite of appearances to the contrary. It is entirely inappropriate to treat these projections of what ought to be as ontological conditions, origins or outcomes of the world process, whether past, present, or futue.[405]

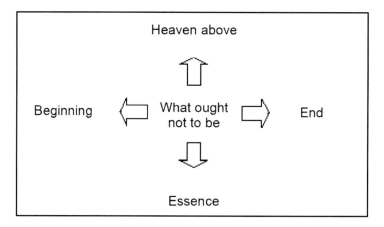

Each of these projections represents the vision of what ought to be. In more concrete terms, a vision of what ought to be arises from the experience of what ought *not* to be. It is a protest against the existence of human depravity and natural evil in the name of a benevolent God. It affirms the validity of divine benevolence. It denies evil the right to exist in God's world. Because one cannot go back in time to a pristine past, nor up into the sky to conquer a heaven above, nor uncover the hidden essence of reality, there is a growing tendency in the biblical documents to concentrate on the eschatological future.

It should be clear that idealised projections of what *ought* to be cannot refer to actually experienced reality as explored by the sciences. The concept of **creation** in theology is not meant to refer to the same phenomenon as the concept of cosmogenesis in big bang cosmology. It is an anthropomorphic metaphor that expresses the ultimate dependence of reality on a transcendent and benevolent Source.

Similarly, **eschatology** is not meant to refer to the same phenomenon as the scientific theory of a big crunch or the continuing dispersion of cosmic energy. It is a metaphor that expresses the thrust of a transcendent and benevolent intentionality towards a dynamic vision of ultimate well-being, which tenaciously confronts the destructive forces operative in reality. Obviously, **heaven above** has nothing to do with outer space. The **essence** of reality is again something entirely different from the quest of quantum physics for the most fundamental level of reality. Science describes fact, faith affirms validity.

In fact, literal interpretations of these projections are positively falsified by the findings of modern science. Big bang cosmology precludes a perfect reality at the beginning of time. The law of entropy precludes a perfect reality at the end of time. The violent nature of outer space, which is hostile to all forms of life, precludes the existence of a perfect alternative reality (heaven above). A reality that subsists in evolutionary flux, precludes a pre-established and enduring essence of reality.

But then these projections were never meant to be quasi-scientific propositions concerning what reality was, is, or will be. Protological and eschatological statements in the Bible explode the limits of the probable and the possible. They typically and deliberately assume highly improbable features. With that, they indicate that comprehensive well-being is a valid, yet elusive aim. Trust in the power and benevolence of the transcendent Source and Destiny of reality goes beyond what seems to be possible in earthly and human terms. And that kind of reassurance is of decisive importance for individual life, society and the ecological future. Let us discuss the concepts of creation and eschatological consummation in some detail.

Section IV
The concept of creation in a scientific perspective

The science-religion debate has invested immense energy on the relation between the biblical creation narratives and the scientific theory of cosmic (and biological) evolution. But the two discourses simply do not match. Scientists are frustrated by Christian fundamentalists who maintain that, as the eternal Word of God, the biblical account must be literally true, and by creationists who try to prove in scientific terms that the world is only a few millennia old because the Bible seems to say so. Even more enlightened participants in the debate are not entirely free from this preconception.[406]

The creation story found in Genesis 2 is quite obviously a myth. A myth is not fantasy gone wild. It has a message. It discloses the inner meaning of social and natural reality as it is experienced in particular natural, historical and cultural contexts. Myths of creation abound in ancient cultures. Often they are remarkably similar. They typically personify animals and assign supernatural qualities to plants. The divine, the human and the earthly are all part of the same reality.[407]

The later account of creation found in Genesis 1 is a carefully crafted theological construct. Again it is not meant to be a primitive scientific theory. It has a theological agenda, namely, to ascribe the reality we experience to God, the Source and Destiny of reality. The 'light' that dispelled 'darkness' according to Genesis 1 is not meant to be the light produced by the big bang. The canopy made by God to keep out the waters above and below the earth is not meant to refer to cosmic space. The six days of creation are not meant to refer to cosmic time. All three motives praise God, who overcame primeval chaos and established a reliable structure.

The statement that God created the sun after God created the animals is not a sign of primitive ignorance. It is meant to discredit the adoration of heavenly bodies as divinities in the religious environment. The human being as the 'image of God' has nothing to do with the scientific theory of human uniqueness, but refers to the mastery of humanity over plants and animals as representative of God on earth.[408] The statement that the woman was created after the animals has nothing to do with the evolution of *Homo sapiens*. It highlights the priority of the husband-wife relationship over the man-flock relationship in a nomadic family. Such comparisons can only increase the confusion in the science-religion debate.

Faith and theology must get used to the idea that God's creative activity is identical with the world process experienced by us and explored by the sciences.[409] They must abandon the merely *imagined* God of wishful thinking and Platonic abstraction in favour of the *real* God who drives the evolutionary process, the God who is at work in chance, contingency and necessity; the God whose 'activity' is manifest in differentiated regularities at various levels of emergence—probability, causality, propensity, teleology, agency and intentionality.

God's specific interventions (the miracle story)

Scientists have immense problems with the assumption of faith that God intervenes in response to prayer and does so by suspending or overriding the regularities that govern immanent processes. It is essential for the science-faith interface to gain clarity on what believers mean when they speak of miracles.

Science recognises that there are unpredictable or extraordinary sequences of events. There may be uncertainty and probability at the quantum level. There may be chance and random. There may be sensitivity to initial conditions and non-linear processes. But all these instances do not suggest a divine subject who is free to suspend or override regularities in nature, who intervenes in the process in response to prayer, who punishes with natural calamities and rewards with natural blessings.

To link up with scientific insight, faith and theology must sober up. Actual potentiality (rather than myth or vision) is constrained within the parameters laid down by the past. Given the way God constructed the universe we know, surprising things happen all the time, but not everything is possible, not even for God's own creative and redemptive intentionality.

In fact, it is God's creative and redemptive intentionality that precludes arbitrary interference in the regularities found in reality. Natural laws are essential because without them reality would not be able to exist and unfold. They are God's laws. They are expressions of his benevolence. They are valid. If there were no constraints for a powerful and loving God, it would be unintelligible why the expected new world without death and tears should not have materialised a very long time ago.

The idea of God's specific intervention also undermines the internal coherence of faith. Naive spirituality typically claims that desired or beneficial occurrences are due to special interventions of God. But what about occurrences that are equally inexplicable but detrimental or highly undesirable? Are they also due to the direct interventions of God?

And if not, was God asleep when they happened? Or is there perhaps a counter-God, the devil, or "the prince of this world", who is in charge of reality? Has God perhaps abandoned his rule to such a rival? Or is satanic viciousness capable of outwitting and outperforming divine benevolence?

Moreover, once one speaks of God's specific and direct interventions, one presupposes that, normally, the world process functions on its own. It is only occasionally that God sees it fit to intervene, even if one postulates that in principle (or ontologically), the world depends on God.[410] This is a form of watered-down deism, the view that God set the world going at the beginning of time and then withdrew to let it function on its own.

As mentioned above, God is not a factor within reality that competes or cooperates with immanent factors. God is, rather, the source of all these factors. Both the creative power and the redemptive intentionality of God operate through the world process as such and as a whole rather than apart from, or in addition to the latter.

In the Bible, the transcendent manifests itself *through* the immanent—a fire, a drought, a victory, a great fish, a desert wind, the wisdom of a king, or the prowess of a military commander. These specific 'interventions' presuppose God's overall creativity, which includes the unexpected, the spectacular, the awe-inspiring aspects of reality—whether positive or negative. It also includes the operation of the laws of nature.

Are miracles then impossible?

'Miracles' do happen. I have experienced numerous miracles during my own lifetime. Miracles are unexpected and extraordinary events within the world process that draw our attention to God's creative and redemptive intentions and actions. They provoke amazement, awe and gratitude.

However, for miracles to happen, God does not *have* to suspend or override the regularities that guide the cosmic process. God can also utilise these processes and regularities. Miracles do not have to explode the limits of what God has made possible within the structures and processes of the universe; they happen because God these possibilities.[411]

In fact, trust in God should be defined as the reassurance that God will do for us and through us whatever can be done under particular circumstances. Trust, even trust in human agency, does not believe in supernatural quick fixes that do not need immanent processes and agencies.

It within these immanent processes that the range of possible and unexpected occurrences is amazingly wide. As the physicist Michio Kaku reminded us, what was deemed unthinkable only a few decades ago has become commonplace today.[412] The miracles of modern technology surpass those expected by our ancient forebears in the faith by far.

There are multiple feedback loops between all levels of emergence, leading to extremely complex networks of downward, upward, whole-part, part-whole, part-part causation. Faith attributes this entire evolving network of forces and relationships to the power and benevolence of God. And it is within this context that extraordinary events happen—all the time and everywhere!

In fact, the whole of reality is a miracle. There is no law of nature that says that the universe as such must exist. The amazing 'fine-tuning' of the universe shows that the probability of having come into existence in its present form approaches zero. Scientists seem to agree that there could be any number of alternative universes. This is also true for each level of emergence. The tiniest variations in successive initial conditions would have led to totally different outcomes.

Take your own life. There may have been roughly 5,000 generations since *Homo sapiens*, as we know the species, first came onto the scene. Counting only twenty generations back you have had more than 2 million ancestors. Over a million successful conceptions had to take place. In each of these cases, a particular sperm cell out of a total of about 40 million had to be in time to unite with a particular egg cell to form a viable embryo, which again had to develop into a healthy human being, who had to survive all threats to its existence from infancy to maturity. Counting all the factors that had to be right from the fine tuning of the universe onwards, the odds that you would have come into existence as a healthy human being at all are virtually zero. And yet you do exist and read these lines. Is that a miracle or not? Does it merit some kind of humility and gratitude? Does it not evoke awe, respect and praise?

Conversely, the regularities built into the system are valid. We can rest assured that God will not provide another planet with another set of regularities after we have messed up this one—not because in 'his' unfathomable grace God would not want to do that, but because God would have to start from scratch with the big bang and it would take another 14 billion years to get us where we are now.

That should certainly sober us up. The idea that nothing is impossible for God is pure speculation if it does not refer to what God has, in fact, made not only possible but real. God's omnipotence is located in the power that the universe actually displays, and in the rules that this power observes, not in some fantasy. As we have seen in the previous chapter, the world that we know is highly ambiguous. We have to live with this fact because God lives with this fact. This brings us to the problem of evil.

A loving God and the presence of evil (theodicy)

The most intractable and irritating problem of the biblical faith is the age old issue of *theodicy* (the justification of God). The problem is simple and pervasive: How can a God with positive intentions cause or allow so much meaninglessness, futility, frustration, injustice, human depravity, suffering and death in a realm ostensibly under his control?[413] It has caused the faith of countless believers to shipwreck, whether they were scientists or not.

The problem of theodicy is caused by a combination of valid and invalid faith assumptions, namely, (a) the valid assumptions that ongoing reality in all its facets and occurrences is the product of God's intentionality and agency, (b) the invalid assumption that God's intentionality and agency are unconstrained by the limitations of time, space, energy and the regularities we experience in this world and (c) the valid assumption that God's intentionality and agency are benevolent.

If all three these assumptions were valid, the pervasive existence or occurrence of natural, personal and social evil would be entirely incomprehensible. God could, with a simple decree, eradicate evil once and for all. Why does he not do so? To relegate it to the sphere of 'divine mysteries' is not a valid solution for a problem created by questionable assumptions.[414] Similarly, to argue that the actions of a sovereign God cannot be questioned does not make sense if this God is believed to have disclosed God's unlimited power and unconditional benevolence.

The impasse surfaced very early in biblical times. It caused the same kind of agony then that believers experience today. It led to a number of widely divergent propositions: (a) The prophetic-deuteronomic stance that God punishes our sins. Its refutation dominates the main body of the Book of Job. (b) The view found in Job 1-2 that God's public prosecutor (*Satan* is the Hebrew word for 'accuser') tempts the sincerity of our faith by inflicting suffering on us. (c) The resigned discovery found in Job 38 ff that God, the Creator of the universe we observe, was just too great for us to fathom. (d) The defiant faith in the justice of God, who would not let the unrighteous get away with it but raise the dead to face the last judgement. (e) The apocalyptic vision of a catastrophic dismantling of the present world and its replacement with an entirely new creation. (f) The view that the incarnation of the *logos* and the crucifixion of Jesus, God's representative on earth, shows that God subjects himself to the suffering of humankind to overcome it from within.

All these propositions tenaciously stuck to the assumption that evil in all its forms had no metaphysical right to exist. The biblical God stood for creativity, justice and loving concern. Any metaphysical and ethical dualism was resisted in principle. Yet again and again, the (originally Persian) dualism suggested that an evil counter-god (the devil) caused all the havoc. He would ultimately be overcome by the true God. Then evil would be eradicated once and for all. The (originally Platonic) idealist dualism also surfaced again and again, namely, that the (good) spirit was imprisoned in (evil) matter and will eventually be released from the latter.

None of these attempts provided a satisfactory solution. Every believer stumbles when faced with personal, societal, or natural catastrophes. How can God allow an innocent infant to be overrun by a car? How can one believe in a God of love after Auschwitz, Dresden, or Hiroshima? How can one account for tsunamis and earthquakes that maim and kill hundreds of thousands of people regardless of their conviction or moral excellence?

The impasse for the Christian faith is increased by the fact that, for the natural sciences, this problem just does not exist. You can explain *how* a road accident happened in purely immanent terms. The question *why* such fateful occurrences happen is off the point. You can put the blame on mechanical failure, bad weather, or human error, but causal sequences of events have no purpose. Why God made a cancerous mutation happen in my body is a wrong question to ask, because there is no evidence that God ever did. You can also argue that the viciousness of humans is a variation of the fight or flight syndrome that is programmed into the human psyche by evolution.

For believers in a powerful and loving God, however, this question is crying for an answer. By far, the easiest answer is that there is no such God. Then the problem seems to disappear. But it immediately raises its head in a new and even more intractable form: metaphysical meaninglessness. The world is pointless, life is pointless, and responsibility is pointless. Eat and be merry because tomorrow we are dead! Humans just cannot live authentic lives on this basis. If faith is to regain its credibility, it must address the problem of theodicy and do so urgently. It can only do so by integrating rather than bypassing or ignoring scientific insight.[415]

The scientific theory of emergence can show us the way out of the impasse. As mentioned in chapter 9, God deals with humans at a personal level because humans are persons. But God is much more than a person, just as humans are much more than persons. God is the Source and Destiny of the whole of reality, including its infra—and supra-personal levels of emergence. Not everything that happens in reality is the direct result of a personal decision of God. Most of it follows from the regularities built into the process—and these are also of God.

Intentionality and agency are indisputable aspects of human reality at the personal level of emergence. But not everything that happens within and with the human being is based on personal intentionality. Much of it runs painfully counter to personal intentionality. I certainly did not intend the cancer in my body to happen. The impersonal infrastructure and supra-structure are accessible to personal influence and agency, but only to the extent that the powers available to the intending agent can outweigh the impersonal factors operative in the situation. These impersonal aspects of my existence are as inextricably part of me as my personal intentionality is. Because our experience of human reality provides the metaphors for our concept of God, it is entirely appropriate to apply this fact to God as well.

If God is taken to be the transcendent Source and Destiny of reality as a whole, the impersonal levels of emergence and their regularities are also of God. The biblical faith has always maintained as much. The natural order is God's order. This means that a tsunami is not caused by the personal wrath of an irate deity, or by the cynical indifference of a cruel fate, but by tectonic shifts in the earth's crust. Physical phenomena are governed by causality rather than purpose, and causality is also of God.

This is a prime example of the way science can help faith to become more precise and appropriate. In which way can we then still speak of a benevolent intentionality ostensibly operative in the global process? The answer is that this intentionality manifests itself precisely in the regularities that govern this process. Causality is more than a lifeless, indifferent mechanism. It is indispensable for the world to exist and to function. It is also necessary for intentionality to translate into effective action. It is part of God's creative power and God's redeeming love. Natural law does not contradict the benevolence of God but confirms and reflects it.[416]

God's sacrificial creativity

Strange as it may seem, the message of God's suffering, transforming acceptance of the unacceptable is applicable here as well. This imperfect world is the world God is busy creating and the world that God loves! If the world were perfect, God's creative and redemptive action would be superfluous. God shares the agonies of the victims of natural, biological and social evil and invites the survivors to join him in his creative and redeeming work.

The very concept of intentionality presupposes not only a discrepancy between what reality has become and what reality ought to become, but also sufficient leeway to do something about it. If that were not the case, believers would be paralysed by the occurrence of catastrophes and predicaments. The precise opposite happens: The Christian faith is the most potent recipe against fatalism and despondency.

To put it in blunt anthropomorphic terms, the law of entropy is the price God pays (and we have to pay) for the operation of energy in the cosmic system, because without it nothing would exist and happen. The occurrence of earthquakes is the price God pays (and we have to pay) for the existence of an earth with a crust that allows for the development of life on earth. The fact that all life is inextricably linked to death is the price God pays (and we have to pay) for the very existence and persistence of life.

In sum, the undesirable state of existing reality is no argument against the creative and redemptive intentionality of God. On the contrary, evil prompts God's benevolence, calls for it, tries to galvanise it. Confronted with injustice, oppression, physical suffering and imminent death, biblical believers evoke God's previous redemptive actions and prophetic promises. They cry for relief and redress. With that, they open themselves up for the prospect of becoming agents or instruments of God's creative and redemptive action.

Section V
Biblical future expectations in a scientific perspective

While it may seem at first sight that the concept of a 'creation from nothing at the beginning of time' can still be brought into line with big bang cosmology, though with considerable hermeneutical summersaults, there is no way modern science can affirm Jewish-Christian eschatological expectations, or even accommodate the possibility that they could ever materialise. Here we are faced with an insurmountable incongruence. To get out of this impasse, we have to explore the existential and historical roots of biblical future expectations. As part of the 'Word of God' they emerged and evolved as redemptive responses to human needs and predicaments.

Abraham needed male progeny to maintain his clan. Hebrew slaves in Egypt needed freedom. Nomads needed land. Dispersed tribes savaged by powerful neighbours needed a centralised authority to build up military strength. Jews in exile needed the permission to go home and rebuild their country and their institutions.

These examples are firmly rooted in immanent reality. The horizons of such expectations gradually widened from the needs of a nomadic clan, to the needs of the tribe, the nation, the international order, the spiritual powers that ruled the world and finally to the universe as a whole. They also gained in profundity, culminating in the expectation of a recreated humanity that would live in righteousness and unimpeded fellowship with God and with each other.

The prophets increasingly used metaphors rather than descriptions of possible futures, for instance, that mountains would be flattened, that lions would lie down with lambs, or that swords would be turned into plough shares. In response to increasing suffering, apocalyptic eschatology radicalised earlier prophecies. Giving up on the possible transformation of the current order, they reckoned that Yahweh would have to destroy the existing world and create a new world without human depravity, natural evil, suffering and death.

A literal interpretation of apocalyptic texts misses their theological agenda. They are meant to reassure believers in desperate situations that their God is still with them and for them and not against them. They deliberately employ unrealistic imagery to indicate the transcendent origins of God's benevolent intentionality.

Just consider the depiction of the 'new Jerusalem' in Revelation 21-22. The background is the complete destruction of a historical city that carried Jewish certainties and aspirations for centuries. Its replacement would descend from heaven in the form of a giant cube with each side about 2,400 kilometres in length, made of pure gold, yet transparent like glass, beautifully dressed like a bride, with high walls, about sixty-five meters thick and made of jasper, with twelve gates, each made of a single pearl and never shut—representing the twelve tribes of Israel—and twelve foundations made of precious stones—representing the twelve apostles.

There is no ocean, no sun, no moon, no night and no time constraint. These metaphors pick up the symbolism found in Genesis 1. There is no temple, because God and the Lamb are its temple; their throne is in its midst; the river of the water of life flows from the throne, and on its banks are trees that give fruit and heal the nations. There is no evil and no curse. There is no human depravity, no death and no tears. All the nations will bring their tributes into the city and walk by its light.

Every detail has symbolic meaning; nothing is meant to refer to empirical reality. Nobody can possibly assume that this is the description of a phenomenon that is expected to come about on this earth. It is a symbolic depiction of comprehensive well-being in the metaphorical language of the ancient Jewish traditions.

Where human and earthly possibilities hit insurmountable barriers to well-being, such as oppressive social structures, extreme suffering and inevitable death, faith changes into protest mode. It does not take no for an answer. It does not succumb to fatalism and despondency. It keeps the future open. But statements about a possible future cannot be *ontological* statements, because the future does not (yet) exist.

Visions of perfection give orientation, but they are not designed to materialise. The vast majority of prophetic scenarios presented in the Bible have never come to pass as announced. Theological statements about the future can only be *intentional* statements, statements about what *ought* to become. They are a *protest* against what ought not to have become. They give us the courage to act and trust in the greater capacity of God to channel the world process into more acceptable directions where human limitations are reached.[417]

Seen in this light, faith in the reconstruction of reality does not compete with the law of entropy, or the inevitability of death. A motivating vision is not the same kind of animal as an analysis of probable futures based on an extrapolation of current trends. Theological eschatology differs fundamentally from scientific futurology.[418]

Can we reconceptualise eschatology?

What then do we make of apocalyptic eschatology today? To begin with, the Christian faith is in no ways bound to the apocalyptic model. From a scientific point of view, it comes as no surprise that history has moved on for more than two millennia without the destruction and reconstruction envisaged by apocalyptic prophesies. The present universe may come to an end in a big crunch or total dispersion a few billion years from now, but without issuing in a perfect world. Such a time frame is beyond any existential, social and ecological relevance anyway.

From a biblical point of view, apocalyptic eschatology was a late, radicalised and entirely improbable version of biblical future expectations. It was never generally accepted by Judaism. In fact, the idea that God could give up on the world he created is out of character with the general trend of the biblical tradition. However far the people of God had moved off course, however unbearable its situation had become, and however hopeless the future might have seemed, Yahweh has always opened up the future again. There was always a new beginning.

From a Christian point of view, one can go a decisive step further. While it affirms the fact that evil has no right to exist in God's world, apocalyptic pessimism fundamentally contradicts the message of God's suffering and redeeming acceptance of the unacceptable as manifest in the Christ-event. There is, therefore, no reason why Christian theology should grant priority to apocalyptic eschatology over earlier and more sober biblical versions. On the contrary, apocalyptic should be understood as an extreme historical response to extreme historical situations and no longer capable of responding redemptively to the needs of today.

Faith and theology must get used to the idea that 'eschatology' has nothing to do with the end of physical time.[419] The concept of God as the transcendent Destiny of reality refers to a divine intentionality that tenaciously sticks to the direction towards comprehensive optimal well-being on its way through exploding stars and devastating earthquakes, extinctions of species, deserts and flooded rivers, frustrations and failures, suffering and death. Again we must give up the merely *imagined* God of wishful thinking, Platonic abstraction and speculative deduction in favour of the *real* God, the God who is willing to suffer and sacrifice on behalf of 'his' creation, the God who is constrained by the preconditions of the world 'he' created.[420]

In no way should Christian faith and theology feel bound to maintain historically and situationally conditioned forms of hope, whether for male progeny, agricultural land, a powerful king, or an apocalyptic reconstruction of the world as a whole. Biblical future expectations were mirror images of an untenable or unbearable contemporary situation.

Therefore they were always *Nah-erwartung*, that is, the expectation of imminent change within the ongoing flux of time. They offered hope and direction. They were supposed to materialise within the time horizons of the listeners or readers. A promise that would take half a millennium to materialise is a useless promise for those to whom it was given. Least of all can a timeless or 'time-full' eternity give direction to a historical, existential, or social process.[421]

So we are not bound to apocalyptic. Positively, it is not difficult to find a viable alternative that reflects main line biblical future expectations and does not clash with scientific insight. That is the notion of *God's vision of comprehensive optimal well-being for reality as a whole*, a vision that translates into God's concern for any deficiency in well-being in any dimension of life and that invites humans to participate in God's creative and redemptive project. It operates at the level of individual and collective intentionality and agency, rather than at the lower levels of emergence.

As the biblical examples show, such a vision of comprehensive optimal well-being can be thought of as a 'moving horizon'. As we approach it, it opens up ever new vistas, challenges and opportunities. The Word of God enlightens, prompts and empowers us. It generates a strong motivation to get involved in the world and face the future with courage and determination. It overcomes fatalism and despondency.

Where we fail, it reassures us that our limitations are not the limitations of God. And that has always been the existential, social and ecological rationale of trust in the God of the future.

Section VI
Gaining comprehensive horizons

Up to recently, it was assumed in virtually all cultures across the globe that humankind was the crown of creation, that one's communal, ethnic, or racial in-group was superior to all other groups, that the earth was the centre of the universe. Correspondingly, God was perceived in terms commensurate with this tiny, primitive, trivial and truncated universe. More than that, individual piety took hold of God and made 'him' the personal companion of humans in the way children derive comfort from their teddy bears.[422]

Having a personal relationship with God is one of the invaluable gifts of the Christian faith. But voices out of traditionalist Africa tell us that 'your God is too small a God'![423] Science agrees, if only by implication. Indeed, our God has become altogether too quaint to do justice to what the concept of God always meant to indicate—the transcendent Source and Destiny of reality as such and as a whole in its unfathomable vastness and complexity.

A theology that does not take account of cosmic scales of time and space and the insignificant share of humans in cosmic reality are naïve at best and foolish at worst. We deprive ourselves of a riches that opens up through the mere click of a mouse. I cannot say it better than Paul Sagan:

> How is it that hardly any major religion has looked at science and concluded, 'This is better than we thought! The universe is much bigger than our prophets said, grander, more subtle, more elegant'? . . . A religion, old or new, that stressed the magnificence of the Universe as revealed by modern science might be able to draw forth reserves of reverence and awe hardly tapped by the conventional faiths.[424]

Human self-absolutisation

In view of modern scientific insight, the anthropocentricity of traditional faith assumptions looks positively ridiculous. The absolutisation of the human being as the peak of the evolutionary process by naturalism and the assumption of mastery and ownership by popular postmodernity is as far off the mark as the potential absolutisation of the human being as the "image of God" by an anthropocentric faith. Science can help faith to regain the dread and respect for the numinous that is typical of all earlier kinds of religion and that both Christianity and modernity have lost.

The earth that for hundreds of thousands of years humans believed was "the world", located beneath "the heavens", is less than a speck of dust in the total cosmic context, "one planet moving around one star, which itself is one of the three hundred billion stars of the Milky Way Galaxy, which in turn is one of a trillion galaxies in the wide universe."[425] The universe is believed to be about 13.7 billion years old. The solar system originated about 5 billion years ago, the earliest forms of life soon thereafter.

Humankind is one species among billions of others on earth. It is one of the very latest evolutionary branches among a host of other mammals. Humans don't like seeing themselves as animals, yet they are. There are a "stunning 99 per cent genetic similarities between humans and chimpanzees."[426] Modern humanity appeared only about 120,000 years ago, which is roughly the last 0.0007 per cent of the estimated time since the big bang. The first substantive cave paintings (manifestations of the 'cultural big bang') date from about 30 000 years ago.

Historical civilisations appeared at the earliest about 10,000-12,000 years ago, which is only the last 10 per cent of the time that modern humans existed. Both the biblical faith and the Greek roots of modern science found their first consolidated expressions during the 'axial period' about 2,500 years ago. Both had antecedents. Both have since traversed an evolutionary trajectory that changed their basic content fundamentally. Modernity began its meteoric rise only about five centuries ago. That is 5 per cent of the history of human civilisations.

Regaining comprehensive horizons is today one of the most critically important issues to be faced by believers and non-believers alike. The Christian faith indeed emphasises the dignity of the individual human being. But today the individual human being is one of more than 6.8 billion of its kind, moving around among billions of other species, on an earth that is just a speck of dust in cosmic space. Do we have any idea what we are doing when we believe that we are the centre of the universe?

The vision of 'comprehensive optimal well-being' that posit as the criterion of an appropriate faith, entails not just a 'natural environment' that is healthy and resilient enough to provide for our needs. It is the entire hierarchy of emergences that constitutes the ancestry and the siblinghood without which we would never have come into being and without which we could not sustain ourselves.

Only if we accord this entire hierarchy of emergences the dignity of God's priceless creation, can we do justice to the ecological problem. Only if all levels of emergence have dignity, can we speak of the dignity of a foetus, an infant, a 'human vegetable,' or a person with incurable dementia. Even where there is no operational consciousness left, there is still biological life, which has a dignity of its own.

In terms of cosmic history, the time span of the predicaments caused by modernity is miniscule. Our earth has seen catastrophes and extinctions before. However, this latest period has seen the rise of the human species to unchallenged prevalence. It has rendered the network of life on our planet, which had grown over billions of years, exceptionally volatile and vulnerable.

The exponential growth of the human population, the explosion of rising expectations, the immense accumulation of technological powers, the acceleration of all intellectual, communicative and social processes, and the globalisation of human interactions have led to problems humanity is hardly equipped to master. Yet what is currently happening within humanity will determine the future of the entire planet and all its inhabitants forever after. And what humans will do, will depend on their assumptions, goals and visions.

The anthropocentricity and self-aggrandisement of traditional faith assumptions as well as their appropriation by secular modernity betray an irrational and arrogant attitude, the aspiration 'to be like God'. This attitude has led humanity into dead end streets from times immemorial. It has caused the intractable crises that modernity has to face today.[427] A realistic self-assessment will not be able to consider the human being as the autonomous master and owner of reality—a reality whose only significance is that of a quarry to be mined for human exploitation and self-gratification.

Relevance

All this is sobering, humbling, in fact humiliating, but we have to concentrate on what human individuals, communities and societies can know, aspire and achieve during their limited lifetimes. Most humans are not even remotely interested in the fate of their progeny beyond three or four generations, otherwise they would not treat the earth the way they do. Cosmic time, cosmic space and cosmic energy are way beyond our reach.

We must learn to distinguish, therefore, between the God of metaphysical theory—the Source and Destiny of the cosmos in its multibillion year history, its unimaginably vast expanses, its subatomic, physical and biological complexity, its unfathomable mysteries at all levels from quanta to collective consciousness—and the God relevant to actually lived human lives here on earth at this juncture of human history.[428]

The former is beyond our observation, comprehension and control. The latter is the God of the Bible, the One whom we encounter in the gift of life, the demand of authenticity and the invitation to be involved in a creative and redemptive project. It is here where the relevance of faith is located. God's project is conducted on our own little earth, within our tiny historical time span, involving our limited aspirations and energies.

Having said that, however, we must also say that the God of our lives can be no other than the God of the universe and cosmic history. That is what the Trinitarian doctrine wants to express—the God who disclosed 'his' intentionality to humans in the Christ-event and invites humans to share in this intentionality, is none other than the transcendent Source and Destiny of reality as such and as a whole.

As the biblical traditions responded to ever new predicaments, challenges and interpretations, their horizons widened. Today our horizons are widening exponentially through the insight of the sciences. We can discern the probable distant past, the distant potential future and the infinite complexity of the evolving universe. In comparison with our biblical and medieval forebears, our responsibility has risen by various orders of magnitude. The contrast between traditional concepts and current needs has become so vast that one can argue that we have reached a new level of emergence within the realm of the spirit.

Our past-oriented traditions are no longer capable of responding adequately to the needs of a rapidly unfolding future.[429] But the evolution of consciousness continues, and we must keep abreast. Unless we embrace this new level of emergence, we will endanger everything that evolution has brought about over the last four billion years.[430]

This is serious! Our petty desires and trivial concerns—whether material, social, or spiritual—must be brought back where they belong—into their global and cosmic contexts. This is why the concept of God as the transcendent Source and Destiny of reality as such and as a whole is of such critical importance. It covers reality in its immense spatial dimensions, its multibillion-year history and its unfathomable profundity.

Yet humans are different

To plough the human being under as just another piece of earthly material takes a valid concern too far. The human being is a creature that has attained a level of personal consciousness, a capacity of observation, comprehension, symbolisation and communication, and a versatility of intentions and actions that surpass those of all other living beings on earth. Being less constrained by natural laws, biological needs and ingrained instincts than other animals, the human being is 'condemned to be free', thus also condemned to be responsible.

In terms of our topic, the most remarkable feature of humans is their capacity to imagine, think and act beyond their personal and communal interests and their particular life world. We are the only creatures that have the capacity to ask where reality comes from, where it is going, and what regularities guide its evolution and its demise. We are the only creatures that can ask what reality ought to become. Together with some handy physical features, such as our upright posture and our versatile thumb, the human mind makes us "a radical new experiment in evolution".[431]

Humans are able to objectify reality, see it from above, 'with the eyes of God' and their own place within a greater whole. Only the human being can contemplate the meaning of the cosmic process in general and of human life in particular. Only the human being has been equipped by evolution with the capacity to have a relationship with an intuited transcendent Source and Destiny of reality, however it may be conceived.

It is this capacity that is not being developed in modernity. It is, in fact, crippled and obstructed by the incessant stoking up of personal desire at the expense of a balanced and fulfilled personal and communal life, a rational allocation of productive resources and a judicious dealing with the ecological prerequisites of life. In this sense, modernity has been a hugely retrogressive step in the evolution of humankind. If we cannot reverse this trend towards senseless consumption and self-aggrandisement, humanity has no future.

Part of this disastrous development is the individualisation and spiritualisation of the Christian faith. Although every single individual is of infinite worth and dignity in the eyes of God, and although each one of us can call upon God as his/her 'heavenly Father', the 'cosmic Christ' of the New Testament cannot possibly be 'my personal Saviour' and nothing more.

What should happen, instead, is that every believer should be personally involved in God's overall creative and redemptive project as manifest in the Christ-event. If Christians cannot learn to see themselves within their concentric contexts, they are denying the foundation of their faith, namely, God's creative and redemptive intentionality for the world as a whole.

Seen in cosmic contexts, the only appropriate status that humans are entitled to claim is that of responsible custodians of the life-worlds entrusted to their care and utilisation. That is the ancient biblical view. To become part of God's creative and redemptive project is the highest calling humans can ever imagine or aspire to. It is a calling issued to every single human being, and every other concern should fall into place within this calling.

Only when humans learn to become part of the great network of life do they truly belong, are they truly 'at home in the universe';[432] will they not destroy each other or the infrastructure of their existence; will they have a chance to survive and prosper in the long run. Revelation in nature and revelation in Scripture are not two different kinds of divine self-disclosure that are in conflict with each other, but part of the general evolutionary thrust towards comprehensive optimal well-being at different levels of emergence.

Section VII
The beginning, end and assessment of human life

Being part of creaturely reality, the human being is subject to the biological processes of emergence, evolution, deterioration and decay. Every individual human being pops into existence through the coincidental convergence of an infinitely complex and evolving network of causal antecedents that goes back at least as far as the big bang. By the same token, every individual human being drops out of existence, leaving behind an immensely complex trail of consequences that will continue at least up to the termination of all life on earth.

Like all biological creatures, humans are mortal. Everything that makes up their human nature, dies together with their bodies. Both the biblical tradition and modern science preclude the (Platonic) separation of an immortal 'soul' from a mortal body. The personal level of emergence depends on the entire impersonal infrastructure of emergence. If the body dies, the spirit dies with it. Every individual human consciousness emerges from the world process and remerges with the ongoing world process.

However, every human being has been 'created in the image of God' (Gen. 1:26ff), that is, as the representative of God on earth. The actual life of a particular consciousness, complete with its antecedents and consequences, can never disappear from the record of cosmic history. Expressed in theological terms, human consciousness is the product of God's creative intentionality that is being lured into God's redemptive project. Having become part of God's project, it can never lose its significance.

It is the degree of correspondence of a human life with the creative and redemptive intentionality of God that determines its quality, acceptability and authenticity. It is not often recognised that this is the most basic motif behind the statements about life after death in the biblical tradition. Let me offer a brief overview of the latter.

The earliest biblical traditions (from the oldest creation story in Gen 2 up to Sirach 41) are remarkably realistic about the inevitability and finality of death. Where the biblical faith ventured to make statements about what might happen beyond death, it invariably did so for a reason—and that reason was invariably its overwhelming concern for divine righteousness and human authenticity.

As the oldest biblical creation narrative has it, we have been taken from the earth and will return to the earth (Gen. 3: 19). The possibility that humans could 'live forever' is explicitly denied (Gen. 3: 22). This text reflects the stance of pre-exilic Israel on life and death: while you live, you have your chance to contribute to the ongoing life of your clan, tribe and nation. When you die, it is your progeny that continues to take the latter forward, while you join your fathers in *sheol*. This is the place of death where you no longer see the light and cannot praise God (Is. 26:14; Ps. 88:10 ff; Ps. 6:5 f; Job 14).

This was as painful a prospect for Israel as for any other human community. God was the Giver of life, so why should God want his people to die? One answer was that humans were sinners; death was the result of God's wrath (Ps. 90: 7-12). Another answer was that death was a simple decree of God, thus an inescapable fate (Gen. 3:22 ff; Sirach 41).

The assumption that we rise from the dead to face the last judgement arose in post-exilic times to reassure believers of the incorruptible justice of God: evil-doers would not escape punishment, and the righteous would not forfeit their reward simply by dying. From the outset, therefore, the primary motivation was not the infinite prolongation of human life, but the justice of God.

Apparently influenced by Egyptian ideas of a divine judgement after death, Wisdom of Solomon seems to have been the first attempt to offer a robust argument in support of bodily resurrection (chapters 1-3). God had created humans in 'the image of his own eternity' and cannot want them to die. If it was God's wrath against sin that caused death, righteousness would prevent death from obliterating the human being. Those who denied that possibility did so merely to get a free ticket to sin.

Another root was the apocalyptic expectation that God would create a completely new reality without unrighteousness, suffering and death. As mentioned above, Apocalyptic responded to the desperation caused by situations of severe suffering and affliction. In New Testament times, these motifs had merged with the messianic tradition of a coming Kingdom of God. It is likely that both John the Baptist and Jesus proclaimed the imminence of the Kingdom of God. After the death of both leaders, the early Christian community eagerly anticipated the speedy return of Christ, who would judge humankind and set up the eternal Kingdom of God to replace the existing world of sin, evil, suffering and death.

But as in Judaism, it was sin and righteousness in this life that would determine one's fate in the last judgment. According to Paul, it was the 'flesh' (that is, sinful human existence) that had to die so that the 'spirit' (that is, righteous human existence) could take over (Rom. 6: 7). Jesus came into the 'sinful flesh', died to the flesh, and rose into the 'spirit' (Rom. 6: 9 f; 8:3 ff; 2 Cor. 5: 21, cf Rom. 1:3f).

Through faith we participate in the transition already now (Rom. 6) and receive the Spirit as first fruits of the life to come. As a result, we know neither Christ nor the Christians 'according to the flesh', but as 'a new creation' that comes about through reconciliation with God (2 Cor. 5: 16-21). Placed into an apocalyptic frame of reference, this means that, after our deaths, we will be transformed from a 'fleshly' existence into a "spiritual body" (1 Cor. 15: 35 ff).

When these expectations did not materialise, the motifs of eternal judgement, the Kingdom of God and Christ as the messianic Son of Man shifted from the outstanding future to the transcendent space of the 'heavenly places'. According to the Deutero-Pauline letters, Christ was already enthroned in heaven above all spiritual powers that rule the world. Believers had already joined him in this exalted position (Eph. 2, Col. 3). Eternal life is acquired or forfeited by accepting to be accepted, or refusing to be accepted, into the new life of Christ here and now (Rom. 6: 1-14; Eph 2: 4-6; Col 3: 1-4).

The Apostle Paul is adamant that the denial of the resurrection would render faith in Christ meaningless (1 Cor. 15: 12 ff). That is certainly true, but the question is what 'resurrection' meant for him on the basis of the apocalyptic worldview, and what it could mean for us today at a time when this worldview has lost its plausibility.[433]

It cannot mean resuscitation of decayed bodies. It can also not mean the immortality of the soul. Neither of the two is feasible in terms of scientific insight. Neither represents the concern of the biblical tradition. When pressed to explain what he meant, Paul makes it clear that it is not a physical body, but a "spiritual body" that is sown (1 Cor. 15: 44), that "flesh and blood cannot inherit the Kingdom of God, nor does the perishable inherit the imperishable" (1 Cor. 15: 50).

There is no question that, operating within an apocalyptic frame work, Paul expected a new ontologically concrete reality when Christ returned in glory (1 Thess. 4: 13 ff; 1 Cor. 15: 23 ff), or after death (Phil. 1: 21-23; 2 Cor. 5: 1-10). However, the apocalyptic framework proved not to be viable as a form of Christian hope. So we either reconceptualise the thrust of Paul's message in forms that make sense in terms of current insight, or we have to abandon it altogether. In the rest of his theology (Rom. 6; cf 2: 1-10), Paul provides an absolutely adequate alternative to apocalyptic eschatology—in faith we 'die with Christ to the flesh' (human life apart from God) and rise into the new life of Christ (human life in fellowship with God).[434]

In John's gospel, 'eternal life' means authentic life rather than endless life. The 'last judgement' takes place wherever people take a decision for or against faith in Christ (John 3: 18, 36; 5: 24; 12: 48.). Christ represented and manifested the Spirit and those who believed in him would share this Spirit (John 6: 63; John 7: 37-39). Christ went to the Father to be present in the Spirit (John 14: 18 ff).

Believers were to remain in Christ, as Christ was in the Father (John 15:1-11). Conversely, the Father was in Christ, and Christ was in his followers John 14: 8-16). All this means nothing other than remaining in the redeeming love of God as manifest in Christ (John 15: 9-13; 1 John 4: 7-21). 'Eternal life' in John thus means authentic life.[435] Its historical manifestations may come to an end, without thereby losing its authenticity (John 11:25 f).

The crude biological terminology of the time should not detract from the fact, therefore, that in the main stream of both Judaism and Christianity, resurrection and eternal life were essentially about existential and communal authenticity, rather than human longevity. On the one hand, biological life was deemed the *consequence* of authentic life. On the other hand, it also served as a *metaphor* for authentic life. The concept of 'eternal' life in the New Testament does not refer to Platonic timelessness or to the endless continuation of biological life, but to a life in unobstructed fellowship with God.

We gather from this overview that a literal interpretation of the 'resurrection of the body' should not be taken as timeless truth. In scientific terms it makes no sense anyway. Death is not the 'enemy' of God that will finally be overcome as Paul believed (1 Cor. 15: 26), but a necessity for the very existence and operation of biological reality.

The Bible is surprisingly realistic in its observation that all things come to an end in some way or another. That includes the life of the human being. Even Christ died; otherwise he would not have been a real human being. Similar to its cosmological counterpart—the apocalyptic transition from this age to the age to come—the expectation of bodily resurrection emerged and evolved in human history in response to particular social and existential needs and their worldview interpretations. It underwent qualifications in various directions. It always had redemptive intentions—the warning against irresponsibility and the reassurance of divine acceptance and belonging. The 'last judgment' always referred to the acceptability of human conduct while alive on this earth. Today this intention must find a conceptualisation that takes account of modern scientific insight.

What God can make of such a lived life is a completely different matter. Falling back on the meaning of the cross, the very core of the New Testament message, we must go an important step further. The cross is a prototypical enactment of God's sacrificial love. Concern about the infinite prolongation of my conscious life seems out of character with the kind of intentionality that manifested itself in the cross of Christ.

This intentionality is not reversed, but affirmed by the image of the elevation of the 'crucified Christ' to the status of a new humanity in which we are allowed to participate. The new life of Christ is a life that acts sacrificially and redemptively in the authority of God. To expect never ending life for oneself may be considered a sign of gullibility that craves for more than what a creature is entitled to expect and what it will ever get, rather than a sign of conformity with the creative, redemptive and sacrificial love of God as it actually manifests itself in the reality we experience.

No human life can possess immortality; but it can participate in the 'life of God'. It emerges from the 'life of God', as it were, and remerges into the 'life of God'. According to the biblical faith, 'God's life' is characterised by creative power and redeeming love, rather than by timelessness, perfection and immutability. It is precisely as a temporary and local representative of God's creative and redemptive intentionality that the particular human existence of Jesus of Nazareth possesses divine authority, unsurpassable dignity and irreplaceable significance.

Being with Christ (Phil. 1:23) is being with God. To be with God 'forever' does not imply the unlimited continuation of one's earthly and bodily life, therefore, but the quality of one's limited life on earth in terms of participation in God's creative and redemptive project.[436] It also means that God in his unfathomable grace is willing to suffer the inadequacies of this life, yet with the intention to overcome them. Every human life can be granted precisely that. Nothing greater can ever be imagined than God's unfathomable gift of grace.

Section VIII
Jesus Christ, the messianic representative of God

I dealt with the concept of God and with the concept of the human being first, because Christ, the pivotal centre of the Christian faith, can only be understood against this background. Was Jesus of Nazareth divine or human? For the dialogue between science and faith, this question is of critical importance. Science is unable to contemplate a transcendent entity as part and parcel of immanent reality. This is a contradiction in terms.

From an experiential point of view, it is imperative, therefore, to emphasise the self-evident, namely, that the historical figure of Jesus of Nazareth, as depicted in the New Testament, was ontologically a human being, who lived on earth during a particular historical time as part of immanent reality, rather than God, the transcendent Source and Destiny of reality as such and as a whole.

The proclamation that God disclosed God's intentionality through the life, ministry and fate of this human being, as the Christian faith assumes, does not mean that this human being was ontologically identical with God. It only means that Jesus identified himself with God and God with Jesus. It also means that his contemporaries encountered the intentionality of God, thus the living God 'himself', in this particular human life. And as far as the responded positively, they were drawn into its dynamic.

It is rather odd that countless Christians believe that Jesus was God himself. With this they undermine the very rationale of the 'incarnation', namely our participation in the new life of Christ in fellowship with God. We cannot become divine, after all, but we can become Christ-like, if 'Christ' signifies authentic humanity.

According to the entire New Testament, Jesus was a human being. He was conceived, born and raised to adulthood; he lived in a particular geographical space, at a particular period in historical time, within a particular religious and cultural context and under particular political circumstances; he recruited followers as a Jewish rabbi, helped people in need within the constraints of space, time and energy; he made enemies, clashed with the Jewish establishment in Jerusalem, was condemned as a heretic and impostor by the Jewish authorities of the time; he was tried, tortured and executed under the Roman governor as a suspected insurgent.

Jesus had a limited human lifetime that began with his conception and ended with his death, a limited geographical reach within a specific political, social, religious and cultural context. What followed after his death falls into a different category to which I will presently return. The Gnostic idea that he was God in a pseudo-human garb was implicitly and explicitly rejected by the authors of the documents found in the New Testament and subsequently declared a heresy by 'orthodox' ecclesial authority. The Jewish authors of the New Testament testified to the relation between God and Jesus in terms of intentionality and agency rather than divine and human ontology. God manifested God's redemptive intentionality through the commitment, insight, teaching and action of his messianic representative.

The Church Fathers expressed a similar view but cast it into the static terms of Hellenistic ontology that were in vogue at the time. The Council of Chalcedon (AD 381) declared that Christ encompassed, in one and the same person, a fully divine nature and a fully human nature, which should **not be separated** but which should also **not be confused** with each other. At the Council of Constantinople (680/1 AD), the same formula was used for the relation between God's will and the will of Jesus. While this ontological expression of the relationship between God and Jesus led to paradoxical formulations, the intention remained clear: the Creator acted **in and through** the actions of the creature. Both subjects retain their ontological integrity.

That Jesus was in the likeness of God, came from God, and was sent by God, does not imply that Jesus was God. According to Genesis 1: 27 ff, the authentic human being is 'created in the image of God'—which means that humans have been entrusted with authority over the rest of creation as representatives of God (Gen. 1:28 ff). In 2 Corinthians 3:17 and 4:4, Paul presupposes that Christ is the authentic form of this image into whose likeness we are to be transformed. In Daniel 7:12 ff, we find the notion of an authentic ruler who was "presented before" God and "given dominion an glory and kingship" in contrast with the pagan despots here depicted as beasts. The rest of the chapter makes it clear that it referred to a collective of the "holy ones of the Most High" (Dan. 7: 22 ff; 27 ff). The "form of God" referred to in the hymn quoted in Philippians 2:6 ff obviously harks back to the same tradition.

The idea that Jesus was God is based on a wrong interpretation of the gospel of John. John 1:1 ff does not refer to Jesus directly but to the creative 'Word' of God (according to Genesis 1) or the creative Wisdom of God (according to Sirach 1:9 ff or Wisdom of Solomon 7:22 ff). It is the loving intentionality of God that was 'before all things' (cf John 1:1) and penetrates all things (cf John 1:3) that found its human manifestation in the life and fate of Jesus, the messianic representative of God. Colossians 1: 15-17 says the same thing in different words. God's intentionality, as manifest in Christ, 'was God' in the sense that it was 'with God' (John 1:2). As far as we are concerned, God is identical with his intentionality because that is all we know and can ever know (1 John 4:16).

God's intentionality 'became flesh', that is, it manifested itself tangibly in the form of a living human being (John 1:14; 1 John 1:1 ff; cf Col. 1:19). But that does not mean that this man was God. "The Son can do nothing of his own accord" (John 5: 19 f); it is the Father that acts through him (John 14: 8 ff). The manifestation of God's redeeming intentionality in an authentic human being is the window as it were, through which we "see the Father" because "the Father is in him" and "he is in the Father". This is why Thomas calls him "my Lord and my God" (John 20: 28). God's intentionality manifested in Christ is "grace and truth" in contradistinction to the law of Moses (John 1: 17 ff). We are supposed to share this 'being in the love of God' (John 15: 5; 9 f; 12 f) and thus to know who God really is, namely, self-giving love (1 John 4: 7-21).

Jesus claimed to be acting in the authority of the 'Son of Man', the expected messianic representative of God. What made this particular human being the pivotal centre of the Christian faith was not the claim that he was God, therefore, but the conviction that God's creative and redemptive intentionality manifested itself in the life and ministry of Jesus of Nazareth. And this happened in a way that differed fundamentally from the Jewish religious tradition within which he operated.

Let us deal with authority first. The pervasive imagery used in the New Testament for the Christ-event is the Ancient Near Eastern assumption that the king was the adopted 'Son of God', that is, God's representative and plenipotentiary on earth (Ps. 2). The king's duty was to maintain the moral, social and cosmic order and to channel divine blessings into the nation. This is the imagery on which the Jewish messianic expectations were based.

All the titles attributed to Jesus by his followers were royal titles: the Son of David, the Anointed (*Mashiach, Christos*), the Son of God, (harking back to Ps. 2) and the legendary priestly king of pre-Davidic Jerusalem (harking back to Psalm 110: 1-4). According to the gospels, he claimed to be the 'Son of Man' who would replace the monsters that held sway over the world (harking back to Daniel 7). However, Jesus redefined royal authority in terms of selfless service (Mark 10: 35-45; cf John 13: 1-17). His followers recognised in the crucified Jesus the image of the suffering Servant of God, who became the victim of the sins of others (harking back to Isa. 53).

Christ represents authentic humanity that is entitled and empowered to act in the authority of God. What makes Jesus special in terms of the Christian faith is not that he was divine, therefore, but that he was authentically human. To be authentically human is to have a relation with God that is not obstructed by the human ambition to seize ultimate sovereignty, exclusive mastery over reality and total independence from God, the Creator. Jesus' intentionality fully reflected and thus manifested the creative and redemptive intentionality of God. He was the prototypical agent of God's dealings with God's world.

Authority is only the formal aspect. What matters is the substantive content, that is, what Jesus actually stood for in the name of God. Jesus preached and enacted God's suffering and transforming acceptance of the unacceptable. He invited sinners, outcasts and lepers into his fellowship. Acceptance, forgiveness and healing were manifestations of God's redeeming love. For the Pauline tradition, 'justification' by grace rather than moral achievement or excellent disposition is the most fundamental aspect of the Christian faith. It is summarised by a disciple of Paul in Ephesians 2.

Why was that stance so revolutionary? Jewish orthodoxy at the time had assumed that divine acceptance or rejection determined one's fate. Blessing was the 'natural outcome' of righteousness. Suffering was punishment for sin (Deut. 28 or 30). Jesus reversed the deuteronomic link between meticulous obedience to the Mosaic Law, on the one hand, and divine acceptance (and thus one's fate in terms of health and well-being) on the other. Acceptance was fundamental; transformation was consequential. As a result, human suffering became the object of God's redemptive concern, rather than the consequence of God's wrath against the sinner.

Jesus also rejected the Priestly link between ritual purity and divine acceptance. God's holiness was located in God's redeeming love, rather than God's perfection. Human sanctity was defined in terms of the response of faith to God's calling, rather than in terms of meticulous ritual observances. The particular intentionality of Jesus and the God he proclaimed was subsequently experienced as the Spirit of Christ that permeated, liberated and empowered the Body of Christ, the Christian community of believers.

The fact that a meticulous fulfilment of the Mosaic Law, in fact any moral or ritual law, was no longer deemed a condition of acceptance had dramatic consequences for divine-human relationships. It also led to fundamentally changed individual and communal relationships. Non-Jews, outcasts, sinners, strangers, enemies, women and slaves were accepted into the Christian fold.

It had the potential of reconstituting human attitudes towards society as a whole. If one takes the statement seriously that God is the Source and Destiny of reality as such and as a whole, it also changes human attitudes towards the natural world. The entire concept of what ought to be is placed on a new foundation. Nothing is the same any more.

In sum, Christians are persuaded that the God represented by Christ is the true God. And Christ is the true human being because he represents the true God. The intentionality of the true human being is a manifestation of the intentionality of the true God.

Section IX
The meaning of the cross of Christ

What exactly happened when Jesus was crucified? It can be argued that the condemnation of Jesus by the Sanhedrin was legitimate in terms of the understanding of Jewish law prevalent at the time. According to Deuteronomy 28 and 30, Yahweh's blessing would rest on those who kept the law and Yahweh's curse on those who transgressed it.

In post-exilic times, it was argued that those transgressing the law would not get away with it, even if they died. Yahweh was a God of justice who would raise the dead to face judgement. And that would happen, expressed in rather abbreviated terms, on the 'Day of the Lord', when the Messiah would come in glory to establish the Kingdom of God for all to see.

So in which sense could Jesus' claim to divine authority be legitimate? His seeming disregard for the Mosaic law—both in its deuteronomic and its Priestly form—and his claim to divine authority as the 'Son of Man' evoked immediate and fierce rejection by the Jewish spiritual and political authorities. When Jesus' triumphant procession into Jerusalem and his cleansing of the temple brought things to a head, he was captured, tried and condemned to death as a heretic and impostor by the Jewish Sanhedrin. How could he claim to act as the messianic representative of God, while undermining the law and the ritual!

The execution of Jesus by the Romans was equally legitimate in terms of Roman law at the time. Although the gospels suggest that Pilate was an uncertain and rather unwilling victim of Jewish incitement, there is no doubt that at least the disciples and his Galilean followers saw in him the messianic king expected by the Jews for centuries and that he was crucified as such. Whoever claimed royal status without Rome's authority was a danger to the stability of the Roman state and the punishment for insurrection was crucifixion. Royal pretensions had to be squashed in the bud.

If Jesus was declared 'sinless' by his followers, this had to imply that Jesus' interpretation of the God of Israel was valid and that of the Jewish leaders was not. He was also not a revolutionary instigator but the 'prince of peace' envisaged in Isaiah 9 and 11. It can be argued that Jesus followed and radicalised a particular strand within the Israelite-Jewish tradition, but in the Jewish religious atmosphere of the time, it was either a blasphemous presumption or an astonishing and entirely new revelation.

His followers were convinced of the latter. What Jesus had proclaimed and enacted in the name of God was that Yahweh was not a God of retributive justice, as Deuteronomy seemed to have defined him, but a God of redeeming mercy. And if Jesus' interpretation of the God of Israel was valid, Jesus must have had divine authority. So his claim to be the messianic representative of God was entirely plausible. In his authoritative preaching, his attitude and his redemptive actions he was experienced as the long awaited 'Son of Man' who would bring salvation to his people.

For the Jewish leaders, the successful condemnation and crucifixion of Jesus and the failure of God to come to his rescue proved God's rejection and condemnation of Jesus and his ministry beyond any reasonable doubt. But for the followers of Jesus, the cross of Christ rather unexpectedly acquired a profound theological significance. This seems to have been informed by the texts on the 'suffering Servant' found in Isaiah 53. The cross manifested the extent to which the redemptive intentionality of God would go in the direction of sacrificial love.

God was willing to forgive. One always 'gives away' what one 'forgives'. If you forsake your right to restitution or revenge, you suffer what the guilty person should have suffered. That was the core of the matter. While Jesus was condemned by the Jewish leaders and executed by the Roman authorities, the New Testament authors acclaim with one voice that Christ had died prototypically "for us", that is, for humankind as such in its estrangement from the creative and redemptive intentionality of God.

In Christ, God's royal 'Son', that is, God's plenipotentiary and representative, God had exposed God-self to human depravity and enmity. This represented a momentous reversal not only of Jewish faith assumptions, but of a general trend in many other religions and worldviews as well. Humans were no longer supposed to sacrifice their first-born sons to God to appease God, but God sacrificed his 'only-born Son' (God-self in God's representative) to a self-seeking, rebellious and hostile humanity to bring them back into his fold.[437]

Entropy and sacrifice

Before we come to the resurrection of Jesus, we have to go a step further in our interpretation of the cross of Christ, a step that is absolutely decisive for our concern about economic and ecological sanity. In terms of an experiential approach to Christian theology, there is a profound, but hardly ever contemplated correspondence between the cross of Christ and the law of entropy. It transpires that the basic message of the Christian faith is not out of character with what happens in the rest of reality.

Millions of seeds are produced and perish so that a few of them find the fertile ground needed for them to germinate. New life only materialises through the death of other life. As individuals, we daily devour large quantities of other life, otherwise we would not exist. But it is also true in terms of entire species. The extinction of the dinosaurs brought mammals to prominence and made the ascendancy of the human species possible. Life on earth again depends on the fact that our star, the sun, is slowly burning up. Construction presupposes deconstruction elsewhere in the system. Evolution happens within the context of the entropic process.

In theological terms, God sacrifices parts of God's creation so that we can live. And we are called upon to participate in God's redemptive project by sacrificing parts of our lives, and ultimately our lives as such, so that other creatures may have the space and the resources necessary to live and prosper. There is no way we can be with God without participating in the sacrifice that God offers to his creation. At the social and spiritual levels of emergence, this sacrifice of God culminated in the cross of Christ. Again and again, believers experience that it is precisely this participation in God's overarching redemptive project that brings the deepest fulfilment and joy.[438] This is where human authenticity is located.

Section X
The meaning of the resurrection of Christ

Science must have the greatest difficulties with a literal interpretation of any notion of the 'bodily resurrection of Jesus from the dead'. It could not have been the spirit of Jesus, released from his body, as Platonism would have argued, because according to emergence theory spiritual reality presupposes a biological infrastructure.

It could not have been the resuscitation of a corpse after two nights and three days in the grave because decay would have set in by then. If indeed it had happened, Jesus would have had to die again. If it had been a new creation, this would have required the existence of an entirely different cosmological infrastructure that nevertheless allowed continuity with the world we know, for which there is no evidence.

There are serious theological objections too. The Gnostic idea of a bodiless spirit was rejected by Christian theologians already during New Testament times and finally by the patristic fathers and ecumenical councils. If the risen Christ was nothing but God, the transcendent Source and Destiny of reality as such, rather than the authentic human being, his new life would not be accessible to our participation, as Christians claim it to be, because humans cannot be divine.

The nature of the New Testament sources compound the problem. They were written decades after the event. By that time, the traditions concerning the resurrection had moved far apart from each other. There is no way they can be harmonised. There are no independent sources that could augment, falsify, or affirm any of them. The authors were also very 'creative' in their stories, notably John, showing that they had a theological, rather than a historical agenda. They used all linguistic means at their disposal to make their point. They did not subject themselves to the Enlightenment criteria of empirical evidence and historical precision. It was the message that mattered.

Did all this happen in Galilee or in Jerusalem? Did the risen Christ appear to Peter first (1 Cor. 15:5), or to a number of women (Luke 23:55, 24:10), or to three women (Mark 16:1), or to two women (Matt. 28:1), or to one woman (John 20:1), or to no women at all (1 Cor. 15:5 ff)? Who precisely were these women? Did they run to the disciples with joy, or were they too afraid to speak? What was the nature of the 'appearance'—a material body, a vision, an audition, or a sudden and overwhelming spiritual certainty?

What is the sense of a body that can eat and be touched, yet walk through locked doors? Why is the risen Christ not immediately recognised by his followers in many of the narratives? What is the literary genre of the various narratives about the 'empty grave'? What does it mean that Christ was 'taken to heaven in a cloud'? What is the nature of the claim that the 'Scriptures' were fulfilled in these occurrences? And if all this is metaphorical language, what precisely is the referent to which the metaphors refer? There is no way historical certainty can be reached on any of these questions.

As in so many other biblical cases, empirical evidence and historical precision were quite obviously not part of the agenda of the sources. If we want to make sense of these narratives at all, we have no choice but to try and dig into the intended meaning the message of the resurrection of Christ wants to convey.

To begin with, the possibility of a resurrection from the dead was not the core of the problem between the followers of Jesus and their Jewish opponents at the time. Although not uncontroversial, the expectation that the dead would rise was by no means a strange idea in Judaism. "Why should it be thought incredible by any of you that God raises the dead?" (Acts 26:8) The question was, rather, whether and for what reason the Messiah had to die and rise from the dead ahead of all the others. There was no precedent for this idea in Judaism.

What was deemed heretical by the Jews was a new interpretation of resurrection, namely, that God had identified God-self with Jesus' proclamation and enactment of God's unconditional transformative acceptance of the unacceptable. For Jewish thought, resurrection was for the last judgement.

For the followers of Jesus, the resurrection of Jesus was a vindication of the claim of Jesus to divine authority. Jesus had proclaimed and enacted God's redeeming grace rather than God's condemning law. He had done so in the authority of the messianic representative of God on earth. The Jewish leaders had condemned him for this double heresy.

In the final analysis, it was not a question of ontology, therefore, but of validity. Jesus had interpreted and enacted the God of Israel as a God of suffering, redeeming, transforming acceptance of the unacceptable. In contrast to the Jewish leaders, the followers of Jesus deemed this proclamation and enactment to be authoritative.

In spite of what the Jewish and Roman leaders believed and decreed, they asserted, Jesus represented and manifested God's creative and redemptive intentionality. In consequence, they proclaimed God's unconditional benevolence as manifest in Jesus Christ as universally valid, accessible and determinative for any vision of a future salvation.[439]

Expressed in terms of Jewish messianic expectations, Jesus was indeed the messianic representative and plenipotentiary of God on earth and had now been installed as such 'at the right hand of God'. As mentioned before, the texts on the 'Suffering Servant' in Isaiah 53 helped the followers of Jesus to come to terms with the apparent failure of Jesus' mission. God had used the death of Jesus to manifest God's own sacrificial love and declared this fact to be universally valid by elevating Jesus, the authentic human being, to the position of universal ruler.

Taken in its historical context, it is hardly possible to fathom the audacity of this proclamation. The real craziness was not the message that a corpse had come to life again, or that a deceased person had been granted a new and different lease of life. The real craziness was much more profound. A bunch of unschooled, disillusioned, intimidated and disoriented followers of Jesus, who had fled in horror on that fateful day, suddenly went public with a truly incredible and seemingly ridiculous message:

Jesus of Nazareth, an itinerant Jewish rabbi, hailing from a remote Roman province, rather than Jerusalem, the designated capital of the world, who had gathered a few fishermen, outcasts and women around himself, who was followed by an excited and seemingly deluded crowd of Galileans to Jerusalem, whose followers had not dared to face the temple guard with armed resistance, who was betrayed by one of his followers, denied by another, misunderstood and abandoned by the rest, condemned as a heretic, impostor and blasphemer by the leaders of his religious community, ridiculed, tortured and executed as an insurgent by the authorities of the mighty Roman Empire, who was by all standards a failed human being, was proclaimed the long-expected messianic ruler of the world, the representative of God, the firstborn of creation, the *kurios* of the cosmos, the Lord over all ideological and political powers that determined human existence at the time and ever after. Who could possibly accede to this claim! As Paul says in 1 Corinthians 1:20 ff, it was a crazy and scandalous idea.

There is no doubt in my mind, therefore, that it is the proclamation of God's affirmation of the validity of the redemptive ministry of Jesus as a reflection of God's intentionality that made the Christ-event the unique criterion of truth for all Christians, nothing more and nothing less. In spite of his apocalyptic frame of reference, therefore, Paul was right when he claimed that without this message, the Christian faith has lost its foundation (1 Cor. 15:12 ff).

The interpretation of the cross of Christ as an act of God on behalf of humankind marked a decisive shift in the assessment of the ministry of Jesus of Nazareth. Authentic humanity, which had manifested itself in this particular human being, in a particular geographical space, during a particular phase in history, within a particular religious and cultural tradition, with its limited range of influence, had now become valid, present and accessible for all humans of all times, places and cultures.

Out of the death of Jesus of Nazareth, the epitome of the new human being, had risen a new, authentic, universally accessible kind of human existence, in which all of humanity was invited to participate. To adapt a picture used variably in the New Testament, the 'seed' that had fallen into the ground had grown into a plant that yielded abundant fruit (John 12:24). All those who identified with Christ in faith became participants of the new life of Christ and got involved in God's creative and redemptive project as Jesus had been before.

Expressed in terms of the theory of emergence, the spirit of Jesus, which had rested upon the impersonal infrastructure of his own body before his death, now became the Spirit of Christ that permeated, liberated, transformed and empowered the 'Body of Christ', the community of believers, which now provided the new biological infrastructure and communal suprastructure.

The Synoptic gospels express this message with the concept of discipleship. The disciples do exactly what Jesus did (compare Matt. 9:35 ff with 10:1 ff) and face the same fate (Matt. 10:16, 24 f). He would be with them where two or three gathered in his name

(Matt. 18:20), where they showed mercy to those in need (Matt. 25:31 ff), and where they would go out into the world to proclaim the message (Matt. 28:16 ff).

If Christ did not go, John's gospel says, the Spirit would not come (John 16:7). Once he were 'elevated', he would draw all people to himself (John 12:32). In sending the Spirit, he would be among them, and there would be a much more profound immediacy in the relationship between God, Jesus and the believers (John 14:18-24).

This insight also characterises the theology of Paul. There is a new way of being human—no longer the 'first Adam', but the 'second Adam' (Rom. 5:12 ff). "We are daily given into death with Jesus, so that the life of Jesus may be made visible in our mortal flesh. So death is at work in us, but new life in you" (2 Cor. 4:11). Believers are united with Christ in his death so as to share in his resurrection (Rom. 6).[440]

The message of the resurrection of Jesus from the dead found surprising and increasing acceptance. The inherent power of this message shone through its seemingly ridiculous form. It was the most powerful assertion of the validity of the message of *God's suffering, transforming acceptance of the unacceptable* that one can think of. It appealed immediately to the marginalised, despised, poor, sick and outcast.

It proclaimed the message that God had identified with Jesus' interpretation, proclamation and enactment of the God of Israel as a God of creative and redemptive love.[441] In God's 'Son' (that is, God's messianic representative on earth), God had exposed himself to the vulnerability, depravity and enmity of humanity—and that to the extreme of his death on the cross.

The distinction between theological meaning conveyed by means of a metaphor and the expectation of an ontological reality to back it up was not as clear at the time as it is for us today. One of the reasons for the confusion was that a major biblical tradition interpreted mortality as the consequence of *sinfulness* (Ps. 90), which could only be avoided through righteousness and remedied only through the supreme sacrifice of life. Jesus had borne our sinfulness, Paul argued, and therefore had to die. But being righteous God could not abandon him to death and decay (Rom. 4:25; 5:21; 8:3 f; 2 Cor. 5:21).

Apocalyptic theology proclaimed the imminence of a real reconstruction of the universe, including our human bodies. Paul believed that our 'fleshly' bodies would be transformed into 'spiritual bodies'. In Paul's thought, flesh is the 'old human being' in conflict with God and 'spirit' the 'new human being' in fellowship with God. The apocalyptic expectation did not materialise and, as far as we know, it cannot materialise. But the underlying concern for human authenticity and the insight that authenticity is a gift of God rather than a human achievement does not thereby lose its validity.

Paul's theological approach allowed him to come up with a viable experiential alternative to the apocalyptic framework. We once knew Christ 'according to the flesh', Paul says, but no longer. Christ is a new creation (2 Cor. 5:16). In which sense? The Lord is the Spirit (2 Cor. 3:17 f). Christ rose in the form of a "spiritual body" (1 Cor. 15: 44). We know by now that, to become real and effective in this world, the spirit needs a biological infrastructure. According to Paul, this infrastructure is provided by the bodies of believers (1 Cor. 6:13 ff), who form members of the 'Body of Christ', the community of believers (1 Cor. 12:12 ff, 27).

Today we know that biological death is not the consequence of sin, but a biological necessity, which cannot be overcome. So we have to abandon a quasi-biological interpretation of the bodily resurrection of Christ. But what then? If one wanted to give an experiential expression to the 'new creation' that Paul speaks about, or the 'new birth' that John speaks about, one would have to speak of an emergent reality at the spiritual level of reality.

Where God's creative, redemptive and sacrificial love manifests itself in a human individual or community, this is indeed something extraordinary, something unexpected and novel. A new level of emergence develops its own regularities. But it still presupposes the entire infrastructure of the impersonal levels of emergence. Spirit is not an entity that can hover off into a realm of its own, as in Platonism.

The risen Christ stands for the new humanity in fellowship with God, which we are all invited to share. As Paul says, believers are 'in Christ', or 'in the Spirit'. As a notion of the new life of Christ, it is inextricably linked to the ministry of the historical person of Jesus of Nazareth as a biological reality. But after his death, it is the ongoing spiritual presence of a biologically deceased person among biologically living people that matters.[442]

Cosmic projections of divine validity

Understood as the destiny of reality, the concept of God implies a vision of what ought to be. In Section III, I have discussed four biblical projections of a reality that was deemed 'very good'—the mythological 'beginning' of time, the eschatological 'end' of time, an alternative 'heavenly' space, and the authentic 'essence' of reality hidden under its inauthentic appearances. If the new life of Christ represented the ultimate intentionality of God, it was only natural to identify it with these projections of what ought to be.

In terms of time, Christ was identified with God's creative act at *the beginning* of time—whether in terms of the creative decree of God (Gen. 1, cf John 1:1-18; Phil. 2:6 f) or the creative wisdom of God (Sirach 1; Wisdom of Solomon 7; cf John 1:1-18; Col. 1:15 ff; 2:3). He was also identified with God's envisaged transformation of reality at *the end* of time—whether as the Judge of the last judgement, the King of God's coming kingdom, or the Principle of the new creation (Rev. 21-22).

In terms of space, he was identified with God's superior 'position', being seated 'at the right hand of God'. This was a messianic metaphor for *divine authority* (Eph. 1:20 ff; Matt. 28:18; Mark 16:19; Acts 7:55). He was similarly identified with the *essence* of reality, that is, with what ought to be—which was currently obscured by what ought not to be (Col. 3:4; Eph. 1:9; 3:5, 10). As such, he was the 'new man' (Eph. 1:10; 2:13 ff), in opposition to the 'old Adam' (Rom. 5:12-21). And we are supposed to move out of the one into the other (Rom. 6).

The development of the notion of the 'cosmic Christ' clearly demonstrates, therefore, what the New Testament believers wanted to communicate. They affirmed the divine self-identification of God's intentionality with the proclamation and enactment of Jesus by attributing the 'characteristics' and 'functions' of God to the 'risen Christ'. The God of Israel was a God of suffering, liberating, transforming and empowering love.

Although scientists may find it difficult to enter into the metaphorical language of an ancient and pre-scientific discourse, once they do so they will find nothing that contradicts scientific insight. They will only find a potent source of meaning, a definition of acceptability and an allocation of legitimate authority.

Getting rid of ideological abuses of faith

If God's intentionality has been defined by what happened in the ministry of Jesus of Nazareth, culminating on his crucifixion, we can no longer condone the abuse of the Christian faith to legitimate oppressive and exploitative pursuits and social structures whether in the church or in society. The same is true for any kind of self-aggrandisement by an ethnic, national, racial, or social group at the expense of other such groups.

The use of religion for the ideological legitimation of the pursuit of power and privilege at the expense of others has been a problem in all cultures throughout human history and the world over. This scourge did not bypass the biblical tradition. I have repeatedly referred to the fact that, in the Ancient Near East, the king was believed to be the "Son of God", that is, God's representative on earth. As such, he had the right and the duty to subdue humankind. I have on various occasions referred to Psalm 2 as a case in point.

Psalm 2 seemingly reflects the enthronement ritual of a Davidic king. Taken at face value, it is a disconcerting example of the religious legitimation of oppressive rule. All the earth is given to the king as his rightful heritage. The heavenly king himself scoffs at any rebellion against his earthly representative. Biblical scholarship traced the origins of this paradigm to ancient Egyptian examples. The Egyptian Pharaoh was deemed the son of the sun-god, Re or Ra.

In biblical history, this ideology was progressively undermined and ultimately turned on its head.[443] Jesus declared the oppressive model of Psalm 2 "pagan", which indeed had been in its original form. He defined messianic leadership as humble service (Mark 10:35-45; cf John 13:1 ff). The Apostle Paul applied this principle to his apostolic authority (2 Cor. 1:24). It is a prime example of how God enters human reality to change it from within.

The model of Psalm 2 should have lost its legitimacy within a Christian context from the very outset. Alas, it was just too alluring to be left alone. The gospels speak of rivalry between the disciples of Jesus. The letters of Paul witness to intense power struggles among the apostolic leadership of the church. With Constantine, at the very latest, leadership again assumed the features of Psalm 2.

Art works of the time depict Christ as a majestic and stern emperor. What happened in the state was paralleled by developments in the church. The Roman bishop claimed to be the *vicarius Christi*, the sole representative of the heavenly king and the bearer of universal authority. Christian emperors claimed the same status for themselves, leading to bloody and drawn-out conflicts between the two offices of pope and emperor.

When feudalism began to crack, princely absolutism took its place. The ruler claimed to have received his authority "by the grace of God". Western dictatorships secularised the legitimating authority but not the model. Examples from recent history abound. Hitler and Stalin are only the most frightening outgrowths of a common trend.

A similar trajectory can be traced with regard to ethnocentrism.[444] The Old Testament presents us with a classical example of ethnic exclusiveness and hostile in-group-out-group attitudes. According to Deuteronomy 7 and Exodus 23, the Canaanites had to be driven out of the country because they were not part of the covenant people of God and did not keep the law of Yahweh. Moreover, the law codes of the Priestly Source give ample examples of how anything considered ritually impure had to be excluded (if not exterminated) from the people of God—by violent means if necessary.

According to the texts, these attitudes occasionally culminated in genocide (1 Sam. 15). But such examples are found in most cultures and religions throughout history and the world over. Again ethnocentrism was turned on its head within biblical history. According to Ephesians 2, the dividing wall of the law (that is the body of all conditions of acceptance) was demolished and a new humanity was formed out of former enemies.

In modern times, the urge towards self-absolutisation and exclusion has been democratised. It manifests itself in the pursuit of individual and collective self-interest regardless of the costs to others and to nature.

Wherever this tendency appears within the church of Christ, we need to recognise it as what it is, namely, an aberration of the Christian faith and a denial of its very rationale. However, like the gospel itself on which it is based, it is a lesson that does not easily penetrate our instinctual preference for self-preservation, our insistence on conditions of acceptance, and the exclusion or elimination of potentially threatening competitors and strangers.

Wherever God's suffering acceptance of the unacceptable becomes the basis of human attitudes, actions and institutions, exclusion, discrimination and marginalisation cannot survive. It must be exposed as such and overcome, rather than given out as a legitimate facet of the Christian faith.

Section XI
The community of believers and the divine Spirit

In the New Testament, the new life of Christ was identified with the life of the believers in as far as it was a life 'in Christ' or 'in the Spirit'. The New Testament ascribed whatever was said about Jesus to those who came to participate in the new life of Christ in communion with God through faith.

This is important for our attempt to reconceptualise the New Testament message in terms of an experiential approach to reality. It shows that there is no *ontological* difference between Christ, the authentic human being, and other human beings. The difference lies in the new relationship between God and Christ in which we are invited to participate.

Believers were 'born of God' like Jesus (John 1:12; Gal. 3:25-4:7); they were called to proclaim and enact the same message (compare Matt. 9:35 with Matt. 10:7 f); they would encounter the same resistance; they would suffer the same fate of being cast out, persecuted, imprisoned and condemned (Matt. 10:16-25; John 15:18-21); they die with Christ to be raised with him and enthroned with him 'in the heavenly places' (Eph. 2:4-6; cf Rom. 6:3-11); they were to act as judges of the last judgment (1 Cor. 6:2); they would carry the death of Christ in their bodies, so that the new life of Christ would also manifest itself in their bodies (2 Cor. 4:10-12; Rom. 6).

The Spirit of Christ permeated, liberated, transformed and empowered the 'Body of Christ', the community of believers (1 Cor. 12). With their concrete bodies, they were the 'members' that constituted the ongoing bodily manifestation of the risen Christ on earth (1 Cor. 6:13-17; Eph. 2:14 ff; 4:4) or the 'temple' of God (Eph. 2:15-22). In John, God identified with Christ and Christ identified with those who believed in him. Believers identified with the God of creative and redeeming love that manifested itself in the ministry of Jesus.

These observations link up with my contentions on the nature of the 'spirit' in Chapter 6. There is no question that the concept of the 'Spirit' of God that manifested itself in the Christ-event and continues to do so in the 'Body of Christ' is an example of structured and oriented individual and communal consciousness. It is not something uncanny or otherworldly. It may be in conflict with other such spiritual phenomena, but it is not in conflict with the basic spiritual potential that characterises all humans. The following observations clearly point in this direction:

(a) It is based on a narrative that is deemed to have happened in human history (Luke 2:1 ff; 1 John 1:1 ff) and that is deemed of ultimate significance in terms of human authenticity.

(b) It travels through time and space in the form of a tradition and reaches us through human communication (the proclamation of the gospel—Rom. 10: 14-17).

(c) It becomes operational only when heard and accepted in faith (Rom. 10:14 ff).

(d) It is expressed in terms of a particular frame of reference (focusing on the Jewish concepts of divine-human relationships, divine authority, human sin, divine judgement and a redeemed creation).

(e) It must be distinguished from other such phenomena and can be corrupted by false prophets, messiahs and teachings (1 John 4:1 ff).

(f) Its particular content can easily be lost and believers must do their best to retain it.

(g) According to Paul and John, being 'in the Spirit' is the same as 'being in Christ' which is the same as 'being in God'. Believers must 'remain in' Christ; Christ must 'remain in' the believers, just as Christ remains 'in God' and God remains 'in Christ'. In short, Christ and the believers share the creative and redemptive intentionality of God.

(h) This intentionality is clearly defined. Being in the Spirit is identical with remaining in the truth, which again is identical with remaining in the power of divine love. It refers to God's creative and redemptive project in which we are invited to participate.

According to the Council of Chalcedon, the divine and the human 'natures' of Christ (God's intentionality and its human manifestation) must not be confused, nor must they be separated. By the same token, it is wrong to juxtapose divine and human spirit. The divine Spirit is real, effective and accessible for us only in the form of a human spirit.

We must rather distinguish the 'Holy Spirit' from an 'unholy spirit', a spirit of self-centredness, avarice and arrogance. The 'Holy Spirit' is a spirit that manifests God's vision of comprehensive optimal well-being, that expresses God's loving concern, and that calls us into participation in his creative and redemptive project for reality as a whole.

Section XII
The Trinity

The classical Trinitarian doctrines are not found in the New Testament, but the meaning they want to convey constitutes its very centre. The classical formulations were occasioned by a translation of the Jewish version of the Christ-event into a Hellenistic (ontological) frame of reference. As an attempt to 'become a Greek to the Greeks', as Paul would have expressed his mission to the Gentiles (1 Cor. 9:19 ff), this was entirely legitimate. Given the underlying frame of reference, the decisions of the 'ecumenical councils' were valid expressions of the Christian faith as I see it.

However, in the course of this reconceptualisation, the anthropomorphic metaphors found in the Bible (creator, father, son, word, spirit) were reified and used as building blocks for the construction of a metaphysical edifice, along with the abstract terms typical of Greek thought (being, essence, substance, person, nature, eternity and so on). This is where the difficulty arises for those of us who no longer think in terms of a Hellenistic frame of reference.

The procedure led to a particularly messy picture. Instead of making the gospel message plausible and lucid, these formulations have cast a blanket of obscurity and incoherence over it. Instead of exposing the historical, social and existential roots of the metaphors used in the biblical tradition, they have pinned down the intellectual energies of countless theologians throughout the ages.

Instead of raising the enthusiasm of participating in God's creative and redemptive project, they caused puzzlement among the laity for centuries. Priests and pastors assign them the dignity of 'divine mysteries' that have to be believed and worshipped rather than understood. This sounds very devout, but a lack of comprehensible meaning inevitably undermines the content and credibility of the message itself.

The New Testament exudes a different atmosphere altogether. It speaks of the mysteries of God that had previously been hidden, *but that have now been revealed*, that could be understood, accepted and enacted (Eph. 1:8-10, 17-19 4:12-16; Col. 2:2 f). The self-disclosure of God's creative and redemptive intentionality through the ministry of Jesus and its ongoing manifestation in the community of believers could be understood. No sophistication was needed to "account for the hope that is in us" (1 Pet. 3:15).

It is counterproductive to continue with this ancient rendition of Christian faith assumptions at a time when Hellenistic metaphysics has lost its traction. People informed by modern science just cannot make sense of the classical Trinitarian formulations any longer. In fact, the intended meaning is quite straightforward. In experiential terms, the doctrine expresses three aspects of a living Christian faith:

(a) The primordial experience of derivation, dependence, vulnerability, mortality and accountability to an authority higher than one's own. This experience finds its expression in the notion of God as the transcendent Source and Destiny of reality as a whole.

(b) The authoritative proclamation and enactment of the redemptive intentionality of this God by Jesus of Nazareth in the context of a retributive legal framework, witnessed by his first disciples and reaffirmed by the message of his 'resurrection'.

(c) The ongoing spiritual presence of the historical manifestation of God's redemptive intentionality through the ministry of Jesus of Nazareth in the collective consciousness of the community of believers.

The first two motifs may seem to contradict each other in experiential terms simply because they express the discrepancy between what reality has become and what it ought to become. Faith will throw its weight behind what ought to become in the face of what ought not to have become. As Luther has taught us, faith is the tenacious trust in the benevolent intentionality of God in the face of all experiences to the contrary.

The relation between the second and the third aspect deals with the continuity between what happened in Jesus, the messianic representative of God, and what happens today in the community of believers. On the basis of these observations, the three aspects of the Trinity can be expressed in experiential terms as follows:

(a) the creative and redemptive intentionality of the transcendent Source and Destiny of reality,
(b) that manifested itself paradigmatically in the Christ-event and
(c) that continues to be present and active as the Spirit of Christ, which permeates the life of the community of believers.

All three of these facets of God's creative and redemptive presence are reflected in structures of individual and communal consciousness that are part of the hierarchy of emergences. There is nothing supernatural, mysterious, or uncanny about them.

Those of us who have waded through endless theological speculations and logical somersaults to come to grips with the meaning of the Trinitarian doctrine may be surprised how simple it actually is. Scientists who are open for the spiritual dimension of human existence should have no insurmountable difficulties with this interpretation either.

As a particular concept of the transcendent, the Trinity competes with other such concepts. While indeed it lays claim on human existence and human interaction in a compelling way, it is still subject to human error; it can be critiqued, revised, corrected, augmented, or discarded; it has to prove itself in terms of the criterion of comprehensive optimal well-being; it can be enriched by other 'spirits' and enrich other such 'spirits'.

Let us summarise

Faith must regain its historical dynamic. It must integrate best science. It must listen to its critics. It must be open to transformation. If not, believers lose their integrity and their message loses its credibility. Necessary reconceptualisations can only lead faith closer to its essential core. If faith persists with spurious assumptions, inappropriate formulations and problematic patterns of behaviour, it has no message for the modern world.

The Bible itself is a powerful witness to the dynamic character of the Word of God. It responds redemptively to changing needs and worldviews. Its proclamation and ritual enactment witnesses to God's suffering acceptance of the unacceptable. We must do for our times what the biblical authors did for theirs, rather than absolutising historically relative statements. In this chapter, I have suggested ways in which theology can reconceptualise its traditional worldview assumptions. None of these reconceptualisations clash with modern scientific insights, nor with the essentials of the biblical message. Here they are in a nutshell:

The concept of God refers to the transcendent Source and Destiny of reality as such and as a whole, rather than to a factor among others within immanent reality. As the source of reality, God's 'activity' is coextensive with the world process that we experience and that the sciences explore. As the destiny of reality, God determines what ought to be.

In the Bible, the vision of what ought to be is expressed in four projections—to the mythological beginning of time, eschatological end of time, a 'heavenly' space, and an authentic 'essence' of reality hidden 'beneath' its inauthentic appearances. All four these projections are rendered implausible by current scientific insights if interpreted literally as what reality actually was in the past, is now, or will be in the future.

The concept of creation refers to the derivation, dependence, vulnerability, mortality and accountability of human life within its entire social and natural context, rather than to first beginnings such as the big bang. God is the source of the whole of reality in its evolutionary flux.

The concept of a 'miraculous divine intervention' does not imply the suspension of immanent regularities, but their optimal utilisation to bring about greater well-being. The theory of emergence throws light on the ancient problem of theodicy: while God becomes a person for humans because humans are persons, God's 'action' covers the impersonal levels of reality as much as the level of personal intentionality.

The future expectations of the biblical faith responded redemptively to experiences of what ought not to have become. They are meant to transcend human predicaments and calamities and lure us forward towards a vision of comprehensive optimal well-being.

This vision resembles a horizon that moves on as we approach it, opening up ever new vistas, challenges and opportunities. The apocalyptic promise of a total makeover of reality at the end of cosmic history was a radical redemptive response to seemingly irredeemable situations and should not be absolutised.

The concept of the human being must be de-absolutised and assigned its actual position in the framework of cosmic space, time and energy. An immense network of causes led to the emergence and evolution of an individual human consciousness and an immense network of consequences trail its demise. What matters for faith is not the infinite prolongation of conscious life but the contribution it made to God's creative and redemptive project, thereby participating in 'the life of God'.

The traditional doctrine of Christ becomes plausible if taken to refer to the creative and redemptive intentionality of the true God acting through the intentionality of the true human being, the representative of God on earth. Its substantive content is God's suffering, transforming acceptance of the unacceptable as opposed to a demanding and condemning law.

The Holy Spirit is the Spirit of Christ—marked by self-giving concern for the well-being of the other and the whole—permeating, liberating and empowering the 'Body of Christ', the community of believers.

The doctrine of the Trinity can be understood, therefore, as the redemptive intentionality of the transcendent Source and Destiny of reality manifesting itself in the Christ-event and its continued presence in the community of believers.

Conclusion —
The complementarity of science and faith

If science and faith are to face their common responsibility in regaining a future for suffering humanity and a threatened earth, they have to discover their complementary tasks. Science looks at reality from an immanent perspective, while faith looks at the same reality from a transcendent perspective. Science needs best faith to be responsible; faith needs best science to be credible. Adopting an experiential approach to both pursuits, we have first explored the status of a concept of the transcendent from a scientific perspective and then integrated the findings of science into the perspective of faith. These theoretical reflections gain existential relevance when believers and scientists are confronted with the demand and the gift of an authentic existence, that is, a life that participates in the creative and redemptive project of God.

Summary of the argument

We have come to the end of a long journey, so let us retrace our steps. Beginning in Part I with the incredible achievements and hazardous ambiguities of the modernist revolution, I argued that, in view of the immense dangers that we are faced with, faith and science can no longer afford to be drifting apart. Science needs 'best faith' for an overarching system of meaning that can offer criteria of acceptability, vision, liberation, motivation and orientation. Faith needs 'best science' to pull it out of individualistic spirituality, mystery and fantasy and become engaged with the real world in which humankind is embedded and where its task is located.

I then proceeded to discuss the relation between science and faith in two directions. I looked at the reality that we all experience and that the sciences explore first from an immanent, then from a transcendent perspective.

In Part II, I tried to figure out where faith fits into immanent reality, as examined by the sciences. I called the method that seems to be appropriate for such a perspective 'experiential realism'. It entails a strict confinement to the reality we actually experience, or that we could experience if we had the capacity and the tools to do so. This method avoids the following:

- Sophisticated obscurantism
- Epistemological scepticism
- Reification of biblical metaphors
- Reification of idealised abstractions
- Reification of fantasy and fiction
- Deductions made from untested assumptions
- Postmodern relativity

Experiential realism does not engage in subatomic, physical, or biological reductionism, but treat all levels of emergence as real, including the structures of human consciousness and the structures of society. This approach is wider than crude empiricism because it includes whatever can be described, comprehended, critiqued and transformed, and what has concrete consequences in this world. The theory of emergence allows us to cover all academic disciplines engaged in the quest for an understanding of the reality we experience within the same conceptual scheme.

On this basis, I offered a brief overview of current scientific theories concerning big bang cosmology, entropy and the hierarchy of emergences. The latter included, in principle, all levels of emergence from subatomic physics, through classical physics, inorganic chemistry, organic chemistry, biological evolution, the brain, the subconscious, consciousness, spirit as structured consciousness, historical sequences and social structures and processes. Each of these levels is characterised by its own regularities. Because humans are endowed with intentionality and agency, they are confronted with choices that have consequences and call for accountability.

In Part III, I ventured the suggestion that freedom from the world and responsibility for the world presuppose the capacity to objectify experienced reality, including the reality of the personal self, and subject it to critique and reconstruction. This calls for an (imaginary) vantage point 'above' reality, thus a notion of the transcendent Source and Destiny of reality. Such a notion is capable of providing meaning, criteria of acceptability and authority to existential, communal and social processes.

Is such a notion irrational? I argued that neither humanity nor the world in which it is embedded seems to be self-generated, self-contained, self-sustaining, self-sufficient and self-accountable. A notion of the transcendent Source and Destiny of reality opens up our awareness towards all-inclusive contexts. To be appropriate, it must cover both the transcendence of the human person as well as the transcendence of all impersonal levels of emergence. This is as far as one can get when trying to describe and comprehend the phenomenon of faith from an immanentist perspective.

In Part IV, I looked at the reality that we experience and that the sciences explore from a transcendent point of view. To be relevant, the transcendent must be understood as the transcendent of the *immanent* reality that we experience and that the sciences investigate. I argued that, if God were to be understood by faith as the ultimate Source and Destiny of experienced reality, faith would have to integrate 'best science' in its view of reality. If it could not do that, one would have to suspect that it has hovered off into the sphere of fantasy or speculation. I called this approach 'integrative transcendence'.

Is an experiential approach to Christian theology possible? Is God not infinitely beyond anything humans can experience, describe, comprehend, explore and manipulate? Indeed God is, but there is a difference between the transcendent as such, which is inaccessible, and our notion of the transcendent, which is part of immanent reality and can be described, critiqued and transformed. If that were not the case, theology would be pointless.

However, the Christian faith is persuaded that its concept of the transcendent reflects the self-disclosure of the transcendent as creative and redemptive intentionality, albeit in a very provisional, fragile and problematic sort of way. This conviction is based on the insights gained during the emergence and evolution of the Israelite tradition that responded redemptively to situations of human need over more than a millennium of ancient Israelite history. It culminated in the Christ-event and continues to transform human lives in the context of a community of believers.

With the recognition that the assumption of a self-disclosure of God's intentionality is constitutive for the Christian faith, I entered into the sphere of theology proper without abandoning my commitment to an experiential approach. To explore what can be done, even within a traditional theological frame work, I examined the approach of Martin Luther. This approach is embedded in a pre-scientific worldview, but among classical theologians, Luther follows an experiential approach to faith and theology most consistently. That is what makes his theology more amenable to a discourse centred on faith-science relationships than others of its kind.

The outcome of this investigation was surprisingly positive. Luther distinguished between the creative power and the redemptive intentionality of God. The *creative power* of God is experienced in ordinary life as deeply ambiguous and does not reveal God's redemptive intentions. The proclaimed *redemptive intentionality* of God as disclosed in the Christ-event and paradoxically is manifest most succinctly and paradoxically in the catastrophe of the cross of Christ.

The experience of God's power in ordinary life does not need any special revelation. It can be explored and explained through observation and reason, thus giving the sciences all the leeway they need to do their job. The proclamation of God's redemptive intentionality, in contrast, is a promise that can unfold its transformative power only when it is accepted in faith. As a vision of what ought to be, it stands in contrast to the experience of what ought not to be. It is not based on empirical evidence but on a defiant commitment to what ought to be. It does not take no for an answer but forges ahead in the direction of God's vision of comprehensive well-being.

Seen in this perspective, faith is self-entrustment to, and participation in, the creative and redeeming love of God. This love has manifested itself paradoxically in the catastrophe of the cross of Christ. It becomes effectual when people respond positively to the challenge and the offer of an authentic life—a life that participates in God's creative, redemptive and sacrificial project in the world and that overcomes what ought not to be in the direction of what ought to be.

In the last chapter, I suggested that, to become both plausible and relevant in a world informed by science, faith and theology must reconceptualise their pre-scientific traditions. The biblical faith has been exceptionally versatile in its canonical history. It has responded to ever new situations of need and dominant frames of reference. Contemporary theologians and believers must do for their times what the biblical authors did for theirs. Their worldview must integrate the best insights now available. Without trying to be dogmatic or prescriptive, I offered a few pointers in this direction:

Biblical hermeneutics can no longer afford an abstract and static concept of revelation but must follow the actual historical emergence and evolution of the biblical witness. We cannot continue with a *concept of God* as an inner-worldly factor among others, rather than the transcendent Source and Destiny of reality as a whole. We cannot continue with an understanding of *creation* that confines the activity of God to gaps in the causal network. God is the transcendent Source of the entire evolving network. We cannot continue with an apocalyptic understanding of *eschatology*. God's vision of comprehensive well-being is a 'moving horizon' that opens up ever new vistas, challenges and opportunities.

We should not reify mythological and metaphorical statements about *Christ* but take both the humanity of Christ and the self-disclosure of God's intentionality in the ministry of Jesus of Nazareth seriously. We must not indulge in obscurantist *Trinitarian* speculations but simply state that, according to Christian convictions, the creative and redemptive intentionality of God, the ultimate Source and Destiny of reality, disclosed itself in the Christ-event and continues to be present and effective in the divine Spirit of Christ that permeates, liberates and motivates the community of believers.

We can no longer substitute the *gospel* message of God's suffering transformative acceptance of the unacceptable, as manifest in the Christ-event, with the conditional acceptance based on the fulfilment of a law, divine or otherwise. We can no longer assume that the *Eucharist* is all about mysterious substances but understand it as the identification of an intangible event, the redemptive act of God for us on the cross of Christ, with a tangible medium, bread and wine, so that it can be specifically allocated. We should at all times be alert against the *abuse* of the Christian faith to legitimate the pursuit of individual desire and collective interests at the expense of others, the community, the society and nature.

In short, when dealing with modern people in general and scientists in particular, faith and theology should rigorously apply Occam's razor, confine themselves to the essentials of the Christian faith, eliminate the immense speculative outgrowths that have accumulated over three millennia of interpretation and communication, and aim at as lucid and self-explanatory conceptualisations of the Christian faith as possible.

The existential relevance of the study

So far the overview. In this book, I concentrated on the problem of the disjuncture between a Christianity that has lost its relevance and a modernity that has lost its bearings. Aiming at human mastery of reality, science is the epitome of modernity. In practice, scientists may be more concerned with their daily research and its desired outcomes than with the greater contexts in which it is embedded, the system of meaning it presupposes and its ethical implications in the modern world.

Likewise, believers may be more concerned with their personal spirituality than with the fate of the world of which they are a part.

If the interaction between these two pursuits does not begin to determine our personal lives, they will remain pure theory. If they do not break out of their narcissistic preoccupation with our petty interests and desires, they will not orient our thinking and acting in the direction of the comprehensive well-being of the whole of the reality in which we are embedded. The actual interface between science and faith will have to play itself out in the existential struggle to make sense of the whole of the reality we experience and, at the same time, the struggle to become personally accountable to its assumed transcendent foundations.

Before signing off, therefore, allow me to bring the relation between science and faith, between comprehension and commitment, between explanation and vision back home into the personal sphere. What are the stakes for scientists as persons, and what are the stakes for believers as persons?

Personal stakes for scientists

I have argued that, to become human, science, technology and commerce need to rediscover the transcendent foundations of life. But why bother? The concerns of faith seem to be way beyond the mandate and method of science. What would happen if it could be shown that there was no God? If science, technology and commerce were nothing but tools, nothing would happen at all.

But what if this pragmatic approach translates into the metaphysical assumption that there is no transcendent meaning, criterion, or authority to be concerned about? Then personal desires and collective interests become ultimate motivations. Then science becomes indifferent to faith. Then technology and commerce can simply ignore it and concentrate on their mundane aspirations. Then the consumer culture can jettison all ethical inhibitions.

It is this attitude that makes modernity such a profoundly dangerous mindset. Scientists, technicians, leaders in business and consumers can end up in perpetual denial. They can be just as confined, short-sighted and provincial in their laboratories as prisoners in their cells. They can become victims of demagogues and profit-hunters. Scientists, technicians and business folks cannot afford cynicism to take over, whether in terms of their personal lives, their professions, society, or nature.

Science, technology, commerce and consumer abundance have been a blessing and a curse at the same time. Scientists cannot afford to overlook this ambiguity.[445] They are at the helm of the most potent forces in a globalising society today. Without science, technology and commerce, political and military decision makers would be powerless. The masses of the population have become totally dependent on their work. What these professions do, or do not do, will determine the future of humanity and the planet forever after. They have created or catalysed the development of all typically modern problems. And without them these problems can no longer be resolved.

They must regain the awareness that they are not gods but derived, dependent, accountable, vulnerable and mortal human beings. They cannot shed their responsibility for the well-being of future generations and the earth. They are accountable to an authority higher than themselves. They have to find their bearings within a more comprehensive reality or get lost in the trivialities of life. The immense powers that their professions have unleashed and continue to unleash has raised this responsibility to a higher order of magnitude.[446] All this is not fantasy but fact.

The detonation of nuclear devices in World War II has opened the eyes at least of some scientists to their unimaginably heavy responsibility. Today global warming, the massive extinction of species, the destruction of forests and maritime resources, the depletion of fossil fuels, the dangers of nuclear power, declining food and water security, new and resistant viruses, the dangers of genetic engineering and many other such developments have added urgency to a change in attitudes.

The fact that scientific, technological and commercial developments have "gone far beyond any control of human forces"[447] makes the situation even more precarious. Only the formation of a critical mass of awareness, alertness, responsibility and determination among leaders at many levels and in many fields across the globe can prevent total catastrophe.

This is the reason for my argument that science needs 'best faith' to be responsible. Even the most pragmatic scientists cannot suspend the human quest for the universally most appropriate and most redemptive vision available. Seen in terms of the grounding and orientation of human existence, questions concerning the ultimate Source and Destiny of reality may be the singular most important question humanity faces today, the question of life and death for earthly existence in general and humanity in particular. As stated above, concepts of the transcendent can be hugely inappropriate, dysfunctional and counterproductive. But that makes the struggle for the truth not less but more important and urgent.

I do not claim that the Christian faith is 'best faith', but I do suggest that it has the potential of becoming 'best faith' if it gets its act together. For modern people, it is not very attractive, if it ever has been. I sympathise with those for whom the message of the plausibility and acceptability of naturalist assumptions has been a liberating experience: no more pressure to believe and confess what makes no sense; no twenty-four-hour surveillance by a stern and omnipresent parent or judge; no expectation to be 'meek and mild' when getting hot under the collar; no need for unnecessary frugality and prudish inhibitions; no anxious questions about the hidden designs of a God of love and power who allows atrocities and catastrophes to happen.

But all this presupposes that faith in Christ is an irrational and oppressive state of mind. This assumption flies into the face of what the Christian faith was always meant to be. It is most unfortunate that the disillusionment with an inappropriate faith has led to the abandonment of faith altogether. The concept of God stands for the power underlying reality and the vision of what reality ought to become—the very reality that science explores, that technology manipulates, that commerce exploits, and that consumer hedonism abuses. And the biblical God has offered humanity justice, redeeming love and personal fulfilment.

Christians believe in Christ because in Christ they have been confronted with the demand and the gift of an authentic existence and the prospects of a meaningful life world. In Christ, believers encounter the potential authenticity of their own lives, the lives of others and their entire contexts, defined and empowered by the Source and Destiny of reality as a whole. There is nothing irrational, outlandish, or reprehensible about such an encounter. In fact, believers experience it as the most foundational, the most enlightening, the most liberating, and the most motivating aspect of their existence.

The stakes for believers

Why should believers bother to take the insights of modern science seriously? Faith seems to have survived quite well for three millennia without leaving its own 'symbolic universe'. But this is a deceptive impression. At least in societies affected by modernity, faith has been pushed into the status of an irrational and irrelevant private pastime. The vast majority of the population has voted with their feet. It has become redundant for all practical intents and purposes. It plays no role in dimensions of public life that really matter—whether sexual morality, family life, social construction, economic achievement, or political organisation. Where it tries to do so, it is experienced not as a help, but as an irritant.

If God was indeed the ultimate Source and Destiny of reality, the One that constituted our identity and defined our authenticity, the One that had a vision for reality as a whole and invited us into his creative and redemptive project, how could faith possibly have become a private pastime? If the Christian message really confronted modern humanity with the choice between authentic and inauthentic existence, between a meaningless and a meaningful world, it would be either embraced and treasured or rejected and persecuted. Simply being ignored is the most devastating form of irrelevance and contempt imaginable.

Clearly the Christian faith is under siege.[448] But we must understand what it is that besieges us. It is not simply the godlessness of the world. It is our own godlessness. We do not entrust ourselves to God as we move into the future, but cling to idolised remnants of the past. It is the lack of plausibility of our faith assumptions, the lack of credibility of our commitment, the lack of integrity in our lives that has brought us into this unenviable situation. Our faith has become obsolete, irrelevant and redundant. We have been left behind at the airport while the jumbo jet of modernity has roared off into distant skies.

For believers, the stakes are enormous. They cannot afford to let their faith degenerate into irrationality, superstition and triviality. Theologians cannot continue as if they were living in biblical or medieval times. If their faith cannot integrate the best insight that is currently available, believers cannot hope to be taken seriously. They cannot even take themselves seriously, because in the depths of their hearts, they know better.

They may enjoy the ecstasy of a crowd singing 'hallelujah' on Sunday mornings. But on weekdays, they may sheepishly and awkwardly move about in the modern world with the gnawing suspicion that there is something amiss in their faith assumptions. If that is the case, they will not be pillars of confidence, vision, responsibility and determination in a modern society gone astray. On the contrary, they will betray symptoms of cognitive dissonance and psychological repression.

The certainty of a special covenantal relationship with God has sustained the Jewish faith through most unbearable circumstances. The iconoclastic response of the prophets to an institutionalised religious routine has held the ancient Israelite community of believers on edge. The reconceptualisation of the Jewish faith by an itinerant teacher in a remote province struck the very foundations of the mighty Roman Empire and its Jewish satellite regime in Jerusalem. The reconceptualisation of the Christian gospel offered by the Reformation struck at the very foundations of medieval European society including the institutional church.

It can happen again if we dare to translate the essence of the gospel into modern frames of reference. The conceptual adjustments needed today are no more earth-shattering than the emergence of post-exilic Judaism, the proclamation of Jesus, or the translation of the gospel into Greek metaphysical assumptions that led to the classical dogmatic tradition. We must just allow ourselves to be led out of our emotional comfort zones and inherited doctrinal formulations by the Spirit of Christ. The Spirit of Christ is ever geared to the future; it always responds creatively and redemptively to the challenges that the world poses in every new phase of history.

Only if that happens will the scandal of the cross again present itself to modern humanity. This scandal does not consist of an obsolete worldview. It consists of the invitation to abandon the human quest for autonomy, mastery and ownership of reality in favour of participation in the freedom, the responsibility, the creative authority and the redeeming love of God, the Source and Destiny of reality as a whole. It involves becoming part of God's vision of comprehensive optimal well-being and God's sacrificial intentionality to overcome any deficiency in well-being in any dimension of life on earth.

Bibliography

Albright, C. R. and Haugen, Joel (eds). 1997. *Beginning with the End: God, Science and Wolfhart Pannenberg*. Chicago: Open Court.

Althaus, P. 1966. *The Theology of Martin Luther*. Philadelphia: Fortress.

Altmann, W. 1992. *Luther and Liberation*. Minneapolis: Fortress.

Barney, Gerald O. (ed) for the Council on Environmental Quality. 1980. *The Global 2000 Report to the President*. Washington: US Government Printing Office.

Barth, K. 1968. *Community, State and Church*. Gloucester Mass: Peter Smith.

Barrett, P. 2000. *Science and Theology since Copernicus*. Pretoria: Unisa.

Bayer, O. [2003] 2004. *Martin Luther's Theologie*. Tübingen: Mohr Siebeck.

Berger, P. 1967. *The Social Reality of Religion* (alternative title: *The Sacred Canopy*). Harmondsworth: Penguin Books.

Berger, P., Berger, Brigitte, and Kellner, Hansfried. 1974. *The Homeless Mind: Modernization and Consciousness*. Harmondsworth: Penguin Books.

Beste, D. 2000. Dossier Kosmologie. *Spektrum der Wissenschaft* 2/2000. Heidelberg: Spektrum der Wissenschaft.

Ben-Ari, M. 2005. *Just a Theory: Exploring the Nature of Science*. Amherst, NY: Prometheus Books.

Birch, C. et al. (eds). 1978. *Faith, Science and the Future* (Preparatory readings for a world conference organized by the WCC at the MIT, Cambridge, MA, in 1979). Geneva: World Council of Churches.

Boyer, P. 2001. *Religion Explained: The Evolutionary Origins of Religious Thought*. New York: Basic Books.

Brümmer, V. 2005. *Atonement, Christology and the Trinity: Making Sense of Christian Doctrine*. Aldershot, UK: Ashgate.

Bryson, B. 2003. *A Short History of Nearly Everything*. New York: Broadway Books.

Buddeberg, S. 1962. *Grundformen christlichen Lebensgefühls*. Stuttgart: Ernst Klett.

Cahoone, L. E. 1988. *The Dilemma of Modernity: Philosophy, Culture and Anti-culture*. New York: State University of New York Press.

Clayton, P. [2004] 2006. *Mind and Emergence: From Quantum to Consciousness*. Oxford: Oxford University Press.

Clayton, P. 2008. Toward a theory of divine action that has traction. In Russel, Murphy and Stoeger (eds) 2008:85 ff.

Convey Morris, S. 1998. *The Crucible of Creation: The Burgess Shale and the Rise of Animals*. New York: Oxford University Press (as quoted by van Huyssteen).

Dawkins, R. 1995. *River Out of Eden: A Darwinian View of Life*. New York: BasicBooks.

Dawkins, R. 1998. "Postmodernism Disrobed". *Nature* 394/1998, 141-143.

Dawkins, R. 2006. *The God Delusion*. London et al.: Bantam Press.

Dawkins, R. 2009. *The Greatest Show on Earth: The Evidence for Evolution*. New York: Free Press.

Drees, W. B. 1990. *Beyond the Big Bang: Quantum Cosmologies and God*. La Salle Ill: Open Court.

Drees, W. B. 1998 (1996). Religion, Science and naturalism. Cambridge: Cambridge University Press.

Drees, W. B. *Religion and Science: A Guide to the Debates*. London: Routledge [taken note of the announcement].

Du Toit, C. W. (ed) 2000. *Evolution and Creativity: A New Dialogue between Faith and Knowledge*. Pretoria: Research Institute for Theology and Religion, Unisa.

Du Toit, C (ed) 2010. *Homo transcendentalis? Transcendence in science and religion: Interdisciplinary perspectives*. Pretoria: Unisa Research Institute for Theology and Religion.

Durand, Francois 2010. The rise of the human predator and the transcendence of consciousness. In: du Toit, C (ed) 2010:53-75. Eagleton, T. 2009. Faith, reason and revolution. Yale Univ Press.

Ebeling, G. 1972. *Luther: An Introduction to His Thought*. Philadelphia: Fortress.

Ehrlich, P. E. [1996] 1998. *Betrayal of Science and Reason: How Anti-environmental Rhetoric Threatens Our Future*. Washington: Island Press.

Ellis, G. 2008. 'Scientific issues: Ground covered and horizons unfolding. In Russel, Murphy and Stoeger (eds) 2008:57 ff.

European Union 2005. *Eurobarometer: Social Values, Science and Technology*. http://ec.europa.eu/comm/public_opinion/archives/ebs/ebs_225_report_en.pdf.

Forde, Gerhard O 1972. *When God Meets Man: Luther's Down-to-earth Approach to the gospel*. Minneapolis: Augsburg.

Foster, J. B. 1994. *The Vulnerable Planet: A Short History of the Environment*. New York: Monthly Review Press.

Freeman, Christopher and Jahoda, Marie (eds). 1978. *World Futures: The Great Debate*. London: Martin Robertson.

Friedman, M. [1979] 1980. *Free to Choose: A Personal Statement*. New York: Penguin.

Garner, R. 2000. *Environmental Politics: Britain, Europe and the Global Government*. London: Macmillan Press.

Gillespie, M. A. 2008. *The Theological Origins of Modernity*. Chicago/London: Univ of Chicago Press.

Gitt, W. 2006. *In the Beginning was Information: A Scientist Explains the Incredible Design in Nature*. USA: Master Books.

Greene, B. 2005. *The Fabric of the Cosmos: Space, Time and the Texture of Reality*. New York: Random House (Vintage edition).

Gregersen, N. H. 2000. God: the Creator of Creativity; Theology and the Sciences of Self-organized Complecity; Supervenience Theory and the Brain-culture Problem: Eucharistic Theology in the Framework of a Holist Supervenience Theory. In: du Toit, C. W. (ed), p. 25-129.

Gregersen, N. H. 2008. Special divine action and the quilt of laws: Why the distinction between special and general divine action cannot be maintained. In Russel, Murphy and Stoeger (eds), p. 180 ff.

Guttman, B. S. 2005. *Evolution: A Beginner's Guide*. Oxford: Oneworld Publications.

Habermas, J 1973. *Technik und Wissenschaft als 'Ideologie'*. Frankfurt/M: Suhrkamp, 6th edition.

Habermas, J 2008. Ein Bewusstsein von dem, was fehlt. In: Michael Reder & Joseph Schmidt (eds): *Ein Bewusstsein von dem, was fehlt: Eine Diskussion mit Jürgen Habermas*. Frankfurt/M: Suhrkamp.

Happel, S. 1996. Metaphors and time asymmetry: Cosmology in physics and Christian meanings. In Russel, Murphy and Isham, (eds) 1996:105 ff.

Hawking, S. W. 1988. *A Brief History of Time: From the Big Bang to Black Holes*. New York et al.: Bantam Books.

Hawking, S. and Mlodinow, Leonard. 2005. *A Briefer History of Time*. New York: Bantam Dell (Random House).

Hawking, S. and Mlodinow, Leonard. 2010. *The Grand Design: New Answers to the Ultimate Questions of Life*. London et al.: Bantam Press.

Heeren, F. 2004 (revised edition). *Show me God: What the Message from Space is Telling Us About God*. Olasse, KS: Day Star Publications.

Hefner, Ph. 1993. *The Human Factor: Evolution, Culture, and Religion*. Minneapolis: Fortress Press.

Heisenberg, W. [1958] 1999. *Physics and Philosophy: The Revolution in Modern Science*. Amherst, NY: Prometheus Books.

Heller M. 1996. On theological interpretations of physical creation theories. In Russel, Murphy and Isham (eds) 1996:93 ff.

Isham, C. J. and Polkinghorne, John Charlton. 1996. The debate over the block universe. In Russel, Murphy and Isham (eds) 1996:139 ff.

Jackelén, A. 2006. Relativistic eschatology: Time, eternity and eschatology in light of the physics of relativity. *Zygon* 41/2006, 955-973.

Jones, B. 1982. *Sleepers, Wake! Technology and the Future of Work*. Brighton: Wheatsheaf.

Kaku, M. 2008. *Physics of the Impossible: A Scientific Exploration into the World of Phasers, Force Fields, Teleportation and Time Travel*. New York: Random House (Anchor Books).

Kauffman, Stuart A. 1993. *The Origins of Order: Self-organization and Selection in Evolution*. New York: Oxford University Press.

Kauffman, Stuart. A. 1995. *At Home in the Universe: The Search for the Laws of Self-organization and Complexity*. New York: Oxford University Press.

Kaufman, G. D. 2004. In the beginning . . . creativity. Minneapolis: Fortress Press.

Kauffman, Stuart. 2008. Reinventing the sacred: A new view of science, reason and religion. New York: Basic Books.

Kieffer, G. H. 1979. Bioethics: A textbook of issues. London et al.: Addison-Wesley Publishing Co.

Krauss, L. and Starkman, Glenn. 2000. Das Schicksal des Lebens im Universum. In Beste (ed) 2000:84-91.

Krüger, J. S. 1995. *Along the Edges: Religion in South Africa; Bushman, Christian, Buddhist.* Pretoria: University of South Africa.

Kuhn, T. S. 1970. *The Structure of Scientific Revolutions.* Second enlarged edition. Chicago: University of Chicago Press.

Lockwood, C. 2007. *The Human Story: Where We Come from and How We Evolved.* London: Natural History Museum.

McCarthy, T. and Rubidge, Bruce. 2005. *The Story of Earth and Life: A Southern African Perspective on a 4.6-billion-year Journey.* CapeTown: Struik.

Marcuse, H. [1964] 1968. *One Dimensional Man.* Gray's Inn Road, London, WC: Sphere Books.

Mason, S. F. *A History of the Sciences.* New York: London: MacMillan, 1962. [older editions titled: *Main Currents of Scientific Thought*] I have used the German edition: Geschichte der Naturwissenschaft. Stuttgart: Alfred Kröner, 1961.

McFague, S. 1993. *The Body of God: An Ecological Theology.* Minneapolis: Augsburg Fortress.

McFague, S. 1997. *Super, Natural Christians: How We Should Love Nature.* Minneapolis: Fortress.

Meadows, D. et al. (ed). 1972. *The Limits to Growth.* New York: Universe Books.

Meadows, D. et al. 1977. *Alternatives to Growth: A Search for a Sustainable Future.* Cambridge, Mass: Balinger Publishing.

Meadows, D. and Meadows, H. et al. (eds). 1992. *Beyond the Limits: Confronting Global Collapse, Envisioning a Sustainable Future.* White River, Vermont: Chelsea Green Publishing Company.

Mishan, E. J. 1977. *The Economic Growth Debate: An Assessment.* London: George Allen & Unwin.

Mithen, S. 1997. *The Prehistory of the Mind: A Search for the Origins of Art, Religion, and Science.* London: Thames and Hudson (as quoted by van Huyssteen 2006).

Mitchell, M. 2009. Complexity: A guided tour. New York: Oxford University Press.

Moltmann, J. 1981. *The Trinity and the Kingdom of God: The doctrine of God.* London: SCM Press.

Moltmann, J. 1985. God *in Creation: An Ecological Doctrine of Creation.* London: SCM.

Moltmann, J. 1992. *History and the Triune God.* New York: Crossroads.

Müller, J. and Wallacher, J. 2005. *Entwicklungsgerechte Weltwirtschaft: Perspektiven Für Eine Sozial- und Umweltverträgliche Globalisierung.* Stuttgart: Kohlhammer.

Murphy, N. and Ellis, George Francis Rayner. 1996. *On the Moral Nature of the Universe: Theology, Cosmology, and Ethics.* Minneapolis: Fortress Press.

Murphy, N. 1990. *Theology in the Age of Scientific Reasoning.* London et al.: Cornell University Press.

Murphy, N. 2008. Emergence, downward causation and divine action. In Russel, Murphy and Stoeger (eds) 2008:111-131.

Nancy, J.-L. 2008. *Dis-enclosure: The Deconstruction of Christianity.* New York: Fordham University Press.

Nietzsche, F. 1966. *Friedrich Nietzsche in drei Bänden.* Darmstadt: Wissenschaftliche Buchgesellschaft.

Nürnberger, K. 1987. *Ethik des Nord-Süd-Konflikts. Das Globale Machtgefälle Als Theologisches Problem.* Gütersloh: Güterloher Verlagshaus (p. 333).

Nürnberger, K. 1988. *Power and Beliefs in South Africa. Economic Potency Structures in SA and Their Interaction with Patterns of Conviction in the Light of a Christian Ethic.* Pretoria: Unisa.

Nürnberger, K. 1998. *Beyond Marx and Market: Outcomes of a Century of Economic Experimentation.* Pietermaritzburg: Cluster Publications/London: Zed Books.

Nürnberger, K. 1999. *Prosperity, Poverty and Pollution: Managing the Approaching Crisis.* Pietermaritzburg: Cluster Publications/London: Zed Books.

Nürnberger, K. 2002. *Theology of the biblical Witness: An Evolutionary Approach.* Münster, Germany: LIT Verlag.

Nürnberger, K. 2004. *Biblical Theology in Outline: The Vitality of the Word of God.* Pretoria: CB Powell Bible Centre (Unisa)/Pietermaritzburg: Cluster Publications.

Nürnberger, K. 2005. *Martin Luther's Message for Us Today: A Perspective from the South.* Pietermaritzburg: Cluster Publications.

Nürnberger, K. 2007. *The Living Dead and the Living God: Christ and the Ancestors in a Changing Africa.* Pretoria: CB Powell Bible Centre (Unisa)/ Pietermaritzburg: Cluster Publications.

Nürnberger, K. 2010. *Richard Dawkins' God Delusion: A Repentant Refutation.* London: Xlibris Corporation.

Nürnberger, K 2011. Transcendence as the key issue between naturalism and the Christian concept of God: A dialogue with Richard Dawkins, Stuart Kauffman, and Brian Swimme,' *Ars Disputandi* [http://www.ArsDisputandi.org] 11 (2011), 1-23

Pannenberg, W. 1969. *Theology and the Kingdom of God.* Philadephia: Westminster Press. Edited by Richard John Neuhaus.

Pannenberg, W. 1970. Dogmatische Thesen zur Lehre von der Offenbarung. In *Offenbarung als Geschichte* edited by W. Pannenberg. Fourth edition. Göttingen: Vandenhoeck & Ruprecht.

Pannenberg, W. 1971. *Grundfragen Systematischer Theologie.* Göttingen: Vandenhoeck & Ruprecht, second edition (translations by KN).

Pannenberg, W. (ed). 1984. *Die Erfahrung der Abwesenheit Gottes in der Modernen Kultur.* Göttingen: Vandenhoeck & Ruprecht.

Pannenberg, W. 1988. *Systematische Theologie,* Band 1. Göttingen: Vandenhoeck & Ruprecht.

Pannenberg W. 1989. The Doctrine of Creation and Modern Science. In Peters (ed) 1989, 152 ff.

Peacocke, A. and Clayton, Ph. 2007. *All that Is: A Naturalistic Faith for the Twenty-first Century: Theology and the Sciences.* Minneapolis: Fortress Press.

Peck, S. M. [1978] 1990. *The Road Less Travelled: A New Psychology of Life, Traditional Values and Spiritual Growth.* London et al.: Random House (Arrow edition).

Pedersen, E. M. W. 2002. New spirituality and the crisis in European churches and cultures. *Swedish Missiological Themes* 90/2002, 449.

Peters, T. 1989. *Cosmos as Creation: Theology and Science in Consonance.* Nashville: Abingdon Press.

Peters, T. 1996. The Trinity in and beyond time. In Russel, Murphy and Isham (eds) 1996:263 ff.

Peters, T., Russel, R. J., and Welker, W. (eds). 2002: *Resurrection—Theological and Scientific Assessments.* Grand Rapids: Eerdmans.

Peters, T. 1997. Clarity of the part versus meaning of the whole. In *Beginning with the End: God, Science and Wolfhart Pannenberg*, edited by C. R. Albright and Joel Haugen. Chicago: Open Court.

Plotkin, H. C. 1993. *Darwin Machines and the Nature of Knowledge.* Cambridge: Harvard University Press (as quoted by van Huyssteen).

Polkinghorne, J. 1996. *Scientists as Theologians: A Comparison of the Writings of Ian Barbour, Arthur Peacocke and John Polkinghorne.* London: SPCK.

Polkinghorne, J. and Welker, Michael (eds). 2000. *The End of the World and the Ends of God: Science and Theology on Eschatology.* Harrisburg PA: Trinity Press International.

Polkinghorne, J. 2006. Space, time and causality. *Zygon* 41/2006, 975-983.

Popper, K. 1998. *Conjectures and Refutations: The Growth of Scientific Knowledge.* London: Routledge (as quoted by van Huyssteen).

Proctor, R. 2003. Three roots of human recency: Molecular anthropology, the refigured Acheulean, and the UNESCO Response to Auschwitz. *Current Anthropology* 44 No 2 (as quoted by vanHuyssteen).

Ratschow, C. H. [1957] 1960. *Der Angefochtene Glaube. Anfangs- und Grundprobleme der Dogmatik.* Gütersloh: Gütersloher Verlagshaus Gerd Mohn.

Rescher, N. 1990. *Evolution, Cognition, and Realism: Studies in Evolutionary Epistemology.* Lanham Md: University Press of America (as quoted by van Huyssteen 2006).

Rolston, H. 1996. Science, religion and the future. In *Religion and Science: History, Method and Dialogue* edited by W. Mark Richardson and J. Wildman. New York: Routledge (as quoted by van Huyssteen. 2006).

Rolston, H. 1999. *Genes, Genesis and God: Values and their origins in natural and human history*: The Gifford Lectures, University of Edinburgh, 1997-1998. Cambridge: Cambridge University Press (as quoted by van Huyssteen 2006).

Russel, R. J. 1996. Creation without a beginning: The doctrine of creation in relation to Big Bang and Quantum cosmologies. In Russel, Murphy and Isham (eds) 1996:291 ff.

Russel, Robert J., Murphy, Nancey, and Isham, Christopher. (eds). 1996. *Quantum Cosmology and the Laws of Nature: Scientific Perspectives on Divine Action.* Second edition. Vatican: Vatican Observatory Publications/Berkeley, CA: Center for Theology and the Natural Sciences.

Russel, R. J. 2008b. *Cosmology from Alpha to Omega: The Creative and Mutual Interaction of Theology and Science.* Minneapolic: Fortress Press.

Russel, R. J. 2002. Bodily resurrection, eschatology and scientific cosmology in Peters, Russel and Welker (eds). 2002, 3-15.

Russel, R. J., Murphy, Nancey, and Stoeger, William R. (eds). 2008. *Scientific Perspective on Divine Action: Twenty Years of Challenge and Progress.* Vatican: Vatican Observatory Publications and Berkeley, Cal: Center for Theology and the Natural Sciences.

Russel, R. J., Murphy, Nancey, and Peacocke, A. R. (eds). 1995. *Chaos and Complexity: Scientific Perspectives on Divine Action.* Vatican: Vatican Observatory Publications/Berkeley Cal.: Center for Theology and the Natural Sciences.

Russel, R. J., Murphy, Nancey, and Isham, Christopher. (eds). [1993] 1996. *Quantum Cosmology and the Laws of Nature: Scientific Perspectives on Divine Action.* Second edition. Vatican: Vatican Observatory Publications/Berkeley, Cal.: Center for Theology and the Natural Sciences.

Sachsse, H. 1984. *Ökologische Philosophie: Natur—Technik—Gesellschaft.* Darmstadt: Wissenschaftliche Buchgesellschaft.

Schmid, H. [1875] 1961. *The Doctrinal Theology of the Evangelical Lutheran Church.* Reprint of third translated edition. Minneapolis: Augsburg Publishing House.

Shutte, A. (ed). 2006. *The Quest for Humanity in Science and Religion: The South African Experience. Pietermaritzburg:* Cluster Publications.

Smail, D. L. 2008. *On Deep History and the Brain.* Berkeley: University of California Press.

Stephan, A. 2004. *Emergenz: Von der Unvorhersagbarkeit zur Selbstorganisation.*

Stewart, I. 1998. *Life's Other Secret: The New Mathematics of the Living World.* New York: Wiley. (as quoted by van Huyssteen).

Stoeger, W. R. 2008. Conceiving divine action in a dynamic universe. In Russel, Murphy and Stoeger (eds) 2008:225 ff.

Swimme, B. 1996. *The Hidden Heart of the Cosmos: Humanity and the New Story.* Maryknoll, NY: Orbis Books.

Tappert, T. G. 1959: *The Book of Concord.* Philadelphia: Fortress Press.

Tattersal, I. 1998. *Becoming Human: Evolution and Human Uniqueness.* New Yorm: Harcourt Press (as quoted by van Huyssteen).

Taylor, G. R. 1975. *How to Avoid the Future.* London: Secker & Warburg.

Teske, J. A. 2010. Narrative and meaning in science and religion. *Zygon,* vol 45/2010, 91-104.

Toffler, A. 1970. *Future Shock.* London: Pan Books.

Tomkins, St. (without date). *1½ Cheers for Richard Dawkins.* www.ship-of-fools. com/features/2006.dawkins.html.

Tracy, Th. 2008. Special divine action and the laws of nature. In Russel, Murphy and Stoeger (eds) 2008:249 ff.

Van der Walt, E 2010. The limbic system and the 'religious brain'. In: Du Toit, C 2010:23-39.

Van Huyssteen, J. W. 2006. *Alone in the World? Human Uniqueness in Science and Theology*. The Gifford Lectures. Grand Rapids: William B Eerdmans.

Vincent, J.-D. [1986] 1990. *Die Biologie des Begehrens: Wie Gefühle entstehen*. Reinbeck: Rowohlt.

Von Weizsäcker, C. F. 1990. *Die Tragweite der Wissenschaft*. Stuttgart: Hirzel. Sixth edition. (English: *Philosophy of Modern Physics*).

Wegter-McNelly, K. 2008. Does God need room to act? Theo-physical in compatilism in noninterventionist theories of objective special divine action. In Russel, Murphy and Stoeger (eds) 2008:299 ff.

Welker, M. [1999] 2000. *Creation and Reality*. Philadelphia: Augsburg-Fortress.

Wuketits, F. M. 1990. *Evolutionary Epistemology and its Implications for Humankind*. Albany, NY: SUNY Press (as quoted by van Huyssteen 2006).

Notes

1   For Durand this is no longer a possibility, but a certainty: "The ultimate destruction of nature as we know it is inevitable as a result of the avarice and consumeristic momentum of humankind. Long before the end of this century there will not be enough resources to sustain the human population. Civilisation as we know it, which grew from the Bronze Age civilisations, will cease to exist" (2010:72f).

2   Is the problem not much older than modernity? Durand traces the conquering and devouring behaviour of humankind back to its evolution in prehistoric times, when human survival could only be secured through rapid procreation and, with the rise of technology, by outwitting and mastering nature (2010:56ff). Referring to Gen. 1:28-30, Durant argues that religion and technology are jointly responsible for leading humanity to its demise (2010:73). The biological part of the argument is persuasive, the religious part is not. In whatever inappropriate forms, religion is about dependence on, and accountability to, a higher authority. It is not faith in God that leads humans into self-enslavement, self-torture and self-destruction, but idolatry. Modernity has effectively dismantled the constraints that responsibility towards a higher authority could have imposed on an instinctual collective conditioning. Gen. 1:28-30 made perfect sense at a time when human survival was threatened. Ecologists hostile to religion generally overlook the fact that this text does not see the human being as the sovereign owner, master and beneficiary of nature, but as a representative of God, entrusted with nature and accountable to God. Note that, according to this particular text, humans are not entitled to consume animals!

3   "Science is a dominating force, a universal viewpoint that is reshaping the outlook of humankind everywhere, especially that of the young" (Peacocke 1989:28).

4   Nürnberger 2002 and 2004. According to the biblical witness, the will of God is focused precisely on the existence and well-being of the world, God's glory becomes manifest in his sacrifice for the sake of human redemption, and God's eschatological vision encompasses a new and redeemed world.

5   Looking at the scientific performance of the Japanese in spite of a culture vastly different from that of the West, Ben-Ari concludes that science does not depend on culture (Ben-Ari 2005:124). I disagree. The fact that "science 'happened' only once"—in the West during the last few centuries (2005:116)—says it all. Aspects of all cultures that are not directly involved can hold onto their own by guarding protected spaces against the encroachments of modernity.

6   In a perceptive analysis, Habermas (2008:28ff) derives both science and faith from a much earlier common origin when 'myth' changed into 'logos' in the so-called 'axial period' during the first millennium BC. But this presupposes that faith is rooted in metaphysics. The zoologist and naturalist Francois Durand

goes further back, namely to the Bronze Age when the inception of technology led to a male dominated culture and religion. "Ironically the same culture that gave us our technology also gave us our religions and the combination of the two will destroy the creation of God—which includes humankind and the ecology" (Durand 2010:73).

7 "The relationship between these two claimants on humankind's loyalties, science and religion, is I think one of the most fundamental challenges that faces the mind and the spirit of humanity today" Peacocke 1989:29.

8 The idea of perfection belongs to a Platonic frame of reference. Platonism abstracts the 'essence' of something from time, space and power relations into a timeless, universal, and harmonious idea. The biblical faith is geared not to perfection but to transformation. Where the concept is used at all it usually has moral or ritual rather than ontological connotations.

9 Ben-Ari says that science and religion do not get into conflict with each other where they do not make contradictory claims. There is no conflict about chemical processes, physical optics, or superconductivity (2005:132), simply because religious traditions so far had nothing to say on these matters. That is exactly my point. Once faith understands that it has nothing to say on big bang cosmology or biological evolution, and that mythological presentations of origination have a different agenda, these conflicts should evaporate.

10 In 'post-foundationalist' circles the word 'grounding' may arouse misgivings. But in a scientific discourse, grounding does not have the Platonic connotations of static immutability. The earth rotates and all stable structures on earth rotate with it. But without the force of gravity, operational within the system, no stable structures or processes could materialise.

11 Consonance is suggested by Ted Peters and his colleagues (1989). Complementarity would be a more acceptable term. For a detailed discussion see Drees (1990:26 ff).

12 Drees argues that the search for consonance is based on three assumptions: 1. Science and faith make independent statements about the same world. 2. Some aspects of the world, like its order, dynamics, character, and its origin or ground are subject to both scientific and theological description. 3. Correct descriptions must fit together without contradiction (Drees 1990:28). I disagree with all three these assumptions. Faith statements refer to the transcendent dimension of the immanent world, while science explores the regularities governing this world. If the Bible assumes that the earth is flat and stationary, this is not a theological statement, but the primitive equivalent to a scientific statement. Where such statements convey a theological message, their descriptive content is beside the point.

13 For a discussion see Ben-Ari (2005:134 f).

14 This is the suggestion of van Huyssteen (2006). His attempt to take the findings of the sciences seriously is highly appreciated, but in my view his assumption that science and theology are parallel approaches to reality is misleading.

15 "I will be urging basically that the scientific and the theological enterprises are interacting and mutually illuminating approaches to reality" (Peacocke 1989:29).

16 See e.g. Nürnberger 1998 and 1999 and my webpage *www.klaus-nurnberger. com*.

17 Ben-Ari 2005:116.

18 Christianity "has borne the brunt of the critique of the Enlightenment of its sacred sources . . ." (Peacocke 2008:207).

19 Serious readers should expose themselves to some of the leading ecological analyses available. The video of Al Gore, *An Inconvenient Truth*, is well-known. Although I am critical of his Marxian preoccupations, I recommend Foster (1994) as a short and accessible introduction into the historical genesis and escalation of the ecological problem. Other classics include the studies of the *Club of Rome*, beginning with Meadows et al., 1972, *The Limits to Growth* and Mesarovic and Pestel, 1974, *Mankind at the Turning Point*. See also Freeman and Jahoda (1978), Birch (1978), Barney (1980), Ehrlich and Ehrlich (1996), regular publications of organisations such as *Green Peace*, the *World Watch Institute*, the *World Wildlife Fund*, the World Bank's *World Development Report* of 1992, the *Brundtland Report*, 350.org and many others. A useful summary can be found in Garner (2000, ch. 2). Some of these sources, though dated, are still valid to a considerable extent. Contemporary analyses abound.

20 With the term 'modernity', I do not refer to the subversive cultural movement that emerged in the latter half of the nineteenth century and culminated after World War II, but the long-term development that was triggered by the Enlightenment.

21 "Most of the world's nations now resemble a kind of historical layer cake, in which social groups living side by side embody the lifestyles of different centuries and this layering is largely determined by the extent to which a people has been influenced by Western modernity, either directly or indirectly" (Cahoone 1988:xi). For the relation between traditionalism and modernity in Africa, see Nürnberger (2007, part II).

22 For a detailed analysis see Nürnberger (1999:57 ff, 297 ff, and 377-394).

23 Durand 2010:55ff.

24 Pannenberg argues that secularisation is not derived solely from the history of ideas and emphasises the role of religious wars (1984:19). But the decisive factor was the emancipatory drive of modernity.

25 This argument goes back to Adam Smith, the father of modern economics. Murphy sees modern individualism as "but an instance of the pervasive metaphysical thesis of atomism" (2008:113). But metaphysical speculations are not the main spring of great social movements. Needs, desires, and interests are. The procedure that takes wholes apart to understand and utilize them better is a pragmatic move to serve interests, rather than the observance of a metaphysical principle. It is indeed "methodological reductionism", but for a purpose (2008:118). It was conveniently legitimated and facilitated by an

atomistic metaphysic inherited from ancient Greece that could be elaborated as "causal reductionism" and "ontological reductionism" (2008:118).

26 "During this period a new authority appeared which was completely independent of Christian religion or philosophy or of the Church, the authority of experience, of the empirical fact . . . What we can see and touch became primarily real" (Heisenberg 1999:195 f). The technological and commercial corollaries soon followed: "One was not so much interested in nature as it is; one rather asked what one could do with it" (Ibid. 197).

27 For a historical analysis of the 'theological roots of modernity' see Gillespie (2008). However, one should not overrate the theological contribution. Intellectuals only become influential as far as they are able to articulate and facilitate an emerging mood.

28 "Newton was profoundly religious and wrote more about the history of Christianity and on Christian theology than he did on any other subject" (Ben-Ari 2005:42).

29 "Only in conjunction with the specific, emancipatory anthropocentricity of European modernity did science and technology become the foundation of a world without God" (Pannenberg 1984:13).

30 The idea of the human being as the measure of all things goes back to Greek antiquity (Protagoras). "If humans set the conditions within which they will define themselves, if they set the goal and plan of the *humanum*, then they make themselves the measure of all things—the fixed point in terms of which everything else must be related." And "while humans may be the measure, some are more of a measure than others . . ." (Kieffer 1979:60). Elites consider themselves the measure; lower classes consider elites reference groups and 'significant others'.

31 "In those European countries in which one was wont to follow ideas up to their extreme consequences, an open hostility of science toward religion developed . . . only the ethical values of the Christian religion were excepted from this trend, at least for the time being" (Heisenberg 1999:198).

32 Prominent intellectuals articulated this general mood. According to Friedrich Nietzsche, the quest for power was the natural motivator of the human being. He ridiculed humility, altruism and concern for the weak as a decadent mindset. Sigmund Freud interpreted 'God' and 'conscience' as an oppressive father figure or super-ego, a pathological phenomenon from which adult humans should be released. According to Adam Smith, 'economic man' is by nature a profit and pleasure maximizer. For Ludwig Feuerbach, 'God' was the projection of the capacities of the human being into a non-existent heaven and should be retrieved to reach human fulfilment. According to Marx, faith was an ideology that legitimated oppressive rulers and capitalist exploiters and was bound to collapse once a classless society was established.

33 The 'warfare of science and religion' is by no means a feature only typical of Victorian England (McGrath 2005:140 ff.)

34 It is often argued that secularism characterises only Europe, but not the United States and the rest of the world. But there is much more spiritual and social

disintegration in these countries than meets the eye. Because of their painful history, Europeans may be more honest, more consistent and more outspoken in their rejection of religion than the Americans. Ironically they are also more responsible in terms of social security networks. A spiritualised piety linked with capitalist irresponsibility and great power arrogance is hardly very convincing. Non-Western cultures may pick and choose from the modern package only what suits their economic ambitions, but even there the corrosive effects of modernity are quietly eating away traditional assumptions and convictions.

35 "The isolation of faith from reason and critical argument into the seemingly unassailable bastions of the pure subjectivity of personal convictions and decisions has contributed more to the erosion of its plausibility (than the attempt to reconcile faith and science)" (Pannenberg 1984:14).

36 Pannenberg speaks of the "experience of the absence of God" in modern culture (1984:12). But absence can only be experienced where there is a memory of presence. Taking an explicit stance against faith and for atheism is pointless where it is taken for granted that there is no God.

37 See Nürnberger (2010) for a response to Dawkins' *The God Delusion.*

38 Swimme 1996:57.

39 Foster gives concrete examples (1994:66 ff). Ecological awareness was alive and well as early as the mid-nineteenth century. George Perkins Marsh wrote in 1864: "The earth is fast becoming an unfit home for its noblest inhabitant . . . Another era of equal human crime and improvidence . . . would reduce it to such a condition of impoverished productiveness, of shattered surface, of climatic excess, as to threaten the depravation, barbarism, and perhaps even extinction of the species" (quoted by Foster 1994:72). After a century and a half of warnings of impending disaster, humankind just does not want to wake up. Marsh wrote: " . . . of all organic beings, man alone is to be regarded as essentially a destructive power . . ." (Foster 1994:73).

40 The business motive soon replaced the love of nature (Foster 1994:75 ff).

41 With increasing awareness of the injustices perpetrated by European imperialism and a concomitant appreciation of non-European cultures and identities, there was a tendency to see science and Christian faith as just two aspects of European arrogance that needed to be pushed off its pedestal in favour of a more 'democratic' approach to describing the world (Ben-Ari 2005:117). As a result, both science and Christian faith were "lumped together" with Western imperialism, but Christian faith much more so than science! Science was too important a tool to achieve progress, status and wealth to be dismissed. It was also motivated by an emancipatory agenda, while the enduring symbiosis between religious and political authoritarianism made Christianity an easy target for liberation movements. Especially the Marxist type was fiercely pro-science and anti-religion.

42 Foster 1994:56 ff.

43 Foster 1994:48.

44 Foster 1994:44.

45   An analysis of the structure of the South African economy under apartheid can be found in Nürnberger (1988, chs. 2-5).

46   F. Fukuyama. 1992. *The end of history and the last man*. New York: Avon Books.

47   The typical arguments are found in Milton Friedman's work (1980). He was one of the most influential economists of the neo-liberal school—the mentor of politicians such as Ronald Reagan and Margaret Thatcher. His intentions were noble enough. His proposal of a negative tax for the poor, for instance, is noteworthy. He was also not wrong in highlighting the inefficiencies of a command economy relative to those of a liberal economy. But trashing the former and idealising the latter—a bias that is typical for the ideological justification of the pursuit of collective interests—had serious consequences. The current global financial crisis was a direct result of his teaching being translated into economic policy and financial practice.

48   Garner shows that the 'radical' or doomsday approach of pioneer ecologists such as the Meadows team has made way for the 'reformist' approach. It wants to lure a population addicted to economic growth and growing affluence into ecological responsibility by arguing that the two pursuits are compatible, that we need 'sustainable' and equitable growth, that science, technology, and the market can handle the situation if adjustments are made (2000:10 ff; 35 ff), that ecologically responsible production makes sense in purely economic terms (Gordon Brown in *Newsweek,* Sept. 28, 2009, 21 ff). However, the happy marriage between economic growth and ecological sustainability without substantial sacrifices is wishful thinking.

49   Garner 2000:37 f.

50   Just consider that "the total annual emissions of CO_2 have increased from 1.5 million tonnes as recently as 1950 to almost six billion tonnes in the 1990s, and it is further predicted that CO_2 concentrations will double by the year 2050 unless action is taken" (Garner 2000:27). According to one estimate, "half of the world's tropical forests were cleared in the 1980s alone" (2000:29) and "energy expenditure is now twenty times greater than it was about 150 years ago" (2000:32).

51   Economic growth will end "either against our will, in a succession of famines, epidemics, social crises and wars; or because we want it to . . . in a series of thoughtful, humane, and measured changes" (Edward Goldsmith in his *Blueprint for Survival*, 1972:15 quoted by Garner 2000:36).

52   The Meadows team (1972, 1977, 1992), the Barney team (2000), and many others have developed sophisticated methods of calculation to assess the consequences of 'overshoot'. Although the validity of these projections has been heavily disputed, the overall trend cannot be wished away. A summary of these studies is offered by Garner (2000:36 ff).

53   Garner 2000:15 ff.

54   Alvin Toffler drew public attention to the problem of acceleration in his book *Future Shock* (London: Pan Books, 1983).

55 Sharon Begley in her article "Your Brain Online" (*Newsweek*, Jan. 18, 2010, p. 18).

56 "Most do not fully see this truth that life is difficult. Instead they moan more or less incessantly, noisily or subtly, about the enormity of their problems, their burdens, and their difficulties as if life were generally easy, as if life should be easy" (Peck 1990:13).

57 Swimme 1996:16.

58 Recently 'the good life' had become a regular feature in *Newsweek*, an otherwise serious journal. Countless popular magazines have *no other* agenda.

59 The mood of mass media is encapsulated in the following quotation: "We don't want anything scary . . . (Readers) are tired of bad news, scary facts, and being held responsible for things" (quoted by Ehrlich and Ehrlich 1996:192).

60 "Origin and point of departure of all economic activity are the *needs of human beings*. Therefore production and provision of goods and services to satisfy these needs are the fundamental economic activities . . . While for a very long time the basic needs of the people had determined economic activities, today it is increasingly the economy itself, which tries to create needs through advertising, which means that the basic economic logic has been turned upon its head" (Müller and Wallacher 2005:15—my translation). Of course, it is not the economy that has done this, but people whose priorities have changed from need satisfaction to profit and pleasure.

61 Marxism was basically flawed in that it postulated the messianic, altruistic and ideology-free potential of the proletariat and its avant-garde leadership, misjudged the relation between personal motivation and social structure and justified the totalitarian trap of the model. This is also the weakness of Foster's otherwise informative work.

62 For my assessment of Marxism see Nürnberger (1998: chs. 3, 4, and 7).

63 For a useful discussion of the issue see Garner (2000:65 f).

64 For detail, see Nürnberger (2007: ch. 5).

65 For the underlying mindset and its catastrophic consequences see Taylor (1975).

66 It is significant that one of the most perceptive analysts of modernity, Jürgen Habermas, hardly a believer, has come to similar conclusions. Arguing in terms of the traditional juxtaposition of faith and reason, he is concerned about the 'defeatist' trend brooding in modern reason, whether in postmodernist radicalism or in a science-believing naturalism. Post-metaphysical thought can deal with the theoretical problem, he believes, but not with the concomitant moral fatalism. Modernity has lost a vision for the whole (previously the 'kingdom of God'), as well as an awareness of the global breakdown of solidarity, of that "which cries to heaven" (Habermas 2008:30f). According to Habermas, the rift between science and faith can no longer be healed. Awareness of what has been lost in modernity produces new fundamentalisms and spiritualisms that clash with modernity. Yet in political terms the problem cannot be avoided and must be tackled. When imposing freedom of religion, a democratic state depends on legitimation, which is rooted in convictions. The tension demands

a complementary learning curve in which the secular and the religious sides have to 'entangle each other' (2008:32f). The religious communities must find their own rationale for adopting a post-metaphysical stance and conceding that the pluralistic society is unavoidable, while the secular state must refrain from treating religion as irrational and irrelevant for public policy and law. A weakness of his stance is that, in his dialogue with Catholicism, Habermas identifies faith with medieval metaphysics, which clashes with the approach of the natural sciences. My approach of experiential realism, in contrast, positions faith on the side of the natural sciences against metaphysical speculations. This should make a fruitful complementarity much more plausible.

[67] Durand (2010:62ff) argues that this behaviour us typical of all male dominated religions and goes back to the alpha male behaviour of our primate ancestors.

[68] This was a serious misunderstanding of the original stance of Luther. See Nürnberger (2005:251 ff) for detail.

[69] "Scientists generally are doing what they love; they often work very long hours for relatively small financial rewards . . . scientists are subject to continuous open testing and scrutiny . . . Certainty is a standard commodity for some religious leaders and political columnists, but it is forever denied to scientists" (Ehrlich and Ehrlich 1998:26 f). I do not disagree with these contentions; but they paint a rather idealised picture of science.

[70] "Nothing so reflected the imperial spirit of the age as the outlook of seventeenth-century science, which saw humanity as engaged in a war for the domination of nature." Bacon believed it was the "real business" of the human race to subjugate, enslave and mould nature (Foster 1994:41). According to David Pepper (1986), scientific rationalism displaced faith in God and the human stewardship role, thus separating humans from nature as the masters and beneficiaries of the latter (Garner 2000:39).

[71] Cf. the so-called 'Frankfurt School'. In One Dimensional Man, Herbert Marcuse spoke of "technological rationality and the logic of domination" (1968:120 ff). Jürgen Habermas spoke of "science and technology as ideology", that is, the legitimation of the pursuit of collective self-interests (1973). These analyses expose the domination of humans over humans through the domination of humans over nature.

[72] "The first research laboratory systematically organized for invention was set up by Thomas Edison in 1876 . . . By 1920, there were around 300 such corporate laboratories and by 1940 some 2200" (Foster 1994:110). Today there must be multiples of this figure worldwide.

[73] See Habermas 1976 for detail.

[74] "The distinguishing feature of the many innovations in biology and medicine is that they can contribute mightily to the betterment of the human condition now and in the future, or they can generate massive evil even for generations yet unborn" (Kieffer 1979:345). Kieffer cites the examples of population, environment, genetic intervention, medical engineering, behavioral control technologies etc.

75 'Opportunity costs' give an indication of the potential benefit of making the same investment in alternative pursuits.

76 NASA's human spaceflight programme alone has an annual budget of US$ 18 billion. This is not even enough for returning to the moon in 2020, or even to keep the International Space Station aloft beyond 2015 (*Newsweek* Oct 19, 2009, 45).

77 Ben-Ari 2005:208 ff.

78 "True, the rationality of pure science is value-free and does not stipulate any practical ends, it is 'neutral' to any extraneous values that may be imposed on it . . . Scientific rationality makes for a specific societal organization precisely because it projects more form . . . which can be bent to practically all ends . . ." (Marcuse 1968:129).

79 "(C)hange in the direction of progress . . . would also affect the very structure of . . . the scientific project. Its hypotheses, without losing their rational character, would develop in an essentially different experimental context (that of a pacified world); consequently, science would arrive at essentially different concepts of nature and establish essentially different facts" (Marcuse 1968:136).

80 Nürnberger (1999: chs 6 and 10). An empirical study can be found in Nürnberger, Klaus. 1988. *Power and Beliefs in South Africa.* Pretoria: Unisa.

81 Swimme 1996:16.

82 "We (Americans) are the only country that has written 'the pursuit of happiness' into its founding document, thus guaranteeing that we'll never be satisfied." (Andrei Codrescu in *Newsweek,* Aug. 23 and 30, 2010, p. 54).

83 For a detailed analysis see Nürnberger (1998, chs. 5, 7, and 8).

84 Christopher Dickey in *Newsweek,* Aug. 23 and 30, 2010, p. 41.

85 The ethics of Immanuel Kant, which provided the foundation of the concept of 'human dignity' in the Human Rights culture, expressly distinguished between something that is a means to an end and has a price, on the one hand, and the 'worth' of a human being that is an end in itself, on the other.

86 Foster 1994:13; further explained on pages 36-40.

87 See Meadows et al. (1993:44 ff) (as quoted by Foster 1994:23) and Nürnberger (1999:20 ff, 70 ff, 333 ff).

88 The earth may well be an open system, importing and processing energy from the sun. However, a virtual closure of the system is reached when the self-regenerating power of nature cannot keep pace with human rapaciousness.

89 Ecological awareness, just as economic awareness, seems to presuppose a critical mindset. Leaving aside the obvious flaws of the Marxist analysis, criticality is one of its most important contributions. Friedrich Engels for instance, the friend and co-worker of Karl Marx, displayed extraordinary ecological sensitivity for his time: "The people who, in Mesopotamia, Greece, Asia Minor, and elsewhere destroyed the forests to obtain cultivable land, never dreamt that they were laying the basis for the present devastated condition of

these countries, by removing along with the forests the collecting centres and reservoirs of moisture" (quoted by Foster 1994:65).

90 The human population increased from about 5 million when humans first settled in villages and towns 6000 years ago, to about 500 million in 1600, 600 million in 1700 (**20 per cent** growth), 900 million in 1800 (**50 per cent** growth), 1800 million in 1900 (**100 per cent** growth), and about 6.8 billion today (almost **280 per cent** growth). See also Wikipedia: "list of countries by population". There are now fourteen times as many people on earth than 500 years ago. These figures reflect a typical exponential growth curve. "Human beings now use or co-opt some 40 per cent of the food available to all land animals and about 45 per cent of the available freshwater flows" (Ehrlich and Ehrlich 1996:14). Population growth is uneven due to differentials in the rate of economic development. In developed countries the initial situation of high birth rates and high death rates changed to a situation of lower death rates with continuing high birth rates, and finally to low birth and death rates (Foster 1994:15). But even this balance is unstable. In industrialized countries, increasing numbers of the aged draw on the resources of decreasing numbers of younger generations. Many poor countries are caught up in a "demographic trap" (low death rates and high birth rates) that cannot easily be reduced due to their low development chances in the global system (Foster 1994:16). If world population growth will stabilize by 2035, the world population is projected to reach about 10 billion in 2100; if stability is only attained by 2065, it will double to 14 billion by that date (Ibid. 17). If material expectations rise as they do today, it is hard to imagine how the earth will cope.

91 "World consumption of commercial energy—most of it from non-renewable resources—rose over 60 times between 1860 and 1985" (Foster 1994:19). Industrial production grew sevenfold during the second half of the twentieth century and is still growing exponentially (Ibid. 20). Industrialisation goes hand in hand with the growth of towns and cities that have an immense impact on the ecosystem.

92 "By far the most physical waste is generated by production, not consumption: for every ton at the consumption end of the waste stream there are likely to have been 20 tons of initial resource extraction and 5 tons produced in manufacturing" (Foster 1994:26 f).

93 The discrepancies are immense. "Energy consumption per person in the advanced capitalist countries is *80 times* that in Sub-Saharan Africa" (Foster 1994:19). The Gross Domestic Product per capita ranges from above 40,000 international dollars for countries like Switzerland, the US, and the Netherlands to below 800 international dollars for countries like Sierra Leone, Liberia, and Burundi—a ratio of more than 50:1. The Democratic Republic of the Congo is at the bottom of the pile with 328 dollars (World Bank figures for 2008 quoted in the article "List of countries by GDP (PPP) per capita" in *Wikipedia*. The discrepancies of income and wealth within countries are more drastic still, ranging from billionaires to people sleeping under bridges and living from hand to mouth—provided there is something in the hand.

94   For the relation between population growth, income growth, growth in life chance discrepancies and growth of ecological impact see Nürnberger (1999:70 ff).

95   Nürnberger 1999:73.

96   Nürnberger 1999:385.

97   Ehrlich and Ehrlich note that "the public is fairly well informed" about environmental problems and there is much to be grateful for (1996:45 ff). However, because deterioration happens gradually (and many aspects are rather hidden and motivations are geared to "crucial issues" in their own lives) people remain rather relaxed. The authors also highlight the dismal lack of education and information in America (1996:26).

98   Just consider the following example: "The Japanese have the world's longest life expectancy and the best recovery rates . . . Japan's health insurance system covers everybody, including illegal aliens." And yet "the aging nation spends about $3,000 per person on health care each year; America burns through $7,400 per person and still leaves millions without coverage." Moreover, "it's largely a private-sector system." How is that achieved? Medical fees are low; doctors work long hours. "They are comfortably middle-class, but not in the country club set." There are huge savings in the high-tech realm: An MRI of the neck costs $1,400 in America and $130 in Japan (T R Reid in *Newsweek*, Aug. 23 and 30, 2010, p. 38). All this shows what is possible.

99   Nürnberger 1998 and 1999. The latest instalment was the sub-prime credit crisis. Whole populations were lured into unprecedented spending sprees through easy access to money they did not work for and did not possess. Banks were kept afloat by heavy state subsidies. Many of their unscrupulous chiefs used the money to up their already unbelievable bonuses and that after having run their firms into near bankruptcy. Balances of payments worldwide went out of kilter. The most powerful economy, the United States, indulged in this splurge by borrowing money from emerging economies, notably China. China's phenomenal growth depended on throwing cheap manufactures at the insatiable American consumers. When this unsustainable system cracked, it pulled the rest of humanity into a worldwide recession. Millions lost their livelihoods, their savings, and their homes.

100   International milestones have been Stockholm 1972, the Brundtland Report 1987, Rio de Janeiro 1992, Kyoto 1997 (Garner 2000:77) and Copenhagen 2009. What national governments are able and willing to do depends to a great extent on the question whether their international peers will be part of the party. The rich North cannot demand austerity from the poor while 5 per cent of the population in the developed nations utilize about 70 per cent of the world's energy supplies (Garner 2000:79 ff). The affluent countries of the North have to empower the emerging economies of the South to apply ecological standards to their rapid growth in terms of financial resources and technological expertise.

101   For Britain see Garner (2000:118 ff). For the United States see Ehrlich and Ehrlich (1996:45 ff). Central and Northern Europe have advanced much more in terms of ecological legislation and policy. The previous British prime

minister, Gordon Brown, deserves special mention (see *Newsweek*, Sept. 28, 2009, 21 ff.)

[102] Garner 2000:45 ff. According to Ehrlich and Ehrlich, taking no action because of dissenting voices amounts to "betting our children's futures" (1996:31).

[103] Significantly, the July 2011 conference of the IUGG in Melbourne has the theme—*Earth on the Edge: Science for a Sustainable Planet*.

[104] *Newsweek*, Sept. 28, 2009, 27.

[105] "Vast numbers of people in that secularized world in which scientifically informed technology dominates the scene are rudderless and, basically, hopeless. They do not know what meaning to give to life . . ." (Polkinghorne in Hefner 1993:ix).

[106] Paul and Anne Ehrlich compare the 'anti-science' of this 'brownlash' with the 'anti-science of creationism' (1996:12).

[107] Ehrlich and Ehrlich concede that their previous futurological projections were wrong and that serious scientists disagreed on greenhouse gases, global warming and the consequences for life and human prosperity (1996:33 ff). Not all dissenting voices were "bought", but their views are often exploited. In the meantime, the arguments of the environmentally concerned group have become more generally accepted.

[108] *Newsweek*, Apr. 9, 2007, p. 48.

[109] This is a more precise rendering of Peters' concept of 'hypothetical consonance' (1989:13 ff). My approach is not a simple 'two-language theory' (16), where the two languages have nothing to say to each other.

[110] This is also true for subatomic physics. It operates very close to the frontiers of human observation and empirical evidence is very hard to come by at that level. It has no choice, therefore, but to rely heavily on mathematical construction. Pushing valiantly into the unknown, it ventures highly speculative models. But the ideal of a reliable depiction of objective reality is not ever abandoned, a fact that is often overlooked by postmodern philosophy and pre-modern theology.

[111] For an introduction see Russel (2008:4 ff). See also Drees (2008:7 ff) for a similar view.

[112] Hawking and Mlodinow 2010:7 f.

[113] The principle goes back to William of Occam. The following definition is attributed to Isaac Newton: We are to admit no more causes of natural things than such as are both true and sufficient to explain their appearances (Wikipedia).

[114] Wolfhart Pannenberg's approach can serve as a test case for the scientific plausibility of theological constructs. "Theology must learn that . . . it can no longer utter the word 'God' without giving account of its meaning . . . if it does not want to fall into the hopeless and indeed self-inflicted isolation of a higher glossolalia and draw the entire church with it into this cul-de-sac" (Pannenberg 1971:351). By trying to provide such a meaning, I want to argue, Pannenberg has developed precisely such an indigestible 'glossolalia'. The following example is taken from Pannenberg (1969). His point of departure is the motif of the coming Kingdom of God as proclaimed by Jesus of Nazareth.

The historical evolution of this tradition and its location in the Israelite-Jewish symbolic universe is simply ignored. The argument is entirely deductive. Its logic runs as follows: "Jesus underscored the present impact of the imminent future" (53). This is not simply a left-over from Apocalyptic, Pannenberg maintains. Kingdom means rule; rule means power. God is identical with God's rule, thus with God's power (55). (As the source of power) God is the Creator of the world. So God, as *future power* (the kingdom), creates the world. Pannenberg then flips the sentence around and says that God as *the power of the future* creates the world. To be more concise: divine future creates the world. The divinity of the future is then identified with the future of the world: "God in his very being is the future of the world" (61). The translation of "creative power" into "future power" is logically as invalid as the inversion of "future power" into "power of the future". The argument is completely indigestible. Certainly it is not based either on the biblical tradition, or the existential experience of faith. An experiential realist approach would argue that by any definition the future is *not yet*; so it cannot have power. As pure potentiality it can only have the 'capacity' to be realised. As such it is constrained by the parameters laid down by the past. Any realisation is driven by the behaviour of energy in time and space and subject to the regularities operative in the world we know. A power that is capable of transforming potentiality into reality cannot itself be situated in the future, thus in the sphere of pure potentiality. Pannenberg's concept of the 'future' seems to correspond with my concept of transcendence. Pannenberg may have used it because the future is inaccessible. But so is the past. If so, why use such an inappropriate concept in the first place? The merger of the concept of the "future" with the concept of "eternity" is equally problematic. It seems that "eternity", as Pannenberg uses the term, is the sum total of an endless series of futures. But how can that lead to ultimate revelation? Or to the "end of history"? And if it does, what is left if the end of history means that the end of everything is reached? Who will be the recipients of the ultimate revelation of God when humans have reached the end of their histories? The argument is a series of adventurous logical inferences drawn from a reified biblical metaphor without taking the historical context of this concept and its contemporary lack of relevance into consideration. No scientist will be impressed by such constructs.

[115] An "ontological reality must be assigned to thoughts, emotions, and social human constructions because of their ability to have causal effects on the physical world" (Ellis 2008:67).

[116] "The world-as-a-whole may be regarded as a kind of overall system-of-systems for its very different . . . component systems are interconnected and interdependent across space and time with, of course, wide variations in the degree of coupling" (Peacocke 2008:219).

[117] George Ellis offers a diagram that illustrates the sphere covered by the **natural sciences** (astronomy, geology, chemistry or atomic physics, particle physics, ecology, physiology, cell biology), the **human sciences** (politics, economics, sociology, anthropology, and psychology), and the "**extra-scientific domain**"

(aesthetics, art, ethics, literature, metaphysics, and philosophy as well as pain, joy, fear, hope, tenderness, sadness, frustrations, and loves of everyday life). (2008:81).

[118] Kaku 2008:ix ff.

[119] See the sophisticated analyses of the relation between object and subject in modern science and technology by Marcuse (1968:122 ff).

[120] This discussion is rudimentary. My attempt to depict the contours of the history of epistemology since the Middle Ages quickly ran into eighty pages without achieving a remotely satisfactory result. So I dropped the idea for the purposes of this book. There are hundreds of fine books on the subject at a professional yet accessible level.

[121] E.g. Cahoone 1988.

[122] "Our knowledge is . . . stored as electromagnetic changes in our brain and is obtained from electrochemical impulses transmitted by our senses. Recognizing the futility of achieving absolute knowledge of the universe, the idealist believes that knowledge is synonymous with our mental constructions, and therefore it makes no sense to talk about the existence of the universe. The universe is simply our knowledge. Taking idealism to the extreme, a solipsist . . . concludes that the universe must be a construction of 'my' mind alone" (Ben-Ari 2005:117). This may seem a rather crude assessment of skeptical epistemology and postmodernism lumped into one, yet it reflects the mood of exasperated scientists.

[123] The psychiatrist Peck summarises the scientific package of assumptions as follows: "The universe is real, and therefore a valid object of examination; the universe makes sense—that is, it follows certain laws and is predictable; human beings are poor examiners, subject to superstition, bias and prejudice, and a profound tendency to see what they want to see . . . simple experience is not to be trusted . . . Science is a religion of scepticism . . ." (Peck 1990:208 f).

[124] Hawking and Mlodinow believe that they have found "a candidate for the ultimate theory of everything, if indeed one exists, called M-theory" (2010:8). This illustrates the extraordinary self-confidence and optimism with which physicists go about their work.

[125] "The view shared by most, if not all, scientists is that the universe exists, that it is amenable to scientific research and that science itself is universal, equally accessible to anyone on Earth . . . regardless of gender, ethnicity, or culture" (Ben-Ari 2005:118).

[126] Kuhn 1970.

[127] Ellis offers a profound analysis of the relationships between the large (cosmology), the small (quantum theory), and the complex (life and intelligence) and spells out the tasks these present to interdisciplinary research (2008:57 ff).

[128] Ben-Ari makes the point that going back to the components (reductionism) and placing detail into an overall picture (abstraction) are both indispensable aspects of the scientific enterprise (2005:142 ff). "Analyzing the components and partial actions of the whole in a reductive way is an essential part of

attaining understanding; but claiming this partial view is the same as the whole is a crucial error . . ." (Ellis 2008:78).

129   Model-dependent realism is "based on the idea that our brains interpret the input from our sensory organs by making a model of the world . . . In the history of science we have discovered a sequence of better and better theories or models, from Plato to the classical theory of Newton to modern quantum theories" (Hawking and Mlodinow 2010:7).

130   Ben-Ari 2005:45. "(E)ven theories can also be considered to be facts when no useful purpose would be served by doubting them" (2005:48).

131   The epistemological debate on whether intersubjective verification and falsification are possible (Kant, Hume, Popper, Carnap, Bartley, Albert, von Weizsäcker, Kuhn and others) has almost no impact on most empirical sciences. Scientists routinely verify or falsify their hypotheses. For a discussion see Ben-Ari (2005:63 ff).

132   'Universals' was the name given to Platonic *ideas*, Aristotelian *forms*, or Nominalist *concepts* in the medieval philosophical debates.

133   March 1957:18—my translation.

134   Theologians and philosophers do not always account for the origin of their concepts. How do we come to know that God is eternal, omniscient, or omnipotent? Russel expresses the common view that the universe "is finite; only God is infinite" (Russel 2008:13). But how can we know? Is the mathematics of infinity a valid theological answer, or does it compound the problem by obscuring the Platonic origins of the idea? Ludwig Feuerbach had an experiential answer: attributes of God are idealized abstractions from reality or irrational wishes rooted in the frustration caused by human limitations and projected into a perfect being.

135   In the so-called 'block universe' in cosmology, where time functions as a fourth dimension, time is deemed reversible. Time is also pre-determined. I am not a physicist or mathematician, but as a lay person I cannot help but wonder whether this mathematical construct is not based on questionable assumptions. Is it admissible to translate time into a quasi-spatial dimension within a geometrical edifice? That three-dimensional entities move and change in time is self-evident, but that does not imply that time is a fourth 'dimension'. Significantly, the idea of additional dimensions can only be expressed in mathematical terms. A depiction of the block universe invariably replaces a spatial dimension with the 'time dimension'. Even more adventurous is the idea that wormholes in a curved space-time continuum could provide humans with short cuts to the future or the past (Greene 2004:61 ff).

136   An example is the essay of Jügen Moltmann called 'The Fellowship of the Holy Spirit: On Trinitarian Pneumatology' (1992:57-69). It was preceded by a whole book on *The Trinity and the Kingdom of God* (1981). My comments are in brackets. Moltmann departs from the (modern) assumption that fellowship does not dominate but liberates. So the 'Kingdom of God' cannot be a monarchy. (Certainly the biblical original was a monarchy). Therefore the doctrine of the Trinity must reflect a fellowship in which the three persons have equal

dignity. So, the Spirit cannot proceed from the Father and the Son (*filioque*), as Western theology maintains, otherwise the Son would dominate the Spirit and the reciprocity between them would be disturbed. (If there is complete equality and reciprocity, can the Father then also proceed from the Son?). The ancient concept of *perichoresis* (the three persons are 'walking around within each other') is used to express the special and incomparable fellowship in which the divine persons intimately exist for each other and in each other as a "complete Trinitarian unity". This Trinitarian fellowship is deemed open, so that the Spirit can take true human fellowship into itself so that it corresponds with the fellowship within God. (Just imagine what a natural scientist could make of such a metaphysical construct! Why not simply say that the Spirit of redeeming love that manifested itself in Jesus of Nazareth manifests itself again in the Christian community?).

[137] See Brümmer (2005:6 ff). However, the problem that science has with the metaphors of faith go beyond simple identification between metaphor and referent, selection of particular aspects at the expense of others, denoting relationship and existential commitment, cultural changes that make ancient metaphors incomprehensible, or theoretical-speculative extensions to the existential core of faith. The real problem is that the kind of reality that the sciences explore, reconstruct and conjecture is void of evidence that there is such an entity in the first place. As Brümmer concedes, the theologian "has not that object" to which the metaphor or image refers, only the metaphor itself (2005:14). Well, is this not what fantasy is made of? Can the coherence of a belief system alone underpin its validity? This needs further clarification.

[138] In his work on creation (1985), for example, Moltmann argues that God must be seen as a Trinitarian community that permeates a many-faceted, non-hierarchical, decentralized world of multiple relationships. (As far as such a harmonious world actually exists, God would then just be a name for the vast network of interrelationships that makes up the immanent world and that any naturalist can observe. But what does the *Trinity* have to do with that?) God the Spirit, Moltmann argues, dwells in (material) creation, holding (biological) creatures together and keeping them in life. (How does 'spirit' relate to physical 'matter' and biological 'vitality'?) "Everything that is, exists and lives in the unceasing inflow of the energies and potentialities of the cosmic Spirit . . . we have to understand every created reality in terms of energy, grasping it as the realized potentiality of the divine Spirit". The Spirit is "the sustaining foundation of everything else" (9-11). (Is 'spirit' then nothing but energy realising its potentiality rather than the structure and orientation of consciousness? Or does Spirit here take on the role of the Creator in the original version of the Trinitarian doctrine? One can perhaps detect a similarity of Moltmann's 'Spirit' with my concept of the transcendent Source of reality. But if that is the intention, why use the traditional concept of the Spirit and not the traditional concept of the Creator (in the sense of the *creatio continua*) for that? His terminology makes comprehension almost impossible even in theological terms, let alone in experiential-realist terms. Just imagine the face

of a normal housewife if an electrician would tell her that the sparks that fly from her defective plug are manifestations of the Holy Spirit!).

[139] The Egyptian Pharaoh was deemed the son of the sungod Re or Ra. The intention of the metaphor was to express his divine origin and authorisation. The metaphor was utilised, amongst others, by the Israelite kings (Psalm 2) and eventually applied to Jesus, deemed the messianic king expected by the Jews, but its authoritarian meaning was transformed into its opposite (Mark 10:35-45).

[140] I have offered a more differentiated analysis in Nürnberger 2007, 219 ff.

[141] Lakeland 1997:12. A similar distinction is made between sceptical and pessimistic postmodernists (who believe that we will never get anywhere in the morass of relativity), and optimists who are willing to live with complexity and make moral choices to reach better conditions (Rosenau quoted by Degenaar in Haines and Wood 1998:17).

[142] "There is an underlying paradox in all ideologies that seek to control or contain modernity . . . if one wishes to control modernization . . . one will seek to manipulate the processes of modernization. These very ideas, however, are modern—indeed modernizing *in themselves*" (Berger, Berger and Kellner 1973:157 f).

[143] "These critiques . . . are seen by scientists as uninformed and pernicious in their effects on the real problems that arise in the relationship between science and society" (Ben-Ari 2005:115; see 115-130 for an elaborate refutation).

[144] Ben-Ari 2005:118. See also Dawkins 1998 ("Postmodernism disrobed").

[145] Heisenberg 1999:204.

[146] The concept of 'particularity' (unique among many) must be distinguished from the more restricted concept of 'singularity' (only one in existence) used in physics for the universe 'before' the big bang or a black hole.

[147] A fact is "an observation backed up by such a preponderance of evidence that no useful purpose would be served by doubting it" (Ben-Ari 2005:46).

[148] "That information has an ontological reality apart from the physical system through which it is embodied. It can be encoded in different ways and transmitted from point to point. It is not encompassed in any particular physical representation" (Ellis 2008:64).

[149] "(This) unpredictability derives from the fact that the regularities at lower levels have become strongly affected by regularities emerging at higher levels of organization" (Deacon quoted by Murphy 2008:126).

[150] For a simple description see Gitt (2005:27 ff). For a more detailed discussion see Ben Ari (2005:1-61). The two authors are at opposite ends in their assessment of the theory of evolution. Gitt tries to demolish it on religious grounds by jumping from modern science to the pre-scientific worldviews of the Bible without any historical or substantive mediation.

[151] However, the fact that "one experiment can falsify a well-loved theory while a million successful experiments cannot absolutely confirm it" (Ben-Ari 2005:19) is a truism that plays no significant role in the practice of most sciences.

152 "(I)n science, observation is theory-laden, that is, the theoretical framework within which the observer works is critical to the observation itself" (Ben-Ari 2005:7).

153 Ellis 2008:68.

154 "It is an essential characteristic of scientific theories that they be as universal as possible" Ben-Ari 2005:18.

155 "Given such unpredictability, the only explanation can be an interpretative story that retrospectively retraces the actual changes in dynamics" (Teske 2010:91).

156 Ben-Ari 2005:81.

157 "Often, viewing all fields together, it seems instead a rather ramshackle structure with little coherence among its various parts" (Kuhn 1970:49).

158 See Nürnberger (1999, chs. 6-10).

159 Dawkins believes that only science can hope to answer ultimate questions (2006:56). In an early work, Hawking famously raised these questions without answering them (Hawking 1988:174). In Hawking and Mlodinow 2010:5ff, the authors claim that philosophy, which had traditionally attempted to answer such questions, "is dead" and scientists have become the torch bearers of discovery and the quest for knowledge. But knowledge has always been the task of the sciences. The authors do not answer the 'ultimate questions' of meaning, acceptability, belonging and authority.

160 Kauffman 1995.

161 To assign science to the paradigmatic mode of knowing and faith to the narrative mode of knowing (Teske 2010:92 ff) is misleading. Whatever we know is in flux and prompts the narrative mode of knowing. Paradigmatic abstractions from the flux of reality are found both in science and theology. The parable of the prodigal son, for instance, is a narrative form of the propositional doctrine of justification by grace.

162 I use the term "law of entropy" deliberately to generalise a cluster of phenomena explored by different disciplines under the concept of "entropy". Its basic meaning is the tendency of processes to move from order to disorder, from imbalance to equilibrium, from potent to spent energy. See the entry under "entropy" in the Wikipedia for an overview.

163 Human lives are "complex adaptive systems" that "can only be explained by an interpretive story that retrospectively retraces the actual changes in dynamics" and that "do not permit a reduction of purposive acts to non-purposive elements" (Teske 2010:91).

164 Significantly, the story of the transitions between ancient, medieval, Newtonian, and quantum physics is told again and again by physicists when they explain the 'new physics'. A recent example is Hawking and Mlodinow (2010:15 ff). One can sense the need of subatomic physics for legitimacy in a puzzled population.

165 The so-called 'historical books' found in the Old Testament (Samuel, Kings, etc.) constantly refer to the annals of the kings that existed at the time. It is clear that the canonical books do not want to repeat this historical story. Their function is to give a *theological* interpretation of the reign of the king. To do so

they select their material and subject it to a particular interpretation. Historical precision is not their prime preoccupation.

166 For a demanding, but accessible introduction to modern cosmology, see Greene (2005). A humorous account of the evolution of "nearly everything" is found in Bryson (2003). Hawking and Mlodinow (2005) provide a more popular version of cosmology than Hawking (1988). Drees (1990) offers an overview of the possible links between cosmology and theology. Barrett (2000) summarises the history of science since Copernicus. Kuhn (1970) shows how scientific paradigms change under the impact of new insights. Guttman (2005) offers a useful overview of the theory of evolution. I have benefited greatly from McCarthy and Rubidge (2005). Clayton (2004) offers an accessible treatment of emergence, Mitchell (2009) of complexity theory. Because of its accessibility I have also utilised the Wikipedia.

167 Research conducted in the European Union has revealed an overwhelmingly positive attitude towards modern science and its capacity to bring about a better world. While faith in God is shared by anything between 16 and 95 per cent of the population in various countries, it has dropped to less than 34 per cent in France, the Netherlands and Denmark. The European average is 52 per cent (European Union 2005:7 f). In contrast, 88 per cent of all EU citizens believe that the work of scientists at universities have a positive effect on society. 81 per cent believe that industries that develop new products have such a positive effect (European Union 2005:43).

168 Von Weizsäcker 1990:405 ff.

169 Hawking and Mlodinow 2005:68 ff. McCarthy and Rubidge (2005:62) provide a good summary. For a deconstruction of this view see Greene (2005:272 ff).

170 For an explanation of these complex mechanisms see Greene (2005:272-323).

171 "Despite our good understanding of the development of structure in the expanding universe, a fundamental mystery remains: the complexity of astronomical structure is self-generated by gravitational instability acting on seed perturbations, and this spontaneous structure growth is the basis on which all other complexity in the universe comes into existence" (Ellis 2008:61).

172 A lucid depiction of the relation between cosmic, solar, terrestrial and organic evolution is offered by McCarthy and Rubidge (2005).

173 As mentioned in a previous footnote, I use the generalising term "law of entropy" deliberately as a meta-scientific category. It says that processes in a closed system tend to move from order to disorder, never the other way round. The principle was first formulated by Clausius as the *Second Law of Thermodynamics*. It has since found numerous interpretations and applications in various academic disciplines, which suggests that we are dealing with a universal phenomenon. Emboldened by World Bank economist Hermann Daly, I have applied it to economic structures and processes (Nürnberger 1999:333 ff). The economist and ecologist John Bellamy Foster has taken a similar stance. In a private conversation a physical cosmologist of rank has confirmed the wide applicability of the term. Following a lead from the work of a prominent Old

Testament scholar, Walter Brueggemann, I have also applied it to the biblical tradition (Nürnberger 1999:70 ff.).

174   Kaufman 2004:88-93.

175   This is not always recognised. Peck, for instance, believes that "the flow of evolution is against the force of entropy . . . I state that the process of evolution is a miracle, because insofar as it is a process of increasing organisation and differentiation, it runs counter to natural law" (Peck 1990:283 ff). No, it does not. Peck also believes that, spiritually, the evolutionary thrust can vanquish the power of the entropic process and make us better people (1990:285). He goes so far as to postulate that "we are growing toward godhood" (288). In some sense this expectation is in line with Dawkins' optimistic view of human ethical development, in other senses with Teilhard de Jardin's speculations of an evolution leading beyond current human reality. Single individuals, groups and states may indeed reach superior patterns of spiritual and social organisation. But one should not overlook the vast degree of anomie and misery created among marginalised population groups by the progressive impact of modern civilisation. One should also not overlook the fact that the latter flourishes at the expense of nature as a whole.

176   Sachsse 1984:6-8.

177   What all this *could* mean in cosmic terms is summarised by the following quotation: "Every star that burns, every planet whose orbit is slowly decaying, every breath you take and every calorie you metabolise brings the universe closer and closer to the point when the entropy is maximized, organized movement of any kind ceases, and nothing ever happens again. There is no escape. No matter how magnificent life in the universe becomes or how advanced, the slow increase in entropy cannot be stopped—the universe will eventually die" (Warren Fried 2007). Unfortunately, I lost the bibliographical reference but the quotation is so striking that I want to share it with my readers anyway.

178   Is an end to reality through a big crunch or total dispersion inevitable? My colleague, Detlev Tönsing, suggests that an expanding universe would always provide new energy differentials. "Water will continue to flow if the crevice into which it flows constantly sinks". However, if the speed of the sinking crevice were identical with the speed of the falling water, there would be equilibrium. The question is also how the expansion of the cosmos as a whole affects energy conglomerations *within* the cosmos—which do not necessarily expand at the same speed, if at all.

179   Science calls this condition "high entropy energy" as opposed to "low entropy energy" or energy "far from equilibrium".

180   Some primitive forms of life derive their energy not from photosynthesis but from chemosynthesis, for instance bacteria that thrive in deep ocean vents and who in turn feed worms, shrimps and mussels. Bryson 2003:279.

181   Nürnberger 2002:79 ff.

182   Nürnberger 1999, ch. 13, 333 ff.

183   Murphy and Ellis offer a simple hierarchy of the sciences based on levels of complexity: physics, chemistry, biochemistry, botany/zoology/physiology,

psychology (1996:19; for detail see 22-38). For a more elaborate 'ladder' see Clayton (2004:27f).

[184] For a first introduction see the Wikipedia article on the internet. For a thorough treatment see Clayton (2004) and Kauffman (2008 and his previous works). Russel offers a brief summary of Peacocke's approach to emergence (2008:6 f).

[185] Kauffman 2008:68 and elsewhere in the book.

[186] Kauffman 2008:127.

[187] "As the parameters are changed, a system can pass from simple regular behaviour that repeats itself exactly to the highly complex non-repetitive irregular behaviour described (by chaos theory) as chaotic" (Barrett 2000:120).

[188] High levels of probability seem to indicate causality rather than random. At the subatomic level they may be due to factors emanating from cosmic depths that are not accessible to scientific observation.

[189] A physicist informed me that the simple rules of causality at the macro-physical level (as depicted by Newton) are due to the fact that probability becomes so overwhelming that potential deviations can be neglected.

[190] If oxygen and hydrogen mix, they have the propensity of forming water, but they will do so only under particular circumstances.

[191] I am not using the concept of teleology in the Aristotelian sense (all phenomena have an intrinsic tendency to reach perfection), also not in the anthropomorphic ascription of purpose to impersonal processes, but in the experiential sense that living beings are programmed by evolution to seek survival, well-being and avoidance of pain.

[192] Ben-Ari argues that it is inappropriate to "extrapolate the term evolution from a specific theory within biology to a blanket term covering the origin of everything" (2005:40). I do not think that biology has a monopoly on this term or that it is inappropriate to use it in cosmological or any other scientific context.

[193] Bryson 2003:288.

[194] "Emergence is not . . . the absence of causality; rather, it is a web of causality so intricate that the human cannot grasp it" (Van Huyssteen 2006:55 quoting Ian Stewart).

[195] Quoted by Bryson 2003:291.

[196] For a critique of Dawkins' metaphor of the 'selfish gene' see Nürnberger (2010:106 ff).

[197] McCarthy and Rubidge 2005:174.

[198] For a lucid depiction of the evolution of the human being see McCarthy and Rubidge (2005 ch. 10) and Guttman (2005 ch. 8).

[199] A useful introduction into the field is Guttman 2005. A more demanding depiction can be found in Gould 2002. But the approach of the latter is controversial.

[200] The biologist and mathematician Stuart Kauffman (1995) accepts Darwin's idea of chance mutation and selection of the most adaptable forms, but goes beyond it by positing mechanisms of self-organisation. The emergence of life is not

due to some 'tinkering' but follows regularities. This assumption considerably reduces the time needed for the most complex forms to evolve.

201   Ian Stewart argues strongly against genetic reductionism and believes that there is a higher emergent level of "universal mathematical principles of growth and form that even DNS exploits". (1998. *Life's Other Secret: The Mathematics of the Living World*. New York: Wiley, referred to by van Huyssteen 2006:55). He believes that life still follows mathematical rules, just of a "more subtle, more flexible, more deeply hidden" nature. But this does not yet cover environment and choice.

202   "Contingency is inevitable, but unremarkable. There are not an unlimited number of ways of doing something. For all its exuberance, the forms of life are restricted and channelled" (Conway Morris as quoted by van Huyssteen 2006:58).

203   "For Gould, if the history of evolution were to be repeated, the likelihood of humans evolving is virtually nil, and humankind is just a quirky, evolutionary accident . . . (According to Morris, however,) the reality of convergence suggests that human intelligence would almost certainly always emerge" (van Huyssteen 2006:58). Stuart Kauffman agrees: humans are 'expected' and are 'at home in the universe' (1995).

204   Bryson 2003:279.

205   For the number of cells see Bryson (2003:371); for kinds of cells see Guttman (2005:126).

206   In what follows, I am led by the informative analysis of Vincent (1990), supplemented by various articles on the brain in the Wikipedia.

207   Van der Walt 2010:24f

208   Van Huyssteen 2006:59, referring to Rescher (1990).

209   Henry Plotkin as referred to by van Huyssteen (2006:80). Species adapt to their respective environments, so that "the structure of all organisms is . . . directly informed by their environment", and "living creatures somehow incorporate into themselves those aspects of the world that they actually interact with."

210   There seems to be "only one (evolutionary) line that leads to persons . . . And at exactly this point, evolution by natural selection passed into a higher level of complexity as nature transcended itself into culture" (Van Huyssteen 2006:48, referring to Rolston 1996:69). Culture, of course, follows its own evolutionary, increasingly differentiating trajectories at a higher level of emergence. But we must not overlook the level of emergence that lies between biological and cultural reality, namely personhood as such. That is what I try to unravel in what follows.

211   For more detail see van der Walt 2010:26ff.

212   Van der Walt 2010:27.

213   "Plotkin considers instinct as unlearned and unthinking adaptive behavior, a kind of behavior that is knowledge in the same way as any form of camouflaging coloration of an insect constitutes knowledge of its surroundings" (van Huyssteen 2006:81).

214 "Thus, representational and mental properties are *contextualized* brain states or events . . ." (Murphy 2008:124).

215 As some neuroscientists hypothesize, "learning involves a selective development of some neurological pathways in preference to others" (Tracy 2008:279).

216 "When new and more complicated structures appear in the universe it is the result of extensive sampling of a vast array of alternatives and then keeping and improving on the alternatives that best serve some higher-order need. This is what allows for the spontaneous increase of fitted order and functional complexity in both human designs and in nature" (Murphy 2008:124, referring to Terrence Deacon).

217 Barrett 2000:120 (inset). Smail 2008:135.

218 Kauffman postulates that "with agency, a part of life, comes meaning, doing, and value" so that "we appear to be living in an emergent universe in which life and agency arose with no need for a Creator God" (2008:130). We will later see that immanent 'self-organising processes' do not compete with an assumed transcendent Source and Destiny. The question Kauffman is busy with is how agency arose within immanent reality.

219 I here follow the depiction of van Huyssteen (2006:60 ff).

220 Van Huyssteen 2006:64.

221 Ian Tattersal (1978) as quoted by van Huyssteen (2006:191).

222 Van Huyssteen 2006:192.

223 Van Huyssteen 2006:196 f. referring to Stephen Mithen.

224 Van Huyssteen 2006:76.

225 Van Huyssteen 2006:76. Evolutionary theory tends to speak of information as the basis of all organic systems (2006:84). I would differentiate between the concept of information, which is a system that can be transmitted without personal consciousness, and the concept of knowledge, which presupposes consciousness.

226 Clayton 2004:110.

227 I would tend to disagree with Plotkin, therefore, when he posits the dominance of instinct over intelligence (van Huyssteen 2006:83). This may be a case of unwarranted reductionism. Intelligence presupposes instinct as the lower level of emergence. It is impacted, but not determined by it. In the same way, culture presupposes intelligence but still generates its own dynamic as an emergent reality.

228 Vincent (1990) provides a detailed account of research into this intricate field.

229 During the final phases of the anti-apartheid struggle, black youths invented the so-called 'necklace' murder. A tyre was place around the neck of the victim, filled with petrol and ignited. Youths could often be seen dancing around the scene in ecstatic glee. A similar phenomenon can be observed in some traditional cultures when animals are killed for sacrificial rituals, or when the organs of animals or people are believed to contain magical powers. Here the agony of the victim is perceived to be part of the effectiveness of the ritual.

230   Beehives and ant heaps are sometimes called superorganisms. They presuppose the biological level of emergence. Supra-personal levels of emergence presuppose the personal level of emergence.

231   "As the parameters are changed, a system can pass from simple regular behaviour that repeats itself exactly to the highly complex non-repetitive irregular behaviour described (by chaos theory) as chaotic" (Barrett 2000:120). Or as Stewart put it, "complicated rule-based systems manage to produce robust high-level features that themselves obey their own kind of emergent rules" (quoted by van Huyssteen 2006:55, note 5).

232   "This 'purposive intelligence' could be seen as a novel phenomenon that emerges at a new level of operational complexity" (van Huyssteen 2006:49).

233   "Biologically we do possess some capacity to transcend the here and now and, in fact, humans may be the only animals that can pass the limits of present reality" (Kieffer 1979:347).

234   "Therefore, while animals adapt to their niches, human beings adapt their ecosystems to their needs . . . Other higher animals may also have minds, but the human mind is the only mind that permits the building of complex transmissible cultures" (Van Huyssteen 2006:99 f, referring to Rolston 1999).

235   The philosopher Karl Popper believed "that we do not gain knowledge by induction, but propose theories to the environment, which subsequently falsifies most of these theories and thus provides the criteria for the retention of those it fails to falsify" (van Huyssteen 2006:78). This gives the impression that scientists playfully throw balls at reality to see how they bounce back. Certainly, science and technology are more goal-oriented and aggressive than that. Animal breeding programmes, for instance, purposefully employ the regularities of reality to produce desired forms of life.

236   For a discussion see Murphy and Ellis (1996:24 ff).

237   It may well be that the *scientific* version of the so-called Gaia hypothesis represents pioneering work in this regard. Without going into detail I may refer the reader to the works of James Lovelock, Lynn Margulis and Dorion Sagan.

238   That is the opinion of Durand (2010) and some ecologists.

239   Kauffman 2008:68 and elsewhere in the book.

240   The classical analysis of past, present and future goes back to the *Confessions* of St Augustine. Some authors suggest that there is no storyline from the past through the present to the future. Rather, the past is "re-experienced" and "re-authored" in the present. Presumably then the future is "pre-experienced" and "pre-authored". Obviously this can only refer to what happens in our consciousness; the world process as such does follow a storyline, otherwise we would never have come to exist! The same applies to the idea of a "fusion of horizons" between past and present as espoused by Gadamer. In biblical terms, one can mention the contrast between Luke, who depicts a long storyline from the crucifixion of Jesus to his Second Coming, and John who collapses all these events in Christ being "lifted up". All these alternatives cannot do justice to a past that has objectively happened, which is presupposed by historical-empirical research such as conducted by archaeologists, paleontologists and historians.

The focus is, rather, on what happens in our brains: memory, anticipation and present self-understanding of the human being.

241 A differentiated treatment of chaos theory in a theological perspective is found in Russel, Murphy, and Peacocke (eds) 1995.

242 According to Ian Stewart, mutations "enable the system to explore its phase space" and selection makes "the exploration of this space very efficient . . . If all that happened were just random mutations, 'the system would wander around in its phase space like a drunkard, tottering one step one step forward, two steps back' . . . With selection, however, bits of phase space that do not work . . . are eliminated. Selection thus helps the system to home in on the most interesting regions of phase space where useful things can happen . . ." (quoted by van Huyssteen 2006:56). Although 'usefulness' is here defined as fitness, this notion projects too much intentionality into the system at a level of emergence where intentionality is not in operation.

243 Kauffman 1995:85 ff.

244 Writing on the "reality of consciousness and its causal effects", Ellis finds it "extraordinary that anyone who takes the trouble to write a book or scientific paper should deny this reality, thus undermining the validity of everything they write . . ." Free will is "a central premise at the foundation of the possibility of all the sciences" (2008:67).

245 The physicists Hawking and Mlodinow make a strong case for reductionism and what they call "scientific determinism" (2010:30, 34). The presupposition is that reality is based consistently upon natural law: "given the state of the universe at one time, a complete set of laws fully determines both the future and the past . . . Since people live in the universe and interact with the other objects within it, scientific determinism must hold for people as well" (2010:30). It is clear that this is a deduction from a premise that has not been shown to be valid. "It is hard to imagine how free will can operate if our behaviour is determined by physical law" (2010:32). Well, the authors have no scruples to confront their readers with much less imaginable propositions than that. Complexity theory gives us at least some clues on how the strict sequence of causality can be transcended. But Hawking and Mlodinow see determinism at work even there: "the outcome is determined in such a complicated way and with so many variables as to make it impossible in practice to predict" (2010:32). Indeed, but prediction is not the point, determination is. For such complex phenomena, the authors resort to what they call an 'effective theory': "In the case of people, since we cannot solve the equations that determine our behaviour, we use the effective theory that people have free will" (2010:33). While they seem to arrive at the same outcome as complexity theory, the authors fail to explain anything at all, while complexity theory does.

246 In Darwinian terms, the fitness of a package of meaning in relation to the need structure of the environment determines its chances of survival. In economics, the product that responds best to market demand will constitute the most successful supply.

247   Game theory and complexity theory have recently explored ways in which automatic 'trial and error' sequences work (Mitchell 2009:145 ff). Processes involving collective and individual consciousness will be infinitely more complex than that.

248   For an analysis of culture as outcome of the emergence of human existence see Hefner (1993, part 4).

249   This is recognised by perceptive naturalists such as Stuart Kauffman and Brian Swimme. Kauffman has argued persuasively that even for the physical and mathematical sciences the future is—in general—not predictable. "A radical, and I will say, partially lawless creativity enters the universe . . . And since we can neither prestate, let alone predict, all that will happen, reason alone is an insufficient guide to living our lives forward. This emergent universe, the ceaseless creativity in this universe, is the bedrock of the sacred that I believe we must reinvent" (2008:130). Fair enough, but nature does not, on its own, give us clues concerning the content of what is to be considered sacred. Is the ceaseless and rapacious creativity of the modern liberal economy (discussed in ch. 11 of Kauffman 2008) to be considered sacred?

250   "(A) process of synaptic selection occurs within the repertoires of neural groups as a result of behavioral experience . . . certain synapses within and between groups of locally coupled neurons are strengthened and others weakened without changes in the anatomy. This selectional process is constrained by brain signals that arise as a result of the activity of the diffusely projecting value systems . . . continuously modified by successful output" (Edelman and Tononi quoted by Ellis 2008:67 f).

251   Peck 1990:301.

252   Peck's identification of the individual and collective subconscious with God, the latter characterised by infallibility as opposed to the fallibility of the conscious mind (Peck 1990:301 ff), is clearly off the mark, whether in empirical or theological terms. The subconscious is as subject to evolutionary processes and as problematic a guide to truth and morality as consciousness is.

253   According to Holmes Rolston (1996:69), "in nature information travels intergenerationally through genes, while in culture information travels neurally, as people are educated into transmissible cultures. In nature, the coping skills are coded on chromosomes. In culture, the skills are in craftsman's traditions, in technology manuals, or in religious rituals, texts, and traditions . . . This information transfer on a cultural level can be several orders of magnitude faster than on a genetic level, and can in fact leap over genetic lines" (paraphrased by van Huyssteen 2006:99).

254   Nürnberger 1987:95-117; 1988:137-152; 1999:142-154.

255   Jung 1995:410.

256   Guttman 2005:59 f.

257   Clayton mentions two prerequisites for a successful interaction of the theist with the scientist: first, theology's "description of the various scientific domains (must be) accepted by the specialists in each field . . . and, second, her results

(must be suggestive) of a theological interpretation *that does not do damage to the facts in question"* (Clayton 2008:110).

258 Neo-Marxist social analysis has used the term for possible social and social-theoretical alternatives: "The terms 'transcend' and 'transcendence' are used throughout in the empirical, critical sense: they are designate tendencies in theory and practice which, in a given society, 'overshoot' the established universe of discourse and action toward the historical alternatives (real possibilities)" (Marcuse 1968:10, note 1).

259 Drees 1990:176-178, 193.

260 Ben-Ari is confident that science has already solved most riddles of practical importance, yet mentions instances that may defy scientific enquiry: "The questions are of overwhelming interest, but they may be unsolvable simply because there is no longer enough evidence remaining. At most, we might be able to offer convincing arguments and experimental results that establish their plausibility, but conclusive evidence may never become available" (2005:207).

261 Ellis offers an overview of the serious problems encountered by cosmology today. "The many proposals for quantum creation of the universe have multiplied . . . but the fundamental issue remains the same: there is no way to test our theories about the origin of the universe directly, and indirect tests . . . are very weak and non-unique . . . most of the universe is unobservable and the physical laws underlying the development of the really early universe are untestable . . ." (2008:59 f).

262 Krauss and Starkman 2000:90f.

263 Ellis 2008:77.

264 Hawking and Mlodinow's M-theory ventures especially far into the unknown. Like most theories in subatomic physics, it is based on mathematical constructions in search of evidence, thus on speculation. Leaving aside its high level of sophistication, it is remarkably similar to polytheism where gods responsible for different areas of reality interact to bring about phenomena in this world: "It could be that the physicist's traditional expectation of a single theory of nature is untenable, and there exists no single formulation. It might be that to describe the universe, we have to employ different theories in different situations. Each theory may have its own version of reality, but according to model-dependent realism that is acceptable as long as the theories agree in their predictions whenever they overlap, that is, whenever they can both be applied" (2010:117).

265 Ellis 2008:62.

266 For an accessible introduction to complexity theory see Mitchell (2009).

267 Kauffman 1995:302.

268 I gleaned the examples in this paragraph from a lecture by my colleague Dr Detlev Tönsing at the conference of the South African Science and Religion Forum, Pretoria, 2-3 September 2010.

269 A radical and rather cynical conjecture concerning the origins of religion is offered by Durand (2010). While immensely persuasive at first sight, it is also

quite tendentious. Aggressive behaviour, out-group hostility, genocide, murder, torture, persecution, subjugation and submission of females and youths, glorification of war and heroism, male and female circumcision, celibacy, fasting, self-mutilation, hostility against survival instincts such as enjoyment of good food and promiscuous sexual activity—all such objectionable attitudes and practices are attributed to male dominated religion as such. At the same time they are deemed a reflection of the ingrained behavioural patterns of alpha males among primates, as produced by evolution. This means that they are insurmountable, and yet fatal for humanity and the ecosystem.

[270] Heisenberg 1999:201.

[271] I picked up this telling quotation from a German calendar and translated it back into English.

[272] This fact is widely acknowledged by naturalists (Dawkins 2006:362 ff; Ben Ari 2005:207; Kauffman 2008 deals largely with unpredictability).

[273] "Just how does what you call an invisible, intangible, eternally elusive gardener differ from an imaginary gardener or even from no gardener at all?" (Flew quoted by Clayton 2008:104).

[274] "Through our relative dependence on the world we can experience our absolute dependence on God as Creator of the world . . ." (Russel 2008:14). Schleiermacher, the father of modern theology, spoke of the 'awareness of ultimate dependence'.

[275] Clayton says "it would seem that the event in question must be in some way different than it would have been had divine action not been involved" (2008:104f). I find this "counterfactual principle" rather pointless. The question can only make sense if a particular event is deemed subject to divine intervention, while the whole context in which it occurs is not attributed to divine agency. But this would be an example of the God of the gaps. If, on the other hand, the event and its entire context were the result of divine agency, the event could only be what it is. Without divine agency it would not be different but non-existent. One can also not establish what an event would have been if the fine tuning of the universe, thus its total historical process, had been different. The simple fact is that this particular event would then not have occurred at all. The same is true if there were no transcendent agency that had brought about the event as part of the flow of reality as a whole.

[276] "Step by step, Atkins succeeds in reducing the amount of work the lazy God has to do until he finally ends up doing nothing at all: he might as well not bother to exist" (Dawkins 2006:118).

[277] Richard Dawkins's idea that convictions are "memes", that is, floating mental viruses that have nothing but nuisance value for its carriers (2006:191-201), is certainly one of the weakest parts of his argument. People do not just believe any nonsense. Convictions settle into minds because they provide meaning, acceptability and authority under given social and existential circumstances. They presuppose certain levels of insight. They impose their validity on our consciousness until challenged and replaced by seemingly better options.

278   "We need a willingness to recognize the many-layered character of that experience, so that the same event can be interpreted as a part of physical process, as an occasion for moral decision, as the carrier of beauty, and as an encounter with the divine presence" (Polkinghorn 1996:11).

279   Kauffman's idea that humans have invented the sacred, and now have to reinvent it, misses the point. We cannot invent what we consider to be ultimate truth. Even naturalists cannot. It impresses its validity on our consciousness when genetic predisposition, foetal impressions, toddler socialisation, ongoing experience and new information all fall into place.

280   Kuhn (1970) has argued that the sciences have always been embedded in changing frames of reference. That does not mean, however, that there is no progression of insight as Kuhn apparently believed. See Ben-Ari (2005:101 ff) for a discussion and refutation of this view.

281   "For humans are condemned to choose and to interpret the world so as to choose better . . ." (Cahoone 1988:231).

282   "Moral inquiry can never occur detached from the context of meanings and presuppositions . . . Moral judgments always commit one to a theory of man or a theory of the universe . . . a metaphysics" (Kieffer 1979:60).

283   "The tendency of science to call everything 'subjective' that does not fit into its active construction of the object may deprive us of the last protection (against treating the human being under the aspect of predictability), that is, our readiness at least to hear the voice of our conscience" (von Weizsäcker 1990:425).

284   That is why the suggestion of Cahoone that culture should play the role of religion in a secular society (1988:215 ff) cannot work.

285   The three phenomena that are taken to point towards the transcendent, "the persistence of mystery in an intelligible world, values in a world of facts, and meaning in a material world" (gleaned from the announcement of Drees 2010) are valid, but they are no substitute for transcendence proper. I agree with Drees that theology must go all the way in its recognition of science as a valid and superior quest for explanation—including an explanation of the emergence, evolution, and immanent nature of faith (Drees 2008:3 ff). But transcendence proper cannot be jettisoned without losing the very rationale of faith.

286   See Nürnberger 2007:35ff for detail.

287   According to the psychiatrist, Peck, one's system of meaning determines one's mental health: "But the fact of the matter is that everyone has an explicit or implicit set of ideas and beliefs as to the essential nature of the world. Do patients envision the universe as basically chaotic and without meaning so that it is only sensible for them to grab whatever little pleasure they can whenever it is available? Do they see the world as a dog-eat-dog place where ruthlessness is essential for their survival? Or do they see it as a nurturing sort of place in which something good will always turn up and in which they need not fret much about the future? Or a place that owes them a living no matter how they conduct their lives? Or a universe of rigid law in which they will be struck down and cast away if they step even slightly out of line?"(Peck 1990:200).

288   "Isn't a suspension of belief more rationally justified in the face of this diversity of great religious-metaphysical traditions?" (Clayton 2008:95). No, it isn't. What follows, rather, is an honest and profound struggle for the truth in view of the consequences of alternative traditions in terms of the vision of comprehensive optimal well-being.

289   "Even the most important decisions in life must always contain this inevitable element of irrationality . . . Without such a firm stand our own actions would lose all force. Therefore, it cannot be avoided that some real or apparent truths form the basis of life . . ." (Heisenberg 1999:205).

290   As Jean-Luc Nancy not only admits but envisages, indeed urges, that consistent deconstruction must lead to a radical void that goes even beyond nihilism. Nothingness, it seems, must be faced honestly, courageously and defiantly. Everything else is self-deceit. Nihilism is the most radical assumption of transcendence one can think of, a transcendence that swallows up the immanent (Nancy 2008:3). Is that genuine freedom? How can it possibly lead to responsibility? Through a defiant reaffirmation of its mastery by the human subject in the face of futility and meaninglessness, as some existentialists proposed?

291   In his famous passage on the murder of God by humankind, Nietzsche says the following: "What did we do when we unleashed this earth from its sun? Where does it now move? Where do we go? Away from all suns? Are we not constantly falling? Backwards, sideways, forward, to all sides? Is there still an 'above' and a 'below'? Have we not lost our way in an infinite nothingness? Do we not feel the breath of the empty space? Has it not become chilly? Does the night not constantly approach us—and more and more night?" (Nietzsche 1966, Band II, page 127. Original to be found in *Die fröhliche Wissenschaft* 1886, 2nd ed. par 125).

292   Classical theology rejected this notion as heretical under the name of Pelagianism.

293   "The epic of evolution has renewed the emphasis on God as the eternal Creator sustaining in existence processes which are endowed by God with an inherent capacity of generating new forms and so with possessing a derived creativity . . . God gives existence to each instance of spacetime with all forms of matter-energy themselves dynamically and continuously and creatively being metamorphosed into new entities, forms and patterns" (Peacocke 2008:201 f).

294   According to Tracy, the 'zero-sum' picture is a potential objection to the search for openings in the causal structure of nature: "If God acts, the creature must be passive, and if the creature is to act, then God must forbear . . . in creating the world, God must at least partially withdraw from it, restricting divine activity so that creatures may exercise the powers that God had given them. The problem then arises about how to get God back into the world in a way that does not displace natural causes and agents." Referring to the work of William Placher, he says that "the boundaries between divine and created agencies resulted from the loss of an earlier tradition that appreciated more deeply the theological principle that God's action transcends all the forms of

agency with which we are familiar. This enabled the earlier tradition . . . to recognize that God is active in all events". He quotes Placher as saying: "Since God was not one agent among others, but operated at a different level of agency, it made no sense to ask which things God had done and which things had been done by someone or something else." Tracy concludes that "it is precisely the modern truncated understanding of God's agency that leads to search for openness in structures of nature" (Tracy 2008:254). I could not agree more! Unfortunately, going through Thomism and Calvinism, Tracy's reasoning leads him to the conclusion that in this approach, although "God's determination of the creature's action is fundamentally different than determination by finite efficient causes . . . we are left with determinism after all" (Tracy 2008:260). This conclusion is hardly appropriate because it builds on an idealised and static concept of God that theology inherited from Greek metaphysics. The biblical faith, in contrast, remains existentially open towards the 'historical movements' of the transcendent.

295 "If God interacts with the world-System as a totality then God, by affecting its overall state, could be envisaged as being able to exercise influence upon events in the myriad sub-levels of existence of which it is made without abrogating the laws and regularities that specifically apply to them" (Peacocke 2008:220). While Peacocke comes close to my view, the phrase "able to exercise influence" is hopelessly inadequate to characterize God as transcendent Source of all of reality. In fact, it does not abandon the two-reality frame of reference—one divine, the other material—that he quite explicitly rejects in the form of a spirit-matter dualism.

296 See Russel, Murphy & Stoeger (eds) (2008) for a summary of the project.

297 Wildman enumerates seven challenges to this approach (2008:135) and therefore abandons the idea of specific divine action within immanent reality altogether. But his list does not yet include the greatest challenge of all: it actually presupposes a God of the gaps.

298 If that were not the case, "nature could do something without God doing it (whereby nature is no longer thought of as created by God), or God could do something in creation without doing it mediated by created nature (whereby one would posit a creation outside of creation, which makes no sense). The only theological candidate for truth is to argue that nature is never devoid of divine activity: nature always equals divine-activity-in-and-through, nature" (Gregersen 2008:194).

299 Stoeger 2008:225 ff.

300 If God is outside of reality, rather than its ongoing transcendent Source, God has to come back into reality to achieve a particular purpose. Then God is basically worldless, and the world is basically godless, while God's intervention is akin to magic making. See the insightful reflections of Tracy on the topic (2008:253 ff) as well as Gregersen (in du Toit 2000:30 f).

301 The Persian religion distinguished between a good god, *Ahuramazda*, and his army of angels, on the one hand, and an evil god, *Angra Mainyu*, and his army of demons on the other. World history is characterised by the ongoing struggle

between the two forces. Caught in the crossfire, humans have to choose on which side they want to throw in their lot. The drama will end in a gigantic showdown, the battle of *Armageddon*, when the evil forces will be vanquished and thrown into a fiery pool. Then only good will remain. It is a very powerful mythological statement that only the good has a right to exist in the world.

302 A more elaborate treatment of the subject can be found in Nürnberger 2011.

303 As a committed atheist, Richard Dawkins accuses agnosticism of dishonesty and opportunism (Dawkins 2006:46 ff). From a scientific point of view, however, it is the only position really consistent with the methodological restriction of science to whatever appears to be within the boundaries of human knowledge and the honesty to own up to the fact that there are such boundaries.

304 I should make it clear that the concepts of the openness or closedness of reality as a whole are different from the concepts of openness or closedness of the universe discussed by cosmologists. The question whether the universe will end up in a big crunch or infinite dispersion refers to immanent reality and not to its transcendent Source and Destiny.

305 This is a point readily conceded by Drees 2008:11 f.

306 Marcuse describes the modern stance as follows: "Outside this (scientific-technological) rationality, one lives in a world of values, and values separated out from the objective reality become subjective. The only way to rescue some abstract and harmless validity for them seems to be a metaphysical sanction (divine and natural law). But such sanction is not verifiable and thus not really objective. Values may have a higher dignity (morally and spiritually) but they are not *real* and thus count less in the real business of life—the less so the higher they are elevated *above* reality" (Marcuse 1968:122).

307 "I shall argue that the existence of God is a scientific hypothesis like any other . . . God's existence or non-existence is a scientific fact about the universe, discoverable in principle if not in practice" (Dawkins 2006:50). "If science cannot answer some ultimate questions, what makes anybody think that religion can?" (2006:56). "Traditionally these (ultimate questions) are questions for philosophy, but philosophy is dead. Philosophy has not kept up with modern developments in science, particularly physics. Scientists have become the bearers of the torch of discovery in our quest for knowledge" (Hawking and Mlodinow 2010:5). Both sources presuppose that 'ultimate questions' are about observation, explanation, and prediction. It is no surprise that the answers they come up with are, again, explanatory theories, rather than sources of existential and social grounding, direction, and vision.

308 Durand 2010. Durand used the term 'beefed-up baboon' in a private conversation.

309 Even Dawkins concedes that: "We are built as gene machines and cultured as meme machines, but we have the power to turn against our creators. We, alone on earth, can rebel against the tyranny of the selfish replicators." Richard Dawkins, *The selfish gene* p 200f as quoted by McGrath 2005:46f.

310 "The God Hypothesis suggests that the reality we inhabit *also contains* a supernatural agent who designed the universe . . ." (Dawkins 2006:58—my

emphasis). "I shall suggest that the existence of God is a scientific hypothesis like any other . . . God's existence or non-existence is a scientific fact about the universe, discoverable in principle if not in practice." (2006:50).

[311] For a more detailed argument, see Nürnberger (2010:37 ff).

[312] "If logic alone somehow required the universe to exist and to be governed by a unique set of laws . . . then perhaps we'd have a convincing story. But, to date, that's nothing but a pipe dream" (Greene 2004:310).

[313] Hawking and Mlodinow 2010:7f.

[314] Hawking and Mlodinow 2010:171ff. "Spontaneous creation is the reason there is something rather than nothing, why the universe exists, why we exist. It is not necessary to invoke God to light the blue touch paper and set the universe going" (Hawking and Mlodinow 2010:180).

[315] Hawking and Mlodinow 2010:9, 171.

[316] Hawking and Mlodinow 2010:173.

[317] Swimme 1996:97.

[318] In this particular sense, the postmodern urge to explode all 'encasements' may seem to overlap with the Christian concept of the transcendent. But a panentheistic solution, if it were intended, is not a theological option.

[319] Note the passionate depiction of this aberration by the naturalist Brian Swimme (1996:8-20).

[320] "It is reasonable to ask who or what created the universe, but if the answer is God, then the question is merely deflected to that of who created God" (Hawking and Mlodinow 2010:172). This argument of infinite regress is popular among naturalists. Dawkins is particularly attached to it (2006:77, 120, 143). But theism only gives a name to an open question that naturalism cannot escape and that it cannot answer. Hawking and Mlodinow say that "Why are the fundamental laws as we have described them? . . . We have seen that there must be a law like gravity . . . the theory must have what is called supersymmetry between the forces of nature and the matter on which they act" (2010:180 f). But what is the origin of such laws? Where does energy come from? Why its symmetric behaviour? The authors believe that they can close the chapter on "why there is something rather than nothing, why we exist, why this particular set of laws and not some other" (2010:171). In fact, all they say is 'this is how what exists functions, so it must function like this for it to exist'. Any believer could have said that too. They have opened their initial questions further without providing any answers. "Well, at least with our current level of understanding that's asking for too much . . . Even if a cosmological theory were to make headway on this question (why there is a universe at all), we could ask why that particular theory—its assumptions, ingredients, and equations—was relevant, thus merely pushing the question of origins one step further back" (Greene 2005:310). "The chain of referring to 'the person in the next office' ends . . . with the cosmologist and the elementary particle physicist . . . Physicists and cosmologists cannot refer to 'the person in the next office'. Due to this particular situation they sometimes engage in philosophical and theological speculation with much less embarrassment than scientists from

other disciplines, though not necessarily with greater competence" (Drees 2008:17). Indeed!

321 Dawkins offers a rather adventurous explanation: Evolutionary mechanisms "sometimes misfire" and the persistence of compassion and generosity in humanity over millennia is due to a "blessed, precious mistake" (Dawkins 2005:220 f.). If it is a derailment of the all-encompassing, all-defining, all-important and totally meaningless evolutionary process, why should it be 'blessed and precious' in the first place? Why should Dawkins not describe it as a symptom of a sick and decadent mentality, as Nietzsche does?

322 "It seems to require quite a low self-regard to think that, should belief in God suddenly vanish from the world, we would all become callous and selfish hedonists, with no kindness, no charity, no generosity" (Dawkins 2006:227). But (a) why should we regard ourselves in the first place, if we are nothing but the product of a senseless evolutionary mechanism? That Dawkins treats this as an issue of self-regard is characteristic of the modernist creed that elevates the human being to ultimate and lonely significance. Maybe we indulge too much in self-regard at the expense of the rest of creation? (b) Self-regard can also consist of collecting as many scalps as possible and hanging them on the poles surrounding the yard. Dawkins would probably be taken aback by what happens in some societies where kinship solidarity and subservience is mandatory to keep malicious sorcery at bay and where cruelty against outsiders—animal or human—is a source of glee. He would perhaps also be slightly more careful if he could witness what happens in traditionalist societies that are uprooted from their religious assumptions and denuded of their moral inhibitions by the unshielded impact of the forces that drive modernity. It was once claimed that 1/3 of all rapes committed in South Africa (one of the rape capitals of the world) are committed by teachers against their female pupils. One can be stabbed to death in broad daylight for the possession of a cell phone. I think Dawkins, the product of public school socialisation and living in an Oxbridge environment, is not aware of his privileged embeddedness in a long-standing moral and cultural tradition of a particular kind.

323 Nürnberger 2007, Chapter 5.

324 Radical postmodernism seems to be completely oblivious of this total anomie when it considers the "possibility of a religious and hyperreligious upheaval or resurrection" the greatest potential danger in the secularised world (Nancy 2008:3 f).

325 For detail, see Nürnberger (2002 or 2004), Chapter 9 (for royalty) and 10 (for sacrifice) respectively.

326 Dawkins 2006:31.

327 This solipsistic concept of the human subject that confronts an objective world out there is often blamed on René Descartes, the father of the Enlightenment, because he juxtaposed the *res cogitans* (thinking stuff) from the *res extensa* (geometrically extended stuff). But for Descartes, the two were held together in unity by the overarching concept of God. It is only when this canopy was deemed redundant and counterproductive in terms of human autonomy that

the personal subject became the autonomous master of the impersonal object. Descartes was a humble and God-fearing person, who tried to get clarity in his mind about the foundations of certainty. It is under the impact of pragmatic utilitarianism and hedonism that the Platonic dualism between spirit and matter and the Christian dualism between God and world turned into the dualism between the dominating human subject and the dominated secular object.

328 "At the beginning of the 17th century, John Donne . . . lamented the collapse of the medieval synthesis: 'Tis all in pieces, all coherence gone'—but after that century, nothing could stem the rising tide of an individualism in which the self surveyed the world as subject over against object" (Peacocke 2008:207). Friedrich Nietzsche called the absence of transcendence 'nihilism'. Nihilism meant "that the highest values are devalued. There is no aim. The answer to the question of 'what for' is missing." Nietzsche continues: "The presupposition of this hypothesis is that there is no truth; there is no absolute essence of things; there is no 'thing as such' . . . there is no corresponding reality in these values and never was; they are only a symptom of power on the side of (the human being) who posits these values, a simplification for the purpose of life" (III 1966:557).

329 For Freud, faith is self-imposed immaturity; for Nietzsche, it is self-enslavement; for Marx, it is opium for the oppressed. The theologian Dietrich Bonhoeffer picked up the modernist sentiment when he argued that secular humanity has "come of age". "The truly adult view . . . is that our life is as meaningful and as wonderful as we choose to make it" (Dawkins 2006:360). In all these cases Paul's contention that it is faith in Christ that allows us to come of age is overlooked (Gal 3:23—4:7).

330 See Nürnberger (2002 or 2004), Chapter 10 respectively for the historical detail.

331 "Scholars like J. S. Krüger have argued that religion should be seen as the dialectic between the synthesizing faculty of the human mind operating at its most inclusive level and the urge to radically transcend the given" (Van Huyssteen 2006:204).

332 Faith is concerned with what Paul Tillich called, on the one hand, the "depth dimension" of reality (the source of reality in my terminology), and on the other hand, our "ultimate concern" (the destiny of reality in my terminology).

333 Clayton speaks of "traction" in this regard: "Theological language has 'traction' if and only if it makes claims that can be impacted, either positively or adversely, by the results of philosophical critique, historical-critical research, or scientific knowledge" (2008:92).

334 For a detailed analysis of this process see Nürnberger 2002.

335 For my evolutionary approach to biblical theology, see Nürnberger (2002 and 2004).

336 It is amazing that the potential of Luther's theology for the science-religion interface has not been recognised as yet, even among theologians that hail from the Lutheran tradition. Karl Heim, Paul Tillich, Wolfhart Pannenberg,

Ted Peters, Antje Jackelén, Philip Hefner, and others are examples. I hope that my contribution will prompt further research and reflection in this direction.

337 Analyses and descriptions of Luther's theology abound. Examples are Elert 1962, Althaus 1966, Ebeling 1972, Forde 1972, Hoffmann 1985, Altmann 1992, Bayer 2004. Because I have offered an accessible introduction to Luther's theology with an extensive bibliography (Nürnberger 2005), I will not substantiate my statements with quotations and references in the present context.

338 For this chapter, I utilise some material found in my article 'Martin Luther's experiential theology as a model for faith-science relationships' (*Zygon* Vol 45, No 1, March 2010, 127-148. Copyright 2010 Joint Publication Board of Zygon).

339 Tracy distinguishes between special acts of God that are epistemic or subjective on the one hand and acts of God that are materially or functionally special (2008:250). His third, causal-historical variety, I believe, covers both the other two. While science would see the epistemic category to apply in such cases (believers interpret natural occurrences as actions of God), faith would take them to be materially or functionally special (God intervenes in the world process).

340 It is interesting to note the experience of an avowed atheist, who could find no beauty and joy in life until she subjected herself to psychiatry: "Even though I am still here, living in the same old house and doing some of the same old things, the whole world looks very different, feels very different. It feels warm and safe and loving and exciting and good. I remember telling you I was an atheist. I am not sure I am any more. In fact, I don't think I am . . . I just feel connected, real, like I am a real part of a very big picture." (Peck 1990:224 f).

341 This point has been emphasised by Ratschow 1960:233 ff.

342 This has not always been recognised. See Nürnberger (2005), Chapter 11 for detail.

343 As a biblical scholar, Luther was strongly impacted by assumptions found in the Old Testament. Augustine influenced him in particular ways, but he was critical of classical metaphysics, especially Aristotle. Toying with Nominalism for some time, he developed his own stance, namely *dynamic universalism* (Buddeberg 1962).

344 Langdon Gilkey comes close to Luther when he declares that creation and preservation are "different aspects of the simultaneous activity of God, who continually gives to all that arises existence and form, molding the new as well as preserving the old" (quoted by Russel 1996, 295). Arthur Peacocke's and Ian Barbour's stances are similar (Russel 1996, 296 ff). However, the concepts of "ontological origination" or "pure relationship" for the dependence of reality on God are metaphysical postulates, while Luther's concept is geared to the flow of actual daily experience. In the explanation to the first article of the Creed in his Small Catechism, Luther writes: "I believe that God has created me and all that exists, that he has given me and sustains my body and soul, all

my limbs and senses . . . he provides me daily with all the necessities of life" (Tappert 1959, 345).

345 Stuart Kauffman's and Gordon Kaufman's concepts of "creativity" are close to that of Luther, with the difference that in the former case, creativity is an abstraction from a verb that is deliberately deprived of its subject, because a creative God does not exist, while for Luther it is the creativity of God.

346 God forms humans from soil like a potter (Genesis 2). God creates the cosmos out of the primeval chaos by means of an imperial decree (Genesis 1), or through a violent conflict with primeval monsters that represent life-threatening forces in the world (Ps. 74:12 ff, Ps. 89:9 f; Is. 51:9), God creates by means of God's wisdom (Job 38-42; Prov. 8-9; Wisdom of Solomon 7; Sir 1), or by means of the divine logos, the ultimate rationale that structures the universe (John 1). God brings about a new creation through God's Spirit that manifests itself in the human spirit (e.g. in 2 Cor. 3:17-18; 4:6; cf Rom. 5:18 ff).

347 This true both for theists like Heeren who argues that the findings of modern science point to a Creator God (2000:127) and atheists like Hawking who argues that "it is not necessary to invoke God to light the blue touch paper and set the universe going". (2010:180)

348 For detail, see Nürnberger (2002:204) and the index.

349 "The rationality of the Creator is reflected in the created world" (Heller in Russel, Murphy, and Isham 1993, 103).

350 "Unbridled self-indulgence on the part of one generation without regard to future ones was and still is the modus operandi of biological evolution and may even be regarded as rational behavior" (Kieffer 1979:346). I agree, but in nature, self-indulgence is not unbridled! It is directed by the need of a species to survive and prosper within specific environments. Therefore evolution also leads to multiple synergies and symbioses between species. What Kieffer also does not say in this context is that even traditional cultures have bridled self-indulgence fairly effectively. They did not consider the raw endowment of instinctual self-preservation bestowed on animals by evolution beneficial to, and appropriate for, human communities. Of course, this did not prevent them from collective particularism and in-group-outgroup attitudes.

351 For a critique, see Nürnberger (2010:106 ff).

352 Pannenberg 1989, 153.

353 Classically depicted in Kuhn 1970.

354 Erasmus of Rotterdam and Philip Melanchthon were the most prominent humanist scholars at the time.

355 Powerfully depicted by von Weizsäcker 1990, 405 ff.

356 Nürnberger 2007:29 ff.

357 "Humans are almost certainly the only creatures who ask 'why-'questions . . . Modern science explicitly and emphatically rejects teleology" (Ben-Ari 2005:23 f). But this is true only for the subpersonal levels of emergence.

358 Kauffman 2008:72 ff.

359 "It makes no sense to ask about the function of granite, or the purpose of a galaxy. Though the atoms composing the heart muscle or a neurotransmitter

molecule have no function in themselves, the particular configurations of the heart . . . beg for some sort of teleological assessment, some function" (Deavon quoted by Murphy 2008:127).

360 In the oldest creation narrative found in the Bible, God forms the human being out of soil (matter) and then blows the "breath of life" (not a soul) into this lump of clay. As a result, it becomes a "living soul" (Gen. 2:7). Once God withdraws the breath of life, the human being reverts to soil (Gen. 3:19). The same is true for other forms of life (Ps. 104:29). *Nephesh*, the Hebrew word for soul, denotes a living body as opposed to a corpse, rather than a bodiless spirit as in Platonism.

361 See 1 Corinthians 15:44. Paul juxtaposes human reality without God (the flesh) and human reality in fellowship with God (spirit). Both are bodily concrete. Gnosticism was later rejected as a heresy by the 'orthodox' majority in the church.

362 Pannenberg proposes the scientific 'field' concept as analogous to God's Spirit that permeates the whole of reality (1989, 165). With that he confuses the sub-atomic level of emergence with the personal level. He also ignores its New Testament meaning, which is inextricably linked to the Christ-event.

363 The most radical and most controversial development of this thought is found in Luther's treatise against Erasmus of Rotterdam called *De servo arbitrio* (the enslaved will).

364 See the *Joint Declaration on Justification* issued by the Lutheran World Federation and the Vatican (1999). Subsequently the Methodists identified themselves with this statement.

365 I have discussed this problem in detail in my book *Zuspruch des Seinsrechts* (Münster, Germany: LIT-Verlag 2003, 148-171. Unfortunately the book is only available in German.

366 For instance in my essay 'Can unity be based on consensus? The gospel of God's unconditional, suffering, redeeming acceptance of the unacceptable as the foundation of Christian unity.' *Journal of Theology for SA*, 116 July 2003, 9-25.

367 Drees contrasts assumptions of consonance between theology and science with assumptions of the painful "absence of God" in the world and locates the function of theology in the realm of the latter (Drees 1990, 5 f).

368 I agree with Drees that theology is geared to the conflictual rather than the consonance model (1990:6). However, conflict can only occur where experienced reality is first taken seriously as it is—in its full ambiguity.

369 McGrath rejects the idea that faith is *blind trust* with considerable contempt, repeating again and again that this is a ludicrous invention of Dawkins and that he knows no believer who has such a crazy idea (2005:85 ff). I think he goes way too far in this regard. Most certainly Dawkins is right when he claims that there is a difference between evidence unearthed by the sciences, on the one hand, and the assumptions of faith on the other. Of course, faith is trust, what else? In his Large Catechism, Luther defines faith as trust. And at least to the extent that you have no evidence for the existence and trustworthiness

of the entity you trust, it is indeed 'blind' trust. Yet trust does not have to be arbitrary, infantile or irrational. It can be well founded, critical, responsible and sacrificial trust. I entrust myself to the pilot of an air liner or to the surgeon who operates on my body and I have good reasons for doing so.

370 For a detailed analysis of this process, see Nürnberger (2002).

371 Here liberation theology meshes with Luther's theology, though this is not generally recognised (Nürnberger 2005, Chapter 12).

372 For the worldview background of the Calvinist rejection of this notion, see Nürnberger (2005:223-228). The contrast was articulated in Luther's stance on the presence of Christ in the Eucharist.

373 "(God's right hand) is not a definite location . . . but the mighty power of God that . . . must be present and active everywhere—even in the tiniest leaf of a tree . . . if he is to create and uphold, he must be present . . . in every creature, in its innermost and outermost being, all around, through and through, above and below, behind and in front, so that there can be nothing more present, nothing more intimately connected with every creature than God and his power" (Martin Luther 1527. *That these words "this is my body etc." still stand—against the enthusiasts*. Par 98 f).

374 Utilising the model of Lakatos, Hefner suggests that the problem of evil (theodicy) must be placed "into the hard core of the program, which is never subjected to the abrasion of falsification" (1993, 271). This means that ambiguity cannot be eradicated and must be accepted as part of the core assumption of faith in God. This is indeed what Luther's hidden God stands for. But for Luther, this does not lead to a fatalistic, acquiescent stance. On the contrary, faith is a defiant protest against the ambiguity of experienced reality in the name of a God with redemptive intentions.

375 Gerhard Theissen quoted by Drees 1990, 180.

376 Gregersen distinguishes between causal explanations of divine action and semantic explanations (2008:183). A causal explanation would claim potential evidence for direct and particular divine intervention into causal networks of immanent reality. Luther's approach would not confirm this stance. Semantic explanations "aim to redescribe nature (as already described and partially explained by science) in terms of divine action" arguing that the scientific explanation alone does not fully make sense of experienced reality. This approach approximates Luther's stance but with the proviso that it cannot work without taking account of the cross as a manifestation of God's intentionality *sub contrario*.

377 "(O)nce a metaphysics or worldview is established and recognized, it permeates all the other things that human beings find engrossing" (Kieffer 1979:61).

378 Drees distinguishes between a theology that relates "to a sense of God's presence" and a theology that relates to "a cry because of God's absence" (1990:175 f). The first refers to a hidden dimension of reality that "is expected to shine through in the material world". The second "takes up the experience of imperfection . . . and expresses the concern for justice and love". This comes close to Luther's insight, except that the "sense of God's presence" is

entirely contingent upon the proclamation of the gospel in the face of contrary experiences. Once believed, it extends over the whole of reality. It would appear that Luther posits an existential sequence beginning with the cry for redress and ending with a sense of adoration and gratitude. This kind of experience is typical for many of the Psalms, which Luther treasured highly.

[379] Luther's Small Catechism, Tappert 1959:345 (my translation).

[380] Tipler's version of the Omega Point theory (which posits the ultimate completeness of reality beyond the constraints of space and time) is a metaphysical version of conditional acceptance. If "every point is essential to complete the whole", it follows that "every individual act or decision receives an infinite weight, an absolute yes-or-no character. Either it fits into the universal wave function . . . Or, if it did not fit . . . there would be no Omega Point . . . In that sense there is no freedom, except for the freedom to follow the rules or the freedom to lose everything . . . it would be difficult to incorporate a notion of forgivingness" (Drees 1990, 139). This is the essential character of the 'law' in contradistinction to the gospel as Luther understood it.

[381] Hefner speaks of a "concept of wholesomeness" (1993, 265).

[382] For Calvin, the law is foundational, rather like the rail for a train. Sin is a derailment. The gospel is the emergency measure that puts the train back onto the track where it is expected to continue. For Luther, the gospel is foundational: We are supposed to be free like eagles. When the eagle crashes into a power line, she is put into a cage by a friendly game warden to prevent further damage. Not the gospel but the law is the emergency measure. But when healed, she is released into the open skies where she belongs (Nürnberger 2005, 112 f).

[383] Karl Barth was on track when he demanded that such guidelines must reflect divine intentionality as manifest in Christ, although Lutherans would not call them 'the law' or a 'third use of the law'. See the essay, 'gospel and Law' in Barth (1968, 71 ff).

[384] "The world was changed through the religion of love", namely through the three-pronged struggle between the gods, Christianity and secular reality (von Weizsäcker 1990, 414 f).

[385] For the biblical argument, see Nürnberger (2002), Chapter 10.

[386] Some participants in the science-religion debate have the "desire to minimize, without eliminating, the severity of the theodicy problem. Likewise some (including Ellis) invoked the concept of kenosis . . . to explain why God does not act more often to ease pain and to educate us wayward creatures . . ." (Wildman 2008:173). Some think of God sharing in the suffering of the world, some reject the idea of divine intentionality altogether (Wildman 2008:174). My proposal tightens up the argument. It eliminates both natural inconsistency and divine arbitrariness.

[387] "Entropy will be the price paid for depth" (Peters 1996, 274). Depth is here understood as "growth in organized complexity", that is, evolution.

[388] My theodicy is akin to Leibniz' "the best of all possible worlds". However, the concept of the benevolent Destiny of reality does not refer to what is (thus static and fatalistic), but to what ought to be and could become (thus dynamic).

389   For more detail, see Nürnberger (2002, 280-289).

390   For a refutation of 'pseudo-science' see, for instance, Ben-Ari 2005, 79 ff. and Dawkins 2009, introduction.

391   I have often referred to Nürnberger 2002 in this respect.

392   Due to his boisterous personality and the highly conflictual situation of his time, Luther himself was anything but tolerant. Lutherans cannot defend the rudeness, irrational outbursts, and aggressiveness of the Reformer. Nor can they justify their own notorious intolerance.

393   Note my book on Richard Dawkins's *The God Delusion* (2006) in this regard (Nürnberger 2010), as well as my book on African religion (Nürnberger 2007).

394   "I believe we stand at the brink of a new Reformation . . . If it is to be faithful to its mission, the church can no longer ignore this crisis of meaning—or the opportunity of renewal" (Russel 2008:33).

395   Dawkins argues that theology, "unlike science and most other branches of human scholarship—has not moved on in eighteen centuries" (Dawkins 2006:14). There is no way traditional theology can escape that verdict.

396   Wildman speaks of the "specter of theologians vainly fumbling after stable theological terminology", that "has stirred mystical theologians" and has been "one inspiration for the aggressive attacks on the socalled ontotheological tradition in recent decades" (2008:139).

397   A natural scientist is reported to have said that Teilhard de Jardin "can be excused of dishonesty only on the grounds that before deceiving others he has taken great pains to deceive himself" (Peter Medawar as quoted by Dawkins 2006:154). This is how some scientists view what even enterprising theologians like Teilhard are trying to do, namely to bring science and faith into a relationship with each other.

398   "The result of proceeding in this way will be that we generate provisional and exploratory theological proposals, proposals that are offered in the expectation that they will be revised in light of new developments in the sciences and in response to ongoing critical assessments of their theological adequacy" (Tracy 2008:266).

399   The doctrine of inerrancy does not occur in any of the major sources—Paul and his disciples, the Synoptic gospels, John, or Hebrews. The two texts usually quoted (2 Peter 2:20 f and 2 Tim. 3:16) speak of the inspiration of the Old Testament (the Bible of the first Christians), but inspiration does not imply inerrancy. The biblical authors are at great pains to depict the sinfulness and fallibility even of their greatest heroes. Moreover, the argument is circular: if the Bible is not inerrant, these two verses are not inerrant either. For the problematic genesis of the doctrine of inerrancy, see Nürnberger (2002:25-32) or Nürnberger (2004:18 ff).

400   If the sun stood still, this means, in fact, that the earth, rotating at an equatorial speed of 1,674.4 km/h, came to a sudden halt—which would have caused the mother of all tsunamis, sweeping away the Israelite army rather than helping them to win a battle.

[401] For detail concerning the four mentioned changes see Nürnberger 2002 or 2004, chapters 7, 9, 10 and 11 respectively.

[402] The Catholic doctrine of transubstantiation is based on Aristotelian metaphysics. Here 'substance' refers to the essential content or significance of something, rather than the material of which it is composed. Of course, the significance of bread and wine changes from that of foodstuffs to that of God's redeeming love as manifest on the cross of Christ, while their material characteristics remain the same. Today substance means the 'material' of which something is composed. Then transubstantiation implies two miracles—the change of bread into body and the continued appearance of the body as if it were bread.

[403] This remark refers to the Lutheran concept of con-substantiation, which suggests that another substance is given 'in, with and under' the substance of bread and wine. This idea is theologically just as inappropriate and just as indigestible to the modern mind as its Catholic counterpart. See Nürnberger 2005, chapter 10 for detail.

[404] This is the problem pursued by the series of conferences and publications of the Vatican Observatory and the Center for Theology and the Natural Sciences in Berkeley (summarised in Russel, Murphy and Stoeger 2008). The programme explored the possibility of allowing for special divine actions in situations where natural laws leave the space open for such divine interventions. The entire exercise was thus pre-programmed to become a quest for a God of the gaps.

[405] Moltmann and his many disciples, for instance, assume "a movement from the initial creation through historical reconciliation to eschatological consummation . . . The kingdom of nature, the kingdom of grace and the kingdom of glory are not three aspects of the one Kingdom of God, but three stages on the way to its consummation." (Moltmann 1992:xviii). "The reality of the world that can be experienced conceals and shelters within itself traces of creation-in-the-beginning. These traces are at the same time a reflection of the coming glory. All knowledge of this world 'as' creation is hence a metaphorical knowledge of this world as parable of the world to come" (Moltmann 1985:58 f). I detect not the slightest reflection on what all this could mean against the background of big bang cosmology or the law of entropy. Note also the use of the term 'metaphor'. The metaphorical extrapolation of what ought to be is taken to be the true reality that once was and that once will be, while the experienced world becomes the 'metaphor'. Such a 'logic' is difficult for science to understand. In fact, it is inexplicable even in theological terms.

[406] The assumption that God has created the world out of nothing (creatio ex nihilo), for instance, is usually taken as a theological axiom and associated with big bang cosmology, without reference to its cultural-historical origins, metaphorical character and existential-experiential basis. An example is the intricate discussion about the big bang depicted by Russel in 1996 (cf also Russel 2008:12 f). In view of current hermeneutics, these debates have become quite unreal and, in a way, similar to fundamentalist arguments. The idea is not found in the biblical creation stories or in the existential experience of

faith. Paul's use of the motif in 1 Cor. 1:28 ff and 2 Cor. 4:6 ff does not refer to *ex nihilo* creation but to transformation. *Creatio ex nihilo* is a post-biblical deduction from the abstract concept of divine omnipotence, rather than an expression of the derivation of the ongoing world process from, and dependence on, a transcendent Source and Destiny. In the latter sense, every moment is a 'creation out of nothing'. But then the theological concept of the ongoing creation (*creatio continua*) is much more appropriate.

[407] Creation myths often express two typical assumptions: The world was supposed to be a good place, but human injudiciousness has thwarted the original rationale. Here is an example from Malawi: *Chiuta* (the Supreme Being) first lived high above the earth, which was dry and lifeless. One day the heavens opened and *Chiuta* descended to earth in a torrent of rain together with the first human couple, the animals and the first agricultural tools. The earth began to flourish and there was plenty of food. This 'paradise' came to an end when humans invented fire. It caused a devastating veldt fire that drove some animals away from humans in disgust, other animals to humans for protection and, most importantly, *Chiuta* back to heaven. Departing, *Chiuta* decreed that humans had to die and face *Chiuta*. (Parrinder, E G 1962 (1954). *African traditional religion*. London: SPCK, 40 ff). This myth uses African imagery, but its content is remarkably similar to Genesis 2-3.

[408] Van Huyssteen (2006) uses this comparison to build up his argument of 'transversal overlap' between academic disciplines.

[409] "Divine action coincides with the actual outcomes in the actual universe that we inhabit . . . and our probabilistic laws reflect the way God has chosen to create the actual world . . . God is always acting *as God* but *from within the world of creation*" (Gregersen 2008:195).

[410] "The existence of the whole world is contingent in the sense that it need not be at all. It owes its existence to the free activity of divine creation. So does every single part of the world" (Pannenberg quoted by Russel 2008:15). Indeed, but God's freedom does not imply that everything is possible within the world that God created; "science actually both limits and qualifies Pannenberg's assertion about the contingency of the universe and all of its parts" (Russel 2008:15).

[411] Wildman distinguishes between the miracle (a) as a suspension or abrogation of law-like regularities, (b) a way of speaking "of apparently providential events that strike us as important and surprising" and (c) "the activation of latent features of natural objects that do not show up within the theoretical framework of our existing natural sciences" (2008:142). My proposal is similar to the second alternative.

[412] Kaku 2008.

[413] Just in passing and two pages from the end of a book against Dawkins' atheist stance, McGrath mentions "the existence of suffering in the world" as "the point of weakness" in the Christian world-view, on par with the "persistence of faith" as the point of weakness in atheism (2007:63). But the two 'points of weakness' are not commensurable. Suffering is an empirical and existential condition, while the persistence of faith could be nothing but credulity and

superstition. The experience of meaningless fate is the most persuasive reason for adopting atheist assumptions in modern times. Theologians simply cannot afford to treat it as a side issue.

[414] It is one of the weaknesses of Peacocke's otherwise insightful proposal, I think, that it cannot account for the ambiguity of reality, thus the problem of theodicy. While he observes the problem that emerges when God is deemed involved in all (rather than some) quantum events (2008:216), he does not deal with it in terms of his own proposal. Wildman, in contrast, gives up on the idea of intentional divine action within reality altogether: "I hold that the postulate of intentional divine action exacerbates the problem of theodicy to such a degree that we are justified in rejecting it for moral and theological reasons" (2008:134). If that does not only refer to particular divine interventions, but also to an overarching (transcendent) intentionality nothing remains of the biblical concept of God.

[415] Russel is one of the few participants in the science-religion debate who confront the issue of theodicy. On the one hand, he sees an analogy between entropy in thermodynamics and evil in theology; on the other hand, he sees a 'disanalogy' in the sense that entropy is 'part of natural goodness'. He is critical of theologies that "limit the problem of evil strictly to moral evil . . . and view pre-human nature as entirely amoral" and speaks of a "cosmic theodicy". Significantly he views "God's suffering through the cross of Christ to include the history of life on earth" (Russel 2008:10 f). This is close to my own approach.

[416] This is even true for Dawkins's atheist concept of nature: "It follows from the fact of our existence that the laws of physics must be friendly enough to allow life to arise" (2005:141).

[417] The critique of "two separate and distinct domains of knowing" (Einstein and Gilkey as quoted by Peters 1989:16) does not apply in this case because the vision of what ought to be arises in response to the experience of what ought not to be.

[418] Russel draws attention to the fact that the law of entropy precludes the physical transformation of the current universe into a new age—a fact largely overlooked in the science-religion debate (Russel 2002, 3-15.) However, the debate in his circles is curiously fixed to a literal-material interpretation of resurrection and cosmic transformation. "(If) it were shown that the universe is indeed headed for an all-enveloping death, then this might . . . falsify Christian faith and abolish Christian hope" (John Macquarry quoted by Russel 2002:4). "If it is impossible it cannot be true. But if it is true, it cannot be impossible" (Russel 2002, 16). If big bang and entropy cosmology turn out to be valid, Ted Peters says, "We would have proof that our faith has been in vain. It would turn out to be that there is no God" (p 8). No, that does not follow! Biblical future expectations have emerged and evolved as a series of visionary responses of a redemptive faith to deteriorating situations of need. Once unpacked as protest statements against an unacceptable reality in the name of a loving God, they do not clash with scientific insights.

419   The kind of confusion that ensued if that is not done, can be demonstrated in the essay of Moltmann quoted above ('The Fellowship of the Holy Spirit: On Trinitarian Pneumatology' (1992:57-69). Again my comments are in brackets. "Through the presence of his own being, God preserves his creation against the annihilating Nothingness" (96). (Well, as far as we know, God does not! Living organisms die, tsunamis happen and the law of entropy is in full operation. This idea can only belong to what I call the vision of comprehensive well-being.) Indeed, Moltmann switches to that kind of discourse: believers "hope that in the kingdom of glory God will dwell entirely and wholly and for ever in his creation, and will allow all the beings he has created to participate in the fullness of his eternal life" (5). (Is this a regulative vision or a realistic expectation? Given the law of entropy and the necessity of death for life, existential realists will want to know what the reality content of such an expectation might be.) Moltmann then links God's future glory with the Sabbath and the resurrection of the dead. The metaphorical character of the biblical originals and their relation with the hard realities of immanent reality are not spelt out, with the result that the vision of comprehensive well-being changes into crass speculation about the earth's future: The same "transfiguring indwelling" of the Spirit, he argues, leads to the "self-transcendence of the world into its open future and the eschatological sabbath". (What on earth is Moltmann talking about?) The incongruence between metaphor and reality becomes even more apparent when, in strange contrast to his confident assertion of the indwelling Spirit and the certainty with which he seems to expect the realisation of his eschatological vision, Moltmann speaks of the progressive destruction of nature by the industrial nations and the pile-up of nuclear arms that "have brought the age of subjectivity and the mechanistic domination of the world up against their definitive limits. Faced with these limits, we have only one realistic alternative to universal annihilation: the non-violent, peaceful, ecological world-wide community in solidarity" (12). (Sure, but how can this be the only realistic alternative, if the creative and redeeming Spirit of God is at work in all of reality, keeps reality in one piece and leads the world inexorably to self-transcendence and a kingdom of glory? Why should the only hope we have consist in the human achievement of a global community in solidarity, if there is a Spirit at work within it who keeps it from disintegrating anyway? In short, there is an unbridgeable gap between Moltmann's speculative use of traditional theological terminology and his realistic assessment of where modernity is leading us. No scientist can make sense of such an argument. A theological vision is designed to give meaning, orientation and spiritual empowerment to humans; it differs fundamentally from a scientific projection of where reality is heading).

420   Moltmann, Pannenberg, Polkinghorne and Russel are all very inventive when it comes to the possibility of a 'completely different' kind of future: " . . . there is no reason to suppose that in the world of the new creation, there cannot be a new kind of *time*, whose processes will have a different character, one no longer condemned to the thermodynamic drift to disorder that is characteristic

of this present world" (Polkinghorne 2006:65). Yet Polkinghorne insists that the future of the 'new world' is continuous with the past of the 'old world' that we know. He also motivates this view with the difference between bottom-up and top-down causation (which are surely characteristics of the 'old world') as well as with complexity theory, continuous creation etc. So how new is the new kind of time really? How malleable is the old kind of time? Current history and eschatological vision are just not part of the same frame of reference. Moreover, the historical roots of biblical eschatology and its original meanings are completely ignored. Even more astounding, of course, are Moltmann's constructions of 'eternal time' (Polkinghorne 2006:65f), let alone those of Pannenberg, which have been taken up by Ted Peters, Russel and others.

[421] If geared to a timeless eternity, eschatology **cannot** "function for the orientation within the realm of possibilities" (Drees 1990, 151).

[422] Dawkins ridicules God as 'Binker', the imaginary ideal friend of a child's fantasy (2006:347f).

[423] Nürnberger 2007:50 f.

[424] Carl Sagan, quoted in Dawkins 2006:12.

[425] Swimme 1996:80.

[426] Van Huyssteen 2006:47.

[427] For a good summary of the debate on anthropocentrism in ecological circles, see Garner (2000:44-54).

[428] "It is not realistic to relate the future on a cosmological scale, counted at least in billions of years, to perspectives for humankind" (Drees 1990:117).

[429] "There is a very close link between images of the future and ethics . . . However, traditional ethics . . . has almost entirely ignored the claims of long-range posterity . . . (The Club of Rome) submitted that as many as 90 per cent of all decisions made in the commercial and political worlds are predicated on time spans of six to twelve months. The average duration of governments worldwide is only three years . . . It could be argued that evolution did not prepare us to conceptualize in the distant future." (Kieffer 1979:346 f).

[430] It is counterproductive to disparage the human being as just one creature amongst others as some radicals in the green movement tended to do (Garner 2000:44 ff). It is the only creature that has the capacity to become an observing, comprehending, accountable, free and responsible agent.

[431] McCarthy and Rubidge 2005:301. In terms of competitiveness in the "relentless arms race of life", we are not very well endowed. In fact, we are "physically disadvantaged" (McCarthy and Rubidge 2005:300). We have no "in-built self-protection devices such as a set of horns or a thick hide"; no fur to protect us against the cold; walking upright we are more exposed to predators; our young are born weak and defenceless; they need to be nursed and protected for years . . . But all this compensated by the fact that "evolution of the brain as honed intelligence and communication to unprecedented levels of sophistication and coupled these with unique physical abilities", such as upright posture and free hands with opposed thumbs (2005:301).

[432] This is a reference to the argument and title of Kauffman 1995.

⁴³³ Paul offers a series of images that are taken from the physical, biological and social realms of earthly reality (seed and plant, animals and humans, heavenly and earthly bodies, sun and moon, perishable and imperishable, weakness and power, first and last Adam, earth and heaven). These images are not particularly suited for defining the relation between the two kinds of 'ontology' envisaged. They are also not particularly suited to highlight the difference between human biology and human authenticity, which is the primary concern of the biblical tradition.

⁴³⁴ For an account of the development of the biblical tradition concerning the resurrection from the dead, see Nürnberger (2002:330-340).

⁴³⁵ Even in a Platonic frame of reference the 'eternal' is the 'essential', that is, that quality of reality that is not subject to time, space, and energy relations, thus the ultimately valid or authentic.

⁴³⁶ What else could the "otherness of time" (Jackelén 2006:966 ff) signify if not precisely that? The popular concept of eternity is an idealised abstraction from the flux of time. It has no ontological content. This is true whether one conceptualises it as 'timelessness' or 'time-fullness'. To lump past, present, and future together under the concept of the 'fullness of time' only works if the differences between them no longer matter. And that can only be achieved by means of an abstraction from the flow of time. This is Greek metaphysics, not biblically based theology. The concept of eternity has no equivalent in the Hebrew language. *Min olam* means from primeval times, *ad olam* means "into the farthest future times". Between these extremes lies the sequence of significant events. The Greek term *aionios* originally also had the connotation of 'era'.

⁴³⁷ A detailed discussion of this reversal can be found in Nürnberger 2002 or 2004, Chapter 9 respectively. The early Israelites took over the practice of sacrificing the firstborn son, the most precious asset of the clan, from their Canaanite neighbours. It even became a requirement of the Mosaic Law (Ex 22:29f). Perhaps due to prophetic and deuteronomic objections, the law then decreed the substitution of an animal for the son and subsequently a substitution of the blood of the animal for its carcass. This is where the blood symbolism found in the New Testament came from.

⁴³⁸ For Paul, suffering means participation in the death of Christ, which unlocks participation in the new life of Christ (2 Cor. 4:7-12). Psychiatrist can come to similar conclusions: "It is the giving up of self (or bracketing the self) that human beings can find the most ecstatic and lasting, solid, durable joy of life" (Peck 1990:76).

⁴³⁹ Some modern theologians argued that Christ "has risen into the eschatological future of God". This cannot refer to ontology, because the future is not yet, but only to God's redemptive intentionality or, expressed in Luther's terms, to the gospel as a *promise*. It is an affirmation of what ought to be in the face of what ought not to be.

⁴⁴⁰ If that is the case, it is inappropriate to ontologize the resurrection as the "first instantiation of a new law of nature" (Russel 2008:16). This idea is similar

to Pannenberg's argument that the resurrection of Christ cannot be disproved by science because history consists of singular events (W. Pannenberg 1966. *Grundzüge der Christologie*. Gütersloh: Mohn, second edition 1966, page 96.). The argument is unconvincing. It implies that the risen Christ was a potential, but hitherto unrealised part of immanent reality. If it had happened, the 'first instantiation' of this new facet of immanent reality would have to be accessible to public observation and scientific inquiry. The phenomenon would probably have been repeated since then in other specimen. After 2000 years, it would probably have generated a sizable population of similar cases. If not, it would not be open for our participation, thus defeating its very rationale in the New Testament (2 Cor. 5:14-18; Rom. 6). Participation indeed happened at the *spiritual* level and demonstrably so. It certainly was not something that manifested a new form of material reality, or a new law of nature. The proposal confuses ultimate validity with earthly factuality and misses the intention of the statement.

441 Jürgen Moltmann spelt out this insight in his seminal work, *The crucified God*. 1974: London: SCM, 1974, Chapters IV and V.

442 In another context, I highlighted the parallel between the 'risen Christ' and the ancestor in African religion. (Nürnberger 2007:91 ff.) It is not the vitality of the forebears that continues after death, but their status, significance and authority within the patriarchal hierarchy. Similarly it is not the biological body of the earthly Jesus that was transformed into the 'risen Christ', but the epitome of the 'new humanity' that Jesus represented and continues to represent and in whose novelty we are invited to participate. In both cases, it is the ongoing spiritual presence of a biologically deceased person.

443 For the following, see Nürnberger (2002) or Nürnberger (2004), Chapter 9 respectively.

444 For the following, see Nürnberger (2002 or 2004), Chapter 11 respectively.

445 Von Weizsäcker, a prominent scientist and philosopher, has spelt out these ambiguities in no uncertain terms (see the summary in 1990:405 ff). One can augment his observations with many instances taken from the social and the human sciences.

446 According to Heisenberg, they can either take part in the political decision making process or withdraw into their academic world. But at the very least they have to keep political decision makers aware of the immense repercussions of such decisions when it comes to scientific applications (1999:192).

447 Heisenberg 1999:189 f.

448 "Don't we hear the noise of the undertakers that are about to bury God? Don't we smell the stench of divine decomposition? . . . What are these churches but sepulchres and tombs of God?" (Nietzsche 1966, II:128, my translation). When I visited Halle in Eastern Germany, I was told there are more churches per square kilometre than anywhere else in Germany. But hardly anybody bothers to belong to the church, let alone attend church services. In the Netherlands, once the bulwark of Calvinism, 20 per cent of the population is said to be Catholic and 8 per cent Protestant. Church buildings are sold to secular

entertainers, cafes, clubs, ice rinks and so on. One cafe owner put up a crucifix with the body hanging upside down as a decoration. In England, the Anglican Church struggles to maintain its medieval cathedrals, and priests in their robes beg tourists for donations.

Glossary

abstraction—a concept that indicates common characteristics among divergent entities and ignores the differences between them.

acceptability—the quality of satisfying valid expectations.

accountability—being responsible to an authority.

agency—the capacity to act.

anthropomorphic—seemingly having human personal qualities.

authority—the capacity to issue orders and expect accountability.

avarice—covetousness, greed.

biblical history—the time between roughly 1000 BC and 200 AD when the biblical documents were written and compiled.

biblical tradition—the body of faith assumptions that emerged and evolved during biblical history.

big bang—the event that triggered the process of cosmic evolution.

body and soul—the relation between the biological and the spiritual aspects of human reality.

brain—the biological organ that forms the command centre of the nervous system and the essential infrastructure of consciousness.

Christology, Christological—doctrines concerning the person of Christ.

closed universe—the assumption that immanent reality is all there is, that there is no source or destiny beyond the realty we actually experience.

collective consciousness—patterns of thought held in common by a group of people.

collective interests—needs and wants pursued by a group of people.

commerce—the aspect of society dealing with trade in pursuit of profit and wealth.

commitment—the sense of obligation or dedication to a cause.

compete (competition)—trying to outperform a rival.

complementarity—the capacity of two different entities to make up for each other's deficits.

complexity—a condition characterised by involved relationships.

comprehensive optimal well-being—the most balanced realisation of the potentials of all participants, aspects, or entities that can materialise under a particular set of circumstances.

conquest—the act of overpowering and subjugating the inhabitants of a foreign country.

consciousness—awareness of something.

consumer culture—a body of ingrained attitudes and behaviour geared to constant increases in material consumption and sensual pleasure.

continuous creation—the view that God did not create the world once and for all at the beginning of time, but that unfolding reality manifests God's continuous creative activity.

conviction—a firm belief.

cooperate—work together.

cosmology—the science of studying the genesis, characteristics and functioning of the universe as a whole.

creation—the assumption that reality came into being through a deliberate act of God.

creativity—the capacity of bringing about something new.

credibility—the capacity of making sense to critical observers.

Darwinism, Darwinist—the theory of biological evolution according to Charles Darwin.

deduction—the act of drawing out the implications of basic assumptions or established facts.

deism—the idea that God created the world and then withdrew to let it function on its own.

dependence—having to rely on somebody or something else.

derivation—having an origin in something else.

discrepancy—a difference or divergence that should not exist.

ecology—the science dealing with the natural environments of living beings.

economics—the science dealing with the production, distribution, and consumption of the material prerequisites of human life.

emancipation—gaining freedom from authority.

emergence—coming into being.

empirical—making an impact on sense perception and its technical extensions.

empiricism—the assumption that only what can be demonstrated or proved to have a material existence should be considered real.

energy—the power that is able to trigger and drive processes.

entropy—the tendency of processes to move from order to disorder, from potent energy to spent energy, and from concentration to dispersal.

epistemology—the science of human knowledge and understanding.

eschatology—the body of doctrines dealing with the expected end of times.

evolution—the process by which reality unfolds into divergent entities.

experiential realism—a method taking for real whatever can be described, critiqued, transformed and has consequences in this world.

explanation—making something understandable.

faith—trust in, and commitment to, somebody or something that cannot be shown to exist or happen.

field—the way magnetic-electrical forces organise themselves in space and time.

fundamentalism, fundamentalist—a stance that deems religious texts or doctrines taken literally as fundamental and inerrant.

God—a name for the ultimate Source and Destiny of reality.

grounding—the assumption of a valid foundation of reality.

hidden God—the experience of the operation of power in the universe without apparent purpose or intention, but attributed to the agency of a personal God.

high entropy energy—energy that is 'spent' and unable to do work.

identity—what makes a person or group unique.

immanence—whatever is accessible, in principle, to human observation, explanation and prediction.

immanent transcendence—aspects of immanent reality that are not immediately accessible to human experience and explanation.

indulgence—extravagant pleasure seeking.

information—the transfer of reliable knowledge or procedure through symbolic representation.

infrastructure—the material and organisational prerequisites for something to exist or happen.

integration—the meaningful amalgamation of disparate elements.

integrative transcendence—the method that accommodates all human experiences and scientific insights in its perspective from beyond.

intentionality—personal orientation towards certain goals or visions.

justification—making acceptable in terms of a valid code of law, value, norm, or vision.

legitimacy, legitimation—proclaiming or claiming to be lawful or justifiable.

liberal—claiming freedom from unnecessary constraints.

logos (Greek)—word, design, rationale, or wisdom.

low entropy energy—potent energy that can do work.

mastery—being in full control of something.

metaphor—a concept taken from concrete human experience that is used to refer to an intangible entity or event.

mind—functioning consciousness.

modernity—the civilisation that emerged and evolved in Europe after the Middle Ages emancipating its members progressively from all kinds of imposed authority and striving for personal autonomy and mastery.

mortality—being doomed to die.

myth—a narrative using imaginative plants, animals, persons, objects and actions to express a persistent experience or valid truth and projecting it to the beginning of time or an ideal space (heaven) to express its validity for all times and spaces.

narcissism—an attitude that is preoccupied with self.

natural law—regularities that govern the emergence, evolution and operation of material reality.

Naturalism—a worldview according to which the reality we experience is all there is, thus closed in upon itself; there is no transcendent beyond.

neuron (neurological)—brain cell that processes and transmits information through electrical impulses and chemical reactions.

neurotransmitter—chemical substance that conveys signals from one neuron to another across a synapse.

obsolescence (obsolete)—the condition of being outdated.

Occam's razor—the principle of giving priority to the simplest explanation available that is capable of doing justice to what is to be explained.

open universe—a universe that is not closed in upon itself, thus self-generated, self-contained, self-sustaining and self-destructive but dependent on a transcendent Source and Destiny.

panentheism—the view that the world is contained within the greater reality of God.

pantheism—the view that reality as such is identical with God.

Parsism, Parsist (Persian)—the religion of ancient Persia that assumed that this ambiguous age will make way for an age where the force of evil will be overcome and only the force of good will survive.

particles—subatomic components of matter and radiation.

Platonism (Platonic)—an ancient Greek philosophy that deems the essence of a phenomenon more basic than its concrete existence, or that the idea of something is more real than its material manifestation. (Platonic abstraction moves from historical time to eternity, from space to universality, from power differentials to harmony.)

plausibility (plausible)—making good sense.

postmodernism—a variegated group of philosophical schools that deem modernity obsolete and counterproductive.

postmodernity—a current popular mood that spurns lofty ideals, visions and commitments in favour of the immediate satisfaction of all needs and desires.

prediction—a reasonably reliable anticipation of future developments.

pre-scientific worldview—a view of reality that is not informed by modern science, thus not based on demonstrable fact, mathematical stringency, or plausible conjecture.

rationalism (rational)—an approach to reality that is based on the reliability of human reason.

realism—an approach that assumes the existence of an objective reality, which does not depend on human observation and interpretation.

redemption (redemptive, redeem)—the act of liberating from, or overcoming, an undesirable situation or condition.

reference point—the central issue or organising principle of a system of meaning.

referent—the entity, event, or activity to which a concept or metaphor is meant to refer.

regularities—consistent patterns or rules according to which various levels of reality seem to operate, for instance causality. (See also 'natural law').

reification—'making real', that is, attributing objective reality to a concept, idea, or metaphor.

relevance (relevant)—making a difference in a concrete situation.

revelation—the assumed disclosure of God's intentionality through historical mediation.

Source and Destiny—see 'transcendent Source and Destiny'.

speculation—drawing logical inferences from problematic assumptions.

subatomic physics—the science that deals with the most elementary components of reality.

superstition—an unfounded fear, false explanation, or belief based on ignorance.

switch—the deliberate change of direction at a particular juncture.

system of meaning—an integrated body of assumptions that provides consistency, purpose and orientation to human consciousness.

theodicy—the attempt to justify faith in a powerful and loving God in the presence of evil.

transcendence—(a) immanent transcendence: going beyond a particular situation or condition within reality, (b) radical transcendence: the assumption that experienced reality is open towards, derived from, and dependent upon something greater than itself.

transcendent (the notion of, concept of)—intuition concerning the nature of what lies beyond.

transcendent Source and Destiny of reality—the assumed ultimate origin and purpose of the world process as a whole.

vision—the desired and imagined outcome of an activity or process.

volatility—being subject to rapid and unexpected change.

wave—the movement and transmission of energy in the form of oscillating movements similar to waves in the sea.

Index

Lightning Source UK Ltd.
Milton Keynes UK
174764UK00002B/2/P